FUNCTIONAL MODELS OF COGNITION

THEORY AND DECISION LIBRARY

General Editors: W. Leinfellner (*Vienna*) and G. Eberlein (*Munich*)

Series A: Philosophy and Methodology of the Social Sciences

Series B: Mathematical and Statistical Methods

Series C: Game Theory, Mathematical Programming and Operations Research

SERIES A: PHILOSOPHY AND METHODOLOGY OF THE SOCIAL SCIENCES

VOLUME 27

Scope: This series deals with the foundations, the general methodology and the criteria, goals and purpose of the social sciences. The emphasis in the Series A will be on well-argued, thoroughly analytical rather than advanced mathematical treatments. In this context, particular attention will be paid to game and decision theory and general philosophical topics from mathematics, psychology and economics, such as game theory, voting and welfare theory, with applications to political science, sociology, law and ethics.

The titles published in this series are listed at the end of this volume.

FUNCTIONAL MODELS OF COGNITION

Self-Organizing Dynamics and Semantic Structures in Cognitive Systems

edited by

ARTURO CARSETTI
University of Rome "Tor Vergata"

KLUWER ACADEMIC PUBLISHERS
DORDRECHT / BOSTON / LONDON

A C.I.P. Catalogue record for this book is available from the Library of Congress.

ISBN 0-7923-6072-9

Published by Kluwer Academic Publishers,
P.O. Box 17, 3300 AA Dordrecht, The Netherlands.

Sold and distributed in North, Central and South America
by Kluwer Academic Publishers,
101 Philip Drive, Norwell, MA 02061, U.S.A.

In all other countries, sold and distributed
by Kluwer Academic Publishers,
P.O. Box 322, 3300 AH Dordrecht, The Netherlands.

Printed on acid-free paper

Printed in the Netherlands.

TABLE OF CONTENTS

PART III

METHODS OF FORMAL ANALYSIS IN THE COGNITIVE SCIENCES

ACKNOWLEDGEMENTS

This book owes its existence to many sources. First of all I would like to express my deep appreciation to Werner Leinfellner. He encouraged me to edit this book and I have greatly benefited from discussions with him about the general orientation of the book. I am also grateful to Elizabeth Leinfellner for her invaluable help in preparing the final version of the book and to Johann Götschl who examined and approved the first project of the book.

I am indebted to my collaborators Andrea Cataldi and Enrica Vizzinisi for their help at the editorial level. I would like also to thank Anny Burer and Rudolf Rijgersberg of Kluwer for their editorial comments and suggestions which contributed to the quality of the presentation of the book.

My deep thanks to the authors for their co-operation and to my students at the University of Rome "Tor Vergata" for their stimulus and for their patience.

Finally, I would like to express my gratitude to Vittorio Somenzi. It was he who first stimulated my interest in the area of cognitive studies. I will always remember the late Gaetano Kanizsa, a good friend and a great scholar. I learned a lot from him while I was teaching in Trieste.

ARTURO CARSETTI

INTRODUCTION

Our ontology as well as our grammar are, as Quine affirms, ineliminable parts of our conceptual contribution to our theory of the world. It seems impossible to think of entities, individuals and events without specifying and constructing, in advance, a specific language that must be used in order to speak about these same entities. We really know only insofar as we regiment our system of the world in a consistent and adequate way. At the level of proper nouns and existence functions we have, for instance, a standard form of a regimented language whose complementary apparatus consists of predicates, variables, quantifiers and truth functions. If, for instance, the discoveries in the field of Quantum Mechanics should oblige us, in the future, to abandon the traditional logic of truth functions, the very notion of existence, as established until now, will be challenged.

These considerations, as developed by Quine, introduce us to a conceptual perspective like the "internal realist" perspective advocated by Putnam whose principal aim is, for certain aspects, to link the philosophical approaches developed respectively by Quine and Wittgenstein. Actually, Putnam conservatively extends the approach to the problem of reference outlined by Quine: in his opinion, to talk of "facts" without specifying the language to be used is to talk of nothing. "Object" itself has many uses and as we creatively invent new uses of words "we find that we can speak of 'objects' that were not 'values of any variable' in any language we previously spoke."[1] The notion of object becomes, then, a sort of open land, an unknown territory, as well as the notion of reference. The exploration of this land appears to be constrained by use and invention. But, we may wonder, is it possible to guide invention and to control use? In which way, in particular, is it possible, from a realist point of view, to link together program expressions and evolution?

F. Wuketits, in the article "Functional Realism", which introduces the first part of this volume, starts from Putnam's opinion that Realism has many faces and clearly affirms that he intends to discuss a type of realism grounded in the structure of our perceiving apparatus. In particular, he underlines that his principal aim is to revisit the problem of reality-realism and to look at it through the lenses of evolutionary thinking. Joining together, in a dialectical way, the adaptationist view and the non-adaptationist view, he points out, in a very clear manner, that cognition has to be regarded, first of all, as the function of active biosystems and that it results from complex interactions between the organism and its surroundings. Organisms do not simply get a "picture" of the world but develop a particular "scheme of reaction". " In the moment an organism perceives an object of whatever kind, it immediately begins to "interpret" this object in order to react properly to it...It is not necessary for the monkey to perceive the tree in itself, but rather to develop a notion of what we humans call tree...What counts is survival."[2]

A. Carsetti (ed.), Functional Models of Cognition, 7-23.

At the same time the notion *World in itself* becomes obsolete. Actually, we have to ask which meaning a notion like this could have for an organism when this same organism cannot refer to it at all in its real life. These considerations (among others) induce Wuketits to propose a new kind of realism that he calls Functional Realism.

From the point of view of Functional Realism, sensory perception gives only some specific hints with respect to the properties of the external world. The notion of Functional Realism that Wuketits introduces implies that humans are able to identify different objects in their surroundings by the properties these objects seem to possess and/or by the functions they fulfil for them. At the same time, it also implies that organisms learn how to react to the objects in their surroundings in order to survive. An important task for evolutionary epistemology is to reconstruct the evolutionary conditions for the development of sense organs and, in general, of cognitive tools. This evolution is the final result of a complex interplay between organisms and Nature, an interplay governed not by the laws of correspondence but by the laws of coherence.

"Different species act by means of their own construct of (parts of) "reality" and process information according to what they need in order to survive."[3] In this sense, Functional Realism possesses a pragmatic character and confirms Putnam's statement that: "A use of 'exist' inherent in the world itself...is an illusion"[4] as well as Ruse's fundamental assumption according to which the real world exists, but it is the world as we interpret it. Wuketits' conception of cognitive activity explicitly points out that cognitive systems are complex systems. But how are complexity and cognition related? And, moreover, in which way can humans, considered as complex systems, understand systems that should be described as complex? We have just remarked, for instance, that the interplay existing between an organism and its environment is characterized by the conjunction of three particular features: complexity, coherence and evolution. But how can we describe the inner conditions and the global functioning of this strange interplay?

This interrogative constitutes the starting point of the analysis outlined by D. Parisi in the paper "Complexity and Cognition". Complex systems are composed of many parts that interact with each other. Each part or element interacts locally with a limited number of other elements on the basis of specific rules. "However" as Parisi remarks "the system has global properties and behaviors that, although they result from the local interactions among the system's elements, we are unable to predict, to infer from the rules governing local interactions and to explain by identifying in each particular instance their causes"[5].

If we take into consideration a neural network as a classical example of a complex system, we may immediately see that the network's behaviour is a global property of the system represented by the neural network. However, given the high non-linearity of the system, we are unable in general to predict the network's behaviour from knowledge of the local interaction rules.

Actually, the major features that characterize a complex system concern the sensitivity to initial conditions and the adaptive and evolutionary character of the system. Moreover, in many cases complex systems are hierarchically organized. The growth of complexity can be viewed as a growth in organization. But human cognition as well (considered as a form of organization) emerges in a late stage of a long evolutionary process as a complex system (or rather as a system made up of many reciprocally interconnecting systems). On the other hand, human cognition appears also as the peculiar activity of an organism that lives and interacts with an environment that contains other complex evolving systems and, in particular, complex systems such as: artifacts, social institutions, etc. We are faced with a double circuit that increases complexity.

So we may wonder: "If reality is composed of both simple and complex systems, is human cognition as a knowing and understanding tool equally capable of understanding simple as well as complex systems? The answer is: no."[6] However, human brains, according to Putnam's and Wuketits' statements, are characterized not only by the presence of a passive adaptive strategy but also by the presence of an active one. It is at the level of the development of this active adaptive strategy (a strategy that permits one to modify the external environment in order to increase one's survival) that we can recognize the roots of a crucial capacity: i.e., the capacity to predict the consequences of our actions. This capacity is, in turn, at the basis of our knowledge of the world and science. Actually, the experimental method as utilized at the scientific level is nothing but the prediction of the consequences of our actions (plus the use of instruments and quantification).

However, while simple systems can be brought into the experimental laboratories, complex systems cannot. Fortunately, we now have a new form of understanding: computer simulation. "We can simulate a complex system and the rules that govern the local interactions between elements...In a certain sense the observation of a simulated complex system replaces the predictions. If we are able to construct a simulated system that essentially behaves as a real complex system, then we can say we have understood the real complex system even if we may be unable to predict the behaviour of both the real system and the simulated system."[7]

But, as Parisi remarks, the analysis of the phenomena of reality as complex systems is just beginning. The construction of adequate simulation models of cognitive activities, for instance, is a very difficult task. The reality that we intend to simulate presents at the same time many different aspects: it is adaptive, evolving, dissipative, self-referential and so on. It possesses a linguistic structure and it articulates according to a precise semantic context. For instance, we may wonder, in which way can we outline the real behaviour of a semantic neural network?

In any case the first two aspects that characterize this strange kind of reality are dynamics and self-organization. Cognition (as well as life) is, first of all, a self-organizing and dynamic phenomenon. So, in order to outline adequate models of cognitive activities, we have to focus first of all on the analysis of the dynamic forms and of the self-organizing processes existing in Nature.

The article by H. Haken, "From Visual Perception to Decision Making: a Synergetic

Approach", constitutes an important contribution in order to attain this difficult task. It deals with complex systems, and it is, in particular, concerned with those complex systems that may form spatial, temporal or functional structures in a self-organizing way. Such structures are characteristic of plants or animals, but they can also be found in the inanimate world of Physics and Chemistry. Synergetics searches for the general principles that govern the formation of structures. Synergetics in this sense deals with open systems and has developed a specific analysis in order to describe those situations in which qualitative changes occur. For instance, in Physics pattern formation occurs in fluids once they are heated from below. Such a fluid may form structures in the form of hexagons, rolls or spirals, etc. In a situation like this, there is a specific control parameter value at which the old state becomes unstable and the new state with its macroscopic motion forms. This new macroscopic motion is described and governed by one or several order parameters. In Synergetics, the relationship between order parameter(s) and individual parts of the system is called the slaving principle. Between the order parameters and the individual parts, a precise form of causality exists. Moreover: " Because the order parameters describe the behaviour of individual parts, we need not describe the individual parts directly...but it is sufficient to know the order parameters. This implies an enormous information compression."[8] Such information compression is a typical linguistic function. In this sense, a precise linguistic organisation form appears as inherent in the physical and the biological world.

Haken distinguishes two different approaches to the study of cognitive activities and, in particular, to the study of visual perception (this latter study constitutes for him a sort of privileged path in order to explore the wider realm of cognition): the bottom-up approach and the top-down approach. At the level of the bottom-up approach, we start from model neurons and their connections and then derive the macroscopic properties of their network. In the case of the top-down approach, on the contrary, we start with the tasks a macroscopic system has to fulfil and we look for possible realizations by means of networks of elements.

In accordance with Haken, visual perception concerns first of all the recognition of objects (or patterns) through vision. Pattern recognition, in turn, can be interpreted as the action of an associative memory. Such a memory constitutes an instrument for the completion of a set of incomplete data to a complete set, and it can be realized by some dynamics. Haken assumes as a central thesis that pattern recognition by the brain or the computer is nothing but pattern formation. In other words, we can reconstruct the pattern recognition dynamics in a close analogy to the just mentioned fluid dynamics example.

On the other hand, a close analogy also exists between decision making and pattern recognition: "In both cases the prototype patterns or the sets of known complete data may be learned or given. Incomplete data in decision making have their analogue in pattern recognition in the form of incomplete test patterns...In analogy to pattern recognition we may introduce a similarity measure...We then may establish a dynamics that is based on the similarity measure and may also include bias, attention parameters or awareness."[9]

Actually, we know that Boolean neural networks really classify. However, in order to attain a coherent and stable classification, the "knower" should not be chaotic nor should its classification. Hence, the necessity that a knowing system is subject to a solid regime, perhaps near the edge of chaos. In such theoretical landscape, learning may be seen, for instance, according to Kauffman (as well as to Haken's original intuitions), as a fundamental mechanism which converts chaotic attractors to orderly ones.

The contribution by E. Pessa "Connectionist Psychology and the Synergetics of Cognition", intends to explore, in a very clear way, some of Haken's fundamental achievements; it is concerned with the Synergetics of Cognition and, in particular, with the link existing between the bottom-up approach and the top-down approach.

According to Pessa, in the same way "as Statistical Mechanics which does not negate the relevance of macroscopic behaviours but tries only to explain them in a simpler way through a more fundamental level of description, Connectionist Psychology also does not negate the relevance of symbolic information processing in cognitive phenomena but tries only to explain macroscopic cognitive behaviour through the cooperative action of a large number of single cognitive microscopic units."[10] However, the problems that Connectionist Psychology must solve are more difficult than the ones considered by usual Statistical Mechanics. From here the importance of a combined method: the conjunction of the conceptual tools of Synergetics with some methods of theoretical Physics, a conjunction that appears capable of generating a new class of cognitive models to be tested against experimental data.

The essential problems of the Synergetics of cognition concern first of all the possibility of finding the right form of dynamical laws at the microscopic level and of deriving the dynamics of macroscopic structures from the knowledge of microdynamics. At the level of the bottom-up approach, one starts from the existing phenomena and then tries to find the best models able to explain them. At the level of the top-down approach, one starts, on the contrary, from abstract principles (symmetry, simplicity and so on) and then deduces from them the model equations. When we adopt a bottom-up approach, the goal would be that of finding a general algorithm for building the best neural network able to explain a given observed behaviour. But, according to Pessa, it is really impossible to attain this goal. So we are obliged to explore the path indicated by the top-down approach. In it one chooses the dynamical equations starting from some general principles. With respect to net architectures, their choice is generally constrained by particular requirements imposed by the experimenter. In particular, we can optimise the application of the traditional methods of Synergetics if we formulate the dynamics of connectionist models through macroscopic equations obtained by replacing the discretized neural network's behaviour with the one of a continuous neural field.

Pessa points out that we are very far from the building of a true Synergetics of cognition. Actually, even if, on the one hand, the progressive extension of the conceptual tools of Statistical Mechanics really joins up with an innovative exploration of the nature of neural networks and cellular automata as well as with a wider description of the behaviour of known linear dynamic systems and dissipative systems, on the other hand it still meets with many difficulties. We see, for instance, that in order to attain our goal we have to outline first of all new types of Boolean networks not restricted to the propositional level but having a predicative and, in perspective, an intensional and holistic character. We need, in other words, a precise theory of more sophisticated forms of organization.

R. Luccio, in his paper "Self-organization in Perception. The Case of Motion", which concludes the first part of the volume, intends, first of all, to show that Synergetics is demonstrating itself as a very powerful tool in interpreting perceptual phenomena. Synergetics is particularly apt to show the heuristic value of known dynamical linear systems in many domains of perception and, in particular, in the case of motion.

As is well known, Gestalt theorists use the term Prägnanz to mean both a tendency of the perceptual process to assume the most regular and economic course, given the constraints present in each specific case, and a tendency towards the maximum Ausgezeichnetheit in the concrete phenomenal result of the process itself. Utilizing the conceptual tools of Synergetics as well as some well-established experimental results obtained by Kanizsa, Luccio, Stadler etc., Luccio affirms that a tendency to Prägnanz as singularity does not exist at all. In his opinion, the behaviour of the visual system is not characterized by a tendency towards singularity, but by a tendency towards stability. Stability appears as the result of a capacity of self-organization displayed by the visual system. The system regulates itself according to principles that are essentially the ones that Wertheimer specified (proximity, similarity, common fate and so on). "The synergetic or conflicting action of such principles tends to a perceptual result that is better in the sense of maximal stability... and not to the better result in the sense of the prototypical or singular."[11] After all, we well know that the tendency to stability is one of the principal factors of that "perceptual geometry" that, at the moment, (according to the studies by Marly, Petitot etc.) presents itself as the real basis, from a genetic point of view, of Euclidean geometry.

The second part of the volume is devoted to a thorough analysis of three fundamental conceptual tools: complexity, organization and self-organization. The obtained results are then utilized in order to enlighten some theoretical aspects concerning the link between complexity, self-organization and truth and between evolution and self-organization. Actually, in order to outline concrete and functional models of cognitive activities we have to clarify first of all the complex relationships existing between complexity and tractability, between syntax and semantics, between natural language and formal languages, etc. We have, in particular, to outline a more sophisticated theory of self-organization, a theory capable, for instance, of taking into account also the intentional and semantic aspects of that specific continuous growth in organization that characterizes human cognition.

G. Ausiello and L. Cabibbo, in their article "Expressivesses and Complexity of Formal Systems", which constitutes the beginning of the second part of this volume, clearly affirm that the two needs that characterize human nature are: 1) the need to represent and understand physical phenomena through mathematical models; 2) the need to capture concepts through an abstraction process in order to arrive at the classification of objects and the formalization of their interrelationships.

The use of computers in these last fifty years has permitted a series of important discoveries and has brought out the need for formalizing aspects of computing itself: semantics and efficiency of programs, computation models and so on. The study of the formalization process possesses a metatheoretical character, and it is closely linked to the logical and the metamathematical studies concerning the power of computation processes and computability.

The paper considers two particular aspects of the formalization process: expressiveness and complexity. In particular, the issue of the complexity of formal models is described with reference to two fundamental areas of computer science: the theory of formal languages and the theory of database languages. When we construct a formal model, we are faced with two contrasting needs: on the one hand, the need to make the model more expressive in order to capture more and more aspects of reality, on the other hand, the need to keep the complexity of the model sufficiently low in order to preserve the tractability of its properties. Usually, formal models stabilize in an equilibrium point that constitutes an optimal compromise. A classical Boolean network represents a typical example of a good compromise. Another example of an optimal balancing between the need for expressiveness and the need for efficiency is represented by deterministic context-free grammars. For these kinds of grammars the equivalence problem has recently proved to be decidable, while it is undecidable for the general class of context-free grammars. By examining the growing difficulty of deciding equivalence in various families of regular expressions, the authors finally point out that the computational complexity of the properties of the formal expressions used in a formalism closely depends on their succintness and, in particular, on the fact that very short formulas may be used to represent long computations. In this sense, cognition reveals itself again as a sort of classification and as a sort of process of information compression.

The starting point of C.Böhm's paper, "Imposing Polynomial Time Complexity in Functional Programming", is constituted by an analysis of the notion of practically computable function. Such a notion may be identified through its belonging to the class of poly-time functions. This complexity class does not depend on the choice of the computing model or on a given programming language. In his paper, Bohm focuses his attention on the use of some functional programming languages where functions or even higher order functions are defined equationally or by rewriting systems. In general, we say that a function is poly-time if it possesses at least one definition logically implying a poly-time computation.

The two technical questions that constitute the central core of the paper are the following: 1) is there some way to decide, looking at an equation system defining recursively a given function, if that function has poly-time complexity? 2) Can the Bellantoni-Cook method be generalised to equation or rewriting systems defining functions on *any* algebraic data type? In accordance with Böhm's analysis, we see clearly that, when some thirty years ago M. Blum introduced the theory of complexity of computation of the functions over natural numbers and when in the early 1970s Karp, Cook, Hartmanis etc. introduced the definition of class NP, the related P=NP conjecture, the NP-completeness results and so forth, the theoretical frame within which these problems and results were successively presented and discussed was essentially a syntactic one. The semantic aspects were considered only with respect to an extensional semantic theory. Even the algorithmic complexity as outlined by Chaitin possesses an essentially syntactic character. Actually, whereas the classical theory of algorithmic complexity assumes that computer programs are meaningful, it does not account for what constitutes their meaning. The usual measure of their complexity is based only on the possibility to reduce their length without taking into account the more or less intricate aspects of their task. Similarly, the estimates of natural complexity based on probabilistic information theory are measures of statistical unexpectedness and not of any kind of meaningful complexity.

It is precisely in order to cope with these difficulties that Atlan and Koppel introduced in the late eighties a new concept: the concept of sophistication. Sophistication is defined as the meaningful part of the classical complexity of an algorithm by drawing a formal distinction between the program (which defines a structure with a given meaning) and the data. This new concept permits us to link, in some way, the traditional complexity theory and the theory of self-organization when this theory is considered in an intentional sense and when we seek to define self-organizing systems able to generate projects in the sense that they appear capable of setting a goal for themselves and achieving that goal.

In Atlan's opinion, as expressed in the paper, "Self-organizing Networks: Weak, Strong and Intentional, the Role of their Underdetermination", the actual meaning cannot be considered as an intrinsic property of a sentence or an object. It seems, on the contrary, to be generated by an act of interpretation that takes place at the interface between the observer and the observed. But an animal can have a specific capacity of self-observation without necessarily having semantic and intentional cognitive capacities. So, in order to work as an apparently infinite source of interpretations, self-observation must be connected to self-organizing devices able to indefinitely produce novelty that it will interpret with a new meaning. From the point of view of the theory of algorithmic complexity, such a capacity for interpretation could be assigned to a special type of algorithms defined as capable of generating infinite objects with a particular character of sophistication. Sophistication in this sense really appears as a measure of meaningful complexity, as a measure of a deeper form of organization, namely, the intentional organization. And it is precisely this kind of organization that characterizes human cognition.

New knowledge derived from the progressive outline of an adequate theory of self-organization of complex systems constitutes, in Götschl's opinion, as presented in the paper "Self-organization: Epistemological and Methodological Aspects of the Unity of

Reality", not only a real improvement of our knowledge of the functioning of the complex systems, but also the real basis for a new understanding of Nature and Culture and of the interrelationships existing between them. Götschl's attempt concerns first of all a tentative reconstruction of the fundamental conceptual categories of the theory of self-organization. In its core the theory of self-organization contains the principles of emergence, self-production and creation of new forms. In particular, the theory of self-organization clearly shows that all complex systems are open systems. They are constituted by processes that are dissipative, chaotic and self-organizing. Under certain circumstances, complex systems can be self-determining and can have the structure of self-referentiality. Moreover, at the level of complex systems, we are faced with a specific form of co-evolution.

In this sense, Götschl formulates a precise central hypothesis: "As a theory of the complexity and dynamics of reality (TSO) identifies and represents the existence of new affinities between evolution, self-organization and cognition. These affinities lead to progress in overcoming the dualistic mind-matter hypothesis."[12] Thus, from a philosophical point of view, a particular form of monism emerges; a conception that we can also find in some of Atlan's last papers devoted to an analysis of Spinoza's theoretical investigations.

The essence of the theory of self-organization is that complex systems can increase in degrees of freedom by themselves. In this way, a structural interwovenness of object, knowledge of an object and self-knowledge appears to be at the real basis of cognitive activity. In this sense, the theory of self-organization can really supply a new theoretical model of reality. What is central to this model is its process-evolutive character. This makes possible a more unified foundation of material and socio-cognitive structural levels. Moreover, the process-evolutive modeling of reality by means of the theory of self-organization facilitates the constitution and the identification of more adequate links between different sciences.

It is precisely the evolutionary character of the cognitive processes that constitutes the central core of Arecchi's paper, "How Science Approaches the World: Risky Truths vs. Misleading Certitudes". He remarks, first of all, that, at the cognitive level, the attempt to outline a purely deductive science consisting in the logical construction of a set of theorems yielding a complete description of experimental observation fails because of two main drawbacks: undecidability and intractability. The way out of such a crisis consists in an adaptive strategy, that is, in a frequent readjustment of the rules suggested by the observed events. This way the language no longer has fixed rules, hence it appears as semantically open. Semantic openness implies a re-evaluation of the notion of truth as recognition of the essential role of some external features in orienting the cognitive procedures.

Let us assume, as Arecchi remarks, 1) a set < A > of control parameters responsible for successive bifurcation leading to an exponentially high number of final outcomes, 2) a set <T > of external forces that, applied at each bifurcation point, break the symmetries, biasing towards a specific choice and eventually leading to a unique final state. "We are in the presence of a conflict between (i) syntax represented by the set of rules... and (ii) semantics represented by the set of external agents."[13] We define "certitude" as the correct application of the rules and "truth" as the adaptation to reality that is constrained by the external forces.

The adaptationist description of cognition is, in Arecchi's opinion, the reason why we can observe stable features in the world around us. In fact, chaotic dynamics can be seen as a multiple bifurcation cascade folded on itself and allowing for the onset of many regular orbits which, however, survive for a short time only and are then replaced by another one. We can avoid destabilisation and fix any of the orbits if we start with small corrections once we have reached the desired position in bifurcation space. Chaos is stabilized by the interaction with the rest of the system. Cognition is the final result of a continuous compromise between the power of the theory and its limitations by the "irruption" of reality. With regard to this, we can remark, however, that the forms of the irruption, which Arecchi refers to depend, in turn, on the cognitive tools and the cognitive action expressed by the observer.

It is exactly the investigation of this kind of circularity that constitutes the central core of Svozil's paper, "On Self-reference and Self-description". The "Cartesian prison" represents the starting point of Svozil's considerations. Let us imagine that our world is discretely organized and governed by constructive laws. One might think of such a world as an ongoing computation based on a finite size algorithm, as a sort of computer-generated universe. "Any observer who is embedded in such a system is naturally limited by the methods, devices and procedures that are operational therein. Such an observer cannot step outside of this "Cartesian prison" and is therefore bounded to self-referential perception. Can one give concrete meaning to this "boundedness" by self-reference?"[14].

With regard to this, two observations must be made in agreement with Svozil's opinion: 1) one application of the fixed-point theorem is the existence of self-reproducing automata and therefore the existence of intrinsically representable systems containing a complete representation of themselves (blueprint); 2) one can demonstrate that in any case it is impossible for a given system to obtain such a blueprint by mere self-inspection.

An idealized self-referential measurement attempts to attain the impossible. On the one hand, it claims to grasp the true value of an observable while, on the other hand, it has to interact with the object to be measured. Computational complementarity is a source of indeterminism. In this sense, if we assume the existence of a computer-generated universe, then complementarity is one inevitable feature of its operational perception. Thus, uncertainty, cognition and self-reference appear as aspects of a unique, real entanglement. In order to try to unravel (even if only in a partial way) this entanglement we have first of all to outline a representation of our cognitive abilities by resorting to the development of formal methods, abstract models, computer simulations, etc. that reveal themselves as being more and more sophisticated. These formal tools do not just lead us to a form of abstract awareness; they also constitute the very sources of a progressive extension of our cognitive apparatus, especially at the neurological level. They are, in particular, concerned with the semantic and intentional development of cognitive activities, with the continuous emergence of ever-new forms of perception and of new language. For instance, the possible outlining of an adequate semantics for natural language emerges as a form of interactive knowledge of the complex chain of biological realizations through which Nature progressively expresses itself in a consistent way according to linguistic rules and constraints. The simulation work offers, in effect, real instruments to Nature, considered as the original information Source, in order to perform a self-description process and to outline specific procedures of control as well as the possible map of an entire series of imagination paths.

In the third part of the volume we include four papers which present some of the more important directions of research with regard to the tentative definition of new formal methods capable of supporting the construction of more adequate simulation models of some specific cognitive activities. These directions of research concern: 1) the theory of self-organization and computability theory; 2) the possible extensions of the traditional theory of formal languages and rewriting systems; 3) the mathematical models of visual perception; 4) the algebraic semantics of natural language.

The principal aim of the paper by M. Koppel and H. Atlan, "Self-organization and Computability", is the attempt to give a precise mathematical definition of a new measure of meaningful complexity: sophistication, a concept which has already been discussed, in more general terms, in Atlan's paper included in the second part of this volume. Starting from the formal definition of the measure of complexity introduced by G. Chaitin, namely program-length complexity, the authors introduce the definition of a quality called "sophistication" considered as the correct formal analogue of organization. Also this completely mathematical definition uses the distinction between two different parts of the description of an object: the first part consists of the object structure, while the second part consists of the specification of the object from among the class of objects defined by its structure. On the one hand, we have programs and, on the other, data.

Let P be the shortest program which is part of the shortest description of a string S. Then P is called the sophistication of S. If **a** is an infinite string and $\mathbf{a}^n$ is its initial segment of length n, then a given program P is a compression program for the infinite string **a** if it is a part of an almost minimal description of every initial segment of **a**. The sophistication of an infinite string is the length of the smallest compression program for **a**.

The authors suggest, on the basis of their mathematical results, that sophistication is the adequate formalization of the concept of "organization". Thus, a self-organizing system would be a system that gradually develops higher sophistication. When such a system develops higher and higher sophistication ad infinitum, it is called a true self-organizing system. In the latter case it is evident that the state towards which the system converges cannot be simulated. In order to soften the force of this (and other) difficulties in the end the authors introduce a bounded version of self-organization which can be simulatable using standard computational tools and which might serve as a reasonable stand-in for true self-organization.

It is important to remark that the measure of meaningful complexity (the measure of the semantic content of a given biological structure, for instance) outlined by Koppel and Atlan is closely linked to the real existence and definition of a compression program, a kind of program that, as we have just seen, many other papers presented in this volume refer to, a kind of program that effectively reveals itself as one of the real roots of the language of Life and of the language of Cognition.

The paper by G. Longo, "The Difference between Clocks and Turing Machines", is, at the beginning, concerned with the exact distinction between hardware and software, a distinction that possesses many links and analogies with the distinction between programs and data. This distinction permits a critical re-examination of the conceptual bases of some aspects of computability theory, of the theory of Turing machines as well as of the incompleteness theorems and related results. This kind of analysis leads the author to make an important remark: "Even though it is not yet clear how much relevant mathematics lies outside finitistically provable fragments, the independence results hint that mathematics is an essentially open system of proofs, open to meaning, over an ongrowing variety of structures and history."[15]

In this sense, Longo can affirm, revisiting some original suggestions proposed by Chaitin, that the invention of new techniques, including proof techniques within broad contexts of knowledge is, at the same time, part of the "theory building" and "tools building" so typical of mathematical cognition. To this character of openness of mathematics closely corresponds now a critical reassessment of our knowledge in the field of automata theory. Actual computers are in fact no longer representable by closed systems of axioms and rules. In order to obtain linguistic or geometric representations of the world as well as of the physical devices already mentioned, reductions will still be required. But a broader notion of rationality is now needed. The new artificial complex systems are actually open, as Wegner remarks, to a world which is only partially known and which is progressively revealed by interaction or observation. So the asynchronous, concurrent distributed systems, open to interactions, seem to suggest a further qualitative jump whose mathematics is not fully understood. Moreover, the analysis and the simulation of brain functioning and cognitive abilities also require more mathematics, partly known, mostly yet to be invented.

With regard to this, we can also imagine a possible recomposition, at a more sophisticated level, of Turing's fruitful split between hardware and software. In particular, according to Longo's opinion, we need to recompose knowledge and imbed mathematics into the maze of back connections, which relate it to perception and other forms of cognitive activity. So we have to use the rigor and the generality of mathematics in order to outline new open systems of computation and life, and, at the same time, we should investigate actual mathematical reasoning as a broader paradigm (as an object of study). The invention of new forms of knowledge and the recovery of the genetic development of our cognitive tools appear as the two aspects of a unitary intellectual task.

It is well known, for instance, that the progressive emergence of new open systems of computation as well as the contemporary research concerning the cognitive links between perception, language and action have actually transformed in recent years the traditional methods of analysis utilized in the field of formal semantics. They have led us, for example, to outline a sort of perceptive genealogy of cognitive symbolic structures in accordance with the guidelines that characterize that genealogy of logic that Husserl progressively discovered and described in the first thirty years of our century.

As J. Petitot affirms in the paper "Sheaf Mereology and Space Cognition": "For classical elementary logic the atomic formulae 'S is p' stay at the lowest level of complexity. Both their syntax and their semantics are trivial. But it is no longer the case if one takes into account the perceptive situations to which these statements refer. Indeed, these situations compel us to introduce in the formalization of symbolic logical structures, topological and morphological structures of a completely different kind. If one wants to do semantics that way, one is therefore committed to elaborate a mathematically integrated theory of topology and logic."[16]

So, on the perceptive side, we find a geometrical descriptive eidetics of the morphological organization of perception. On the logical side, we find, on the contrary, a logical eidetics of judgement: we are really faced with a formal syntax and a formal semantics. Morphological eidetics depend on the synthetic laws of perception; logical eidetics depend on analytic laws.

Petitot, following some of Thom's original suggestions, remarks that the truth conditions of atomic formulae as "S is p" depend on the way the "intentions of signification" are intuitively filled at the ante-predicative and pre-judicative level. Without an adequate description of this intuitive filling-in one cannot elaborate a correct theory of predication. Petitot shows that it is possible to model in a mathematically correct way Husserl's intuitions about the "intuitive filling-in". If we assume **W** to be the spatial extension of the substrate S, **W** is filled by a quality p belonging to a given genus G (e.g., colours). The filling-in is then described by a particular map f. This map, in turn, can be interpreted as a section of a particular (trivial) fibration.

The morphological analysis of the filling-in of spatial regions by qualities appears to be closely linked to the constitutive role of the qualitative discontinuities in segmenting visual scenes. The noematic content of an object, in this sense, expresses the synthetic relations of foundations between spatial extensions and their dependent moments (colours, contours, etc.). Hence the "categorial genesis" that transforms the perceptive synthesis where a substrate S possesses a dependent moment p, in the predication "S is p". It is exactly the logical typification of the synthetic dependence relations in syntactic categories (the transformation of the substrates and of the dependent moments into subjects and predicates) that Petitot aims to model in his paper. He shows, in particular, that the geometric logic yielded by Topos theory gives adequate tools in order to model

Husserl's ante-predicative genealogy of logic, and that there exists a precise topological geometrical eidetics of the morphological structures of perception, an eidetics that is, moreover, neurologically relevant.

In other words, if we try to move backwards in order to reconstruct the successive steps of the genealogical process, we finally have the real possibility of deciphering the actual texture of the neurological structures as they express themselves at the visual level. What happens at the level of vision understanding can also be recognized as being present at the level of language understanding. In this latter case, however, the tentative reconstruction reveals itself, if possible, as even more difficult. We need mathematical tools that are more and more powerful and sophisticated. In particular the texture of the constraints as well as the hierarchies of the intentional rules appear more and more stratified.

The article by Carsetti, "Linguistic Structures, Cognitive Functions and Algebraic Semantics", seeks to present some aspects of the present attempts to reconstruct not only a genealogy of logic but also a genealogy of language, with logic being considered as a part of linguistics. First of all, Carsetti underlines that, in the past twenty years, Quine and Putnam have revisited, in a very precise way, Husserl's original intuitions concerning the dialectics existing between substrate and dependent moments.

When we affirm, for example, "the apples are red", then, according to Quine, we realize perfectly as speakers of a natural language that the predication constitutes a link which is stronger than the simple conjunction; the predication, in fact, requires the "immersion" of the apples into the red. From a perceptual, linguistic and conceptual point of view, we are really faced with an object that possesses the precise character of being a whole, a character that anticipates even more complex forms of organicity. A book, for instance, is composed of pages which themselves, as far as they are parts depending on an organic whole, do not constitute an object. These same pages, however, become simple objects once they are torn from the book. Husserl was the first to introduce in our century a sort of logical calculus of parts and wholes. He thought that only certain "organic" wholes constitute real objects, and in this sense he introduced the notion of "substratum". The dialectics existing between part and whole really "homes" in the same act of predication, namely in the language considered as action and intervention. We have to discover the secret functional aspects of such dialectics if we want to enter the mysterious kingdom of primary linguistic expressions. But, in accordance with Carsetti's opinion, to face this kind of dialectics means also to face the problems concerning the symbolic dynamics that governs the successive expression of dependency links, the articulated texture relative to the relationships between parts and whole.

If, as we have just seen, we can recognize at the ontological level the existence of an entire sequence of expression levels of increasing complexity, the dialectics part-whole must be compared to the successive constitution of these different levels as well as to the net of generative constraints through which this kind of process expresses itself. It is only within the secret meanders of this symbolic dynamics that we can, finally, find the theoretical instruments capable of showing us the correct paths that we need to pursue in order to obtain a more adequate characterization of the Fregean *Sinn.* At the level of the arising new horizons, precise algebraic structures will graft themselves onto the previous structures and complex functional spaces will unfold with reference to a specific frame of geometric interactions. In consequence of the successive transformation of the dependency links, the same "shape" of real objects, as it is normally recognized by us, will change. The same modalities through which we regiment the universe of the events as well as the modalities through which we induce, in a co-operative way, by coagulum and selection, a specific coherent unfolding of the deep information content hidden in the external Source, will be subject to a specific transformation. So, the ontological discourse connects itself with a precise genealogical and dynamic dimension, a dimension that must be linked, however, to a specific continuous procedure of conceptual recovery if we want this same dimension to continue unfolding freely, performing thus its specific generative function.

At the end of this brief and incomplete presentation of the main guidelines of the book, let us now make just a few final remarks.

As we have seen, cognitive processes can be considered, in the first instance, as self-organizing and complex processes characterized by a continuous emergence of new categorization forms and by self- referentiality. In order to understand the inner mechanisms of these processes we have to outline a theory of more and more sophisticated forms of organization. We need, for instance, to define new measures of meaningful complexity, new architectures of semantic neural networks, new clusters of synergetic structures, etc. In particular, we have to take into consideration the genetic and "genealogical" aspects that characterize the inner development of cognitive symbolic structures.

However, cognition is not only a self-organizing process. It is also a co-operative and coupled process. If we consider the external environment as a complex, multiple and stratified Source which interacts with the nervous system, we can easily realize that the cognitive activities devoted to the "intelligent" search for the depth information living in the Source, may determine the same change of the complexity conditions according to which the Source progressively expresses its "wild" action. In this sense, simulation models are not neutral or purely speculative. The true cognition appears to be necessarily connected with successful forms of reading, those forms that permit a specific coherent unfolding of the deep information content of the Source. Therefore the simulation models, if valid, materialize as "creative" channels, i.e., as autonomous functional systems (or functional models), as the same roots of a new possible development of the entire system represented by the mind and its Reality.

In the light of these considerations we can easily realize that besides natural selection we also have to recognize the existence of an internal selection considered as essentially linked to the successive expression of the organismic constraints as well as to the final unfolding determined by the inner creative channels just mentioned.

Thus, with respect to the possible outlining of new models of cognition, we are really faced, at the moment, with the appearance of a new frontier.

It appears quite clear, for instance, that the emergence of new cognition coincides necessarily with an extension and an inner re-organization of the language in use. This extension will concern, however, not only the extensional aspects of the first-order structures, but also the intensional ones and, in general, the aspects relative to the second-order structures. As a matter of fact, language can go beyond any limit as determined by a particular conception of linguistic formal rules. As Putnam remarks, "Reason can go beyond whatever reason can formalize."[17]

These considerations can be paraphrased in a different way by saying that, at the level of cognitive studies, it appears necessary now to extend the condition of predicative activity, as defined by Quine, by admitting the necessary utilization of certain abstract concepts in addition to the merely combinatorial concepts referring to symbols. For this purpose we must count as abstract those concepts that do not comprise properties and relations of concrete objects (the inspectable evidence) but which are concerned with thought constructions and, in general, with the articulation of the intellectual tools of invention and control proper to the human mind.

The utilization, at the semantic level, of abstract concepts, the possibility of referring to the sense of symbols and not only to their combinatorial properties, the possibility of picking up the deep information existing in things and the extended use of functional models within a dynamic context characterized by the presence of precise forms of co-evolution, open up new horizons at the level of cognitive studies.

In the thirties Goedel suggested that mental procedures might extend beyond mechanical procedures because there may be finite, non-mechanical procedures that make use of the meaning of the terms. He spoke about "the beginnings of a science which claims to possess a systematic method for such clarification of meaning, and that is the phenomenology founded by Husserl."[18]

We have just seen, for instance, that, from Husserl's point of view, a specific kind of reflection enables consciousness to grasp dependency links and to typify the substrates in subjects and the dependent moments in predicates. In accordance with Goedel's opinion, a particularly significant example of our mental capacity to grasp an abstract concept can be found in the unlimited series of new arithmetic axioms, in the form of Goedel's sentences, that one could add to the given axioms on the basis of the incompleteness theorems. We can use the axioms to solve problems that were previously undecidable. We are simply unfolding our innate capacity to grasp a concept or, in Husserlian terms, our capacity to express a categorial intuition.

In accordance with Husserl, we really can "see" the categories once embodied in a categorial intuition. So, we are able to realize the "way" according to which the objects present themselves in our mind. In other words, we can reflect on categories, we can arrive, for instance, at a better understanding of their bounds or their relations. According to a famous example, when we see a chair *and* a table, we can also perceive, in the background of this perception, the "connection" existing between the two different things. We perceive a categorial form with respect to a specific "situation" and this form determines the *way* in which this situation is given. We are dealing with a kind of categorial perception (or rational perception), a perception that does not concern simple data (relative to the inspectable evidence) but complex conceptual constructions. In Husserlian terms, meaning "shapes" the forms creatively.

However, in order to understand how this "shaping" takes place we need more information about the genealogical aspects of this process. At the same time, we need a new theory concerning the link between software and hardware. In fact, a program, in accordance with the perspective advocated in some of the papers included in this volume, must also be considered as the performer of itself. Moreover, we also have to handle in a semantic and dynamic way the processes of information compression as they express themselves at the biological level. In particular, we need, as we have just seen, more and more adequate measures of meaningful complexity also capable, for instance, of taking into account the dynamic and interactive aspects of depth information. In short, we need new models of cognition. Functional and co-evolutive models, not static or interpretative ones. At the level of these kinds of model, emergence (in a co-evolutive landscape) and truth (in an intensional setting) for many aspects will necessarily coincide.

NOTES

[1] Cf.: Putnam, H., Representation and Reality, Cambridge, 1983, p.137.
[2] Cf.: Wuketits, F. M., "Functional Realism", this volume, p.30.
[3] Cf.: Wuketits, F. M., "Functional Realism", this volume, p.34.
[4] Cf.: Putnam, H., The Many Faces of Realism, La Salle, 1987, p.121.
[5] Cf.: Parisi, D., "Complexity and Cognition", this volume, p.39.
[6] Cf.: Parisi, D., "Complexity and Cognition", this volume, p.44.
[7] Cf.: Parisi, D., "Complexity and Cognition", this volume, p.47.
[8] Cf.: Haken, H., "From Visual Perception to Decision Making: a Synergetic Approach", this volume, p.52.
[9] Cf. : Haken, H., "From Visual Perception to Decision Making: a Synergetic Approach", this volume, p.63-64.
[10] Cf.: Pessa, E., "Connectionist Psychology and the Synergetic of Cognition", this volume, p.67.
[11] Cf.: Luccio, R., "Self-organization in Perception. The case of Motion", this volume, p.94.
[12] Cf.: Götschl, J., "Self-organization: Epistemological and Methodological Aspects of the Unity of Reality", this volume, p.156.
[13] Cf.: Arecchi, F. T., "How Science Approaches the World: Risky Truths vs. Misleading Certitudes", this volume, p.177.
[14] Cf.: Svozil, K., "On Self-reference and Self-description", this volume, p.189.
[15] Cf.: Longo, G., "The Difference between Clocks and Turing Machines", this volume, p.218.
[16] Cf.: Petitot, J., "Sheaf Mereology and Space Cognition", this volume, p.233.
[17] Cf.: Putnam, H., Representation and Reality, op. cit., p.134.
[18] Cf.: Goedel, K., "The Modern Development of the Foundations of Mathematics in the Light of Philosophy" in S. Feferman et al. (Eds.) Kurt Goedel: Collected works. Vols. I, II, III, 1986, 1990, 1995, Oxford, pp.374-387.

PART I

MODELS OF COGNITION

FRANZ M. WUKETITS

FUNCTIONAL REALISM

> "It is true that I have a strong impression of an external world apart from any communication with other conscious beings. But apart from such communication I should have no reason to trust the impression".
>
> A. EDDINGTON

INTRODUCTION: REALISM REVISITED

According to a naive understanding we perceive the world – with all its particular objects and characteristics of these objects – as it actually is. From our everyday's point of view stones are *stones*, trees are *trees*, tables are *tables*, and so on and so forth. No mentally healthy person, it seems, can have serious doubts that a stone is *really* a stone, a tree is *really* a tree, a table is *really* a table, etc. Since philosophers have expressed such doubts, there is either something wrong with their mental health or there is some substance to the question whether we perceive the world and all its particular objects and characteristics of these objects as they "really" are.

In their everyday life humans also tend to think that reality has to do with – or is rather an assemblage of – concrete things. "Ask any man", said Eddington (1958, p. 274), "who is not a philosopher or a mystic to name something typically real; he is almost sure to choose a concrete thing". But what about "time", "light speed", "energy", "entropy", and other physical entities? They are not concrete objects, but are nevertheless – at least for the physicists (and most probably for all philosophers) – real. We can state that not only those middle-sized objects like stones, trees, and tables that we perceive in our everyday life are real; there exist many other real "things", although we do not see, hear or smell them. Moreover, the middle-sized (concrete) objects that we *do* see (hear or smell) are maybe not what we think that they are.

Realism has many faces (Putnam, 1987), and there is not one good (philosophical, scientific) argument for the exclusive truth of our everyday commonsense realism. However, this type of realism is quite important for our life. In fact, we have learned to trust our eyes, ears and noses and thus to identify different objects; if something *looks like* a knife, for instance, we usually also know that we should be careful when using it in order to avoid more or less dangerous wounds. This is to say that we have indeed some reasons to believe that our perception of different objects and their characteristics is not completely wrong even though this belief does not say anything about the "world as it really is", but rather says something about the "world as we have to interpret it". The reasons for this belief are not philosophical but biological.

A. Carsetti (ed.), Functional Models of Cognition, 27-38.

In what follows I discuss a particular type of realism which I preferably call *functional realism* and which is grounded in the structure of our perceiving apparatus that was developed during our evolution by natural selection. Systematically, this type of realism is located somewhere between our naive everyday understanding of the world and all those sceptical positions that teach us that reality is a mere construction. My point of view is evolutionary epistemology or, better to say, one version of evolutionary epistemology (see also Wuketits, 1987, 1989, 1990, 1991, 1992, 1995). My starting point is that the very meaning of "getting reality" is – not only for humans, but for other animals too – *survival*. I want to invite the reader to revisit, so to speak, the problem of reality-realism and to look at it through the lenses of biological, evolutionary thinking.

EVOLUTIONARY EPISTEMOLOGY: THE ADAPTATIONIST VIEW

In his *Behind the Mirror* Konrad Lorenz stated that "what we experience is… a real image of reality – albeit an extremely simple one, only just sufficing for our own practical purposes" (Lorenz, 1977, p. 7). Much earlier, in his seminal paper on evolutionary epistemology he had written that "just as the hoof of the horse is adapted to the ground of the steppe which it copes with, so our central nervous apparatus for organizing the image of the world is adapted to the real world with which man has to cope" (Lorenz, 1941, 1982, p. 12). Similarly, Riedl (1984) stated that there is a correspondence between our cognition and nature, and Vollmer (1984) stressed that our sense organs *fit* their environments. I was, years ago, also very much attracted by this stance of evolutionary epistemologists and spoke of the "representation of reality" by the cognitive apparatus (Wuketits, 1984).

Since adaptation is one of the key notions of evolutionary biology, one is easily tempted to apply "adaptationism" to the evolutionary study of cognition and knowledge. Therefore, at least for some time, the adaptationist version of evolutionary epistemology was predominantly advocated by most naturalists who were to study cognitive and knowledge phenomena in evolutionary terms. One of the standard arguments of evolutionary biologists, as far as they advocate a Darwinian theory (particularly the synthetic theory) of evolution, is that natural selection favours those organisms that are comparatively well adapted to their respective environment, so that adaptation is the necessary result of evolutionary processes. For example, Simpson (1963) says that natural selection in the Darwinian sense includes an adaptive trend. From this argument one can easily deduce an epistemologically relevant statement which runs as follows: "Our perceptions do give true, even though not complete, representations of the outer world because that was and is a biological necessity, built into us by natural selection. If it were not so, we would not be here"! (Simpson, 1963, p. 98).

Two points are of particular relevance in the context of an adaptationist version of evolutionary epistemology:

(1) The cognitive apparatus of any organism is adapted to the outer world with which the organism has to cope.

(2) What an organism perceives, is a true, but simplified picture of the outer world or certain structures of its environment.

This means that the advocates of the adaptationist view do not think that we – or other organisms – get a complete image of what is "out there" and are therefore not naive realists. But what they do maintain is that any organism's perception obeys to what really exists in the surroundings to which the perceiving apparatus is adapted after all. Clark (1986, p. 152) comments this stance as follows: "On the plausible assumption that basic cognitive orientation or instinct is as adaptively strategic as gross bodily form, the evolutionary epistemologist brings his selective paradigm to bear on the issue of the relationship between an animal's environment (as we recognise it) and its knowledge of that environment (as expressed in its observable behaviour)". It is for this reason that Simpson (1963, p. 98) could explicitly state that "the monkey who did not have a realistic perception of the tree branch he jumped for was soon a dead monkey – and therefore did not become one of our ancestors". But what does "realistic" mean? Is the monkey supposed to grasp the "tree branch in itself"? One could hardly think so.

Popper (1989, p. 453) writes: "*The thing in itself* is unknowable: we can only know its appearances which are to be understood... as resulting from the thing in itself and from our own perceiving apparatus". If this is true for humans, then it must be even more true for monkeys and other animals. But if the perceiving apparatus (of humans, monkeys, and other living beings) is not adapted to the things in themselves (but rather to their appearances), then any organism's picture of its surroundings cannot be "realistic" in a strict sense: there cannot be a true correspondence between the cognitive apparatus and the outer world. However, we can hardly deny that any organism obtains information about the outer world, so that evolution can generally be described as a cognition process (see, e.g., Lorenz, 1977, Riedl, 1980, Wuketits, 1986) or at least an information-gaining process. The "Proto-Cognitive Model" (Elitzur, 1994) puts forward such a broad generalization; however, this model does not appear to be in tune with a strict adaptationist version of evolutionary epistemology. What would be the alternative to this version of an evolutionary theory of cognition?

EVOLUTIONARY EPISTEMOLOGY: THE NON-ADAPTATIONIST VIEW

Since "organism are not puppets operated by environmental strings" (Weiss, 1969, p. 362), we have also to claim that their cognitive apparatus is not a passive organ waiting for getting impressions from outside. Therefore, as I pointed out in another pubblication (Wuketits, 1989):

– cognition has to be regarded as the function of *active* biosystems (and not of "blind machines" that just respond to external influences);

– it is not a reaction to the outer world, but results from complex *interactions* between the organism and its surroundings;

– it is not a linear process of step-by-step accumulation of information, but a process of continuous error elimination[1].

I have changed somehow my position and have moved, so to speak, from an adaptationist to a non-adaptationist version of evolutionary epistemology. (One should keep in mind, however, that "non-adaptationist" does not mean "*anti*-adaptationist). The former has been criticized for example by Falk (1993, p. 163): "According to it our models of the mind represent the world because our brains have evolved to respond to 'the environment' – as if there exists such an entity, which is distinct and essentially independent of the organism – so as to increase our adaptation".

Now, humans (and other organism) do not simply "portray" their environment(s). Particularly, what does seem clear is that different species perceive different aspects of "the world". A tree is a *tree* for a human being, but for a dog or a cat or a sparrow it is something different – we do not know what these animals exactly perceive when they perceive a "tree", but we might infer from their behavior that for them the object that we call "tree" is not necessarily the same as it is for us.

However, organisms are obviously able to cope with different objects in their surroundings. Evolutionary epistemologists have maintained that animals operate on the basis of inborn "hypotheses" (see, e. g., Lorenz, 1977, Riedl, 1980) that have been stabilized during evolution by a large number of experiences and confirmed expectations. The same would be true also for human beings. Therefore, evolutionary epistemology includes *hypothetical realism* as a particular approach to the problem of reality. Like other living beings, humans calculate the world on the basis of innate hypotheses or "innate teaching masters" (Lorenz, 1977) that can be understood as a system of instructions regarding the "world as it has to be interpreted". And like other species Homo sapiens perceives only a rather tiny fraction of "the world" and lives in a particular *mesocosm* or "world of medium dimensions" (Vollmer, 1984). The implications of this stance have not been clearly seen by those evolutionary epistemologists who have been commited to the adaptationist view.

The are at least two implications to be mentioned:

(1) Organism do not simply get a "picture" of (parts of) the world, but develop a particular "scheme of reaction". In the moment an organism perceives an object of whatever kind, it immediately begins to "interpret" this object in order to react properly to it. Simpson's famous metaphor – "The monkey who did not have a realist perception of the tree branch he jumped for was soon a dead monkey..." (see above) – then has to be modified. It is not necessary for the monkey to perceive the tree in itself, but rather to develop a notion of what we humans call "tree" (or "branch") and calculate, as it were, its own behavior, its own reaction. What counts is survival.

(2) However, the notion *World in itself* becomes obsolete or redundant, if we lay particular stress on an organism's own behavior. Clark (1986) and Ruse (1989) have arrived at this conclusion which some may find a drastic step. But we have to ask seriously which meaning a notion could have for an organism when the organism in its "real life" cannot refer to it at all. There are, of course, many entities beyond an organism's perceiving capacities, but they do not play any role for the organism. This conclusion almost inevitably follows if one takes a strict biological point of view and regards survival as the primary target in each organism's life.

As it is obvious, I hold a view that pays more attention to the organism and its ways of behavior. One point that has been made also by some advocates of evolutionary epistemology representing the "constructivist wing" of this epistemology (e. g. Diettrich, 1989, 1994) is this: A particular environment does not simply prescribe the organisms' reactions to it; different species have different possibilities to cope with their environment. To pick up once again Lorenz' statement, this means that developing hoofs, for instance, is not the only way to cope with the steppe. In fact there are not only horses living in the steppe – there are also snakes, rodents, and many other types of animals living there which definitely have not developed hooves.

FUNCTIONAL REALISM

My central point regarding realism is that when we perceive any object, we do so since all objects – as far as they can be perceived by us at all - show certain properties or have certain functions for us. How do we perceive, for example, a table as a *table*? We have learned that all objects of a particular size and showing a particular form *are* tables, i.e., objects serving special functions or purposes. Of course, one could argue that this is just a matter of contention and not so much a question of realism. So, let me make my point clearer.

First of all, we should abandon the correspondence theory which basically holds that there is an objective world and that our perceiving apparatus is adapted to it – so that, after all, what we perceive is a (simplified) picture of this objective world or some of its objects. Actually, the subject-object distinction that has attracted many philosophers and scientists is not tenable. The supposed correspondence between object and subject, the supposed congruence between the objective world and the subjective perception

$$\text{object} \approx \text{subject}$$

has to be replaced by a broader view expressing the close relations between subject and object and showing that they both are parts of *one* reality:

$$\boxed{\text{object} \leftrightarrow \text{subject}}$$

What we have to face then is rather *coherence* than correspondence (or congruence). This has also implications for discussing the relations between nature and man. I cannot pursue this topic further here, but Götschl (1991), for one, has given an interesting account of what the man-nature relation from this point of view means and which philosophical consequences it implies[2].

Second, speaking of coherence leaves us with the image of "active organisms" and the assertion that any organism's view of reality rests on *success in life*. (See also Oeser, 1987, for similar reflections). Take, for example, a cat hunting a mouse. We have no reason to believe that the cat perceives the mouse in the same way as we do.

For the cat is what we call "mouse" simply a prey and all objects that resemble what we call "mouse" have from the point of view of the cat the same purpose. What is now the "real mouse"? Is it the "mouse" that a human being perceives? Or is it the cat's "mouse"? Well, these questions do not make much sense. Dennett (1991) draws our attention to color vision and to the fact that many humans are red-green color-blind. "Red" and "green" are properties that for those people are not the same as for other people. Moreover, we should keep in mind that many animals are lacking (lacking?) color vision; among them are dogs and cats. Is now the world perceived by a normal (not color-blind) human being truer than the world perceived by dogs or cats? Again, the question makes no sense.

The point is this. There are different types of eyes or, more generally, photoreceptors, and therefore different ways to perceive different objects of the world. In fact, what, for example, the statement "this is a *red* rose" – a typical statement at the level of our common-sense realism – means when translated into a more critical realist position is the following:

– There are *real* objects, called roses;
– and this is one of its representatives;
– which triggers a stimulus in the retina of our eyes;
– and this stimulus causes a special sensation which we call red (or, depending on the language we use, rouge, rosso, rot, and so on and so forth).

In our everyday life it is sufficient to say "this is a red rose", and for the sake of effective communication we should continue using the parlance of naive realists which, however, does not imply anything about the very nature of things, e.g., roses. Certainly, in our everyday life as well as in science we presuppose that there is a real world. But as scientists (or philosophers) we know that the world is not necessarily what we perceive by the means of our sense organs. As Rescher (1986, p. 77) says: "We recognize, or at any rate have no alternative but to suppose, *that* reality exists, but we are not in a position to stake any final and definitive claims as to *what* is is like". We rather have to accept that our sensory perception gives only hints at the properties of the external world (cf. Rensch, 1968).

From an evolutionary point of view it is not necessary to suppose that the perceived objects in the brain of an organism are congruent with the objects in the outer world. Sjölander (1995) uses as a metaphor an orchestra where the players (as active organisms!) are playing according to given notes (inputs from the sense organs). Although the musicians pay attention to the notes they are not acting like puppets operated by the musicbook. Their playing is a self-governing activity; is is not created by the notes, the notes do not even give complete information about how to make music. Similarly, a dog hunting a rabbit does not give the rabbit the complete information necessary in order to know how to run away; its running is a self-governing activity, even though the appearance of the dog does provide the rabbit with important stimuli. The rabbit knows, in a way, that it has to react in a particular manner when a dog appears. Its running serves its own survival. To complete Sjölander's metaphor we can say that like any successful musician who is able to find his own style of playing music (and not just to read notes very well), any successful rabbit has found its ways to escape from dogs (and not just learned to recognize dogs and to obtain a "realistic perception" of them). (See the figure below).

My notion of functional realism therefore has two meanings that, however, are interrelated):

(1) It implies that humans (like other living beings) are able to identify different objects in their surroundings by the properties these objects seem to have or by the functions they fulfill for us (or other animals).

(2) It also implies that organism learn how to "function" when confronted with different objects in their surrounding, i.e., how to react to these objects in order to survive.

TWO POSTULATES FOR EVOLUTIONARY EPISTEMOLOGY

Critics of evolutionary epistemology have lamented that this type of epistemology has failed to understand "how much of what is 'out there' is the product of what is 'in here' (Lewontin, 1982, p. 169). This criticism, however, hits only the adaptationist version of evolutionary epistemology. As far as I can see, most of the advocates of this epistemology today would agree that "much of what is out there is indeed the product of what is in here". This is not because many evolutionary epistemologists have mutated and become constructivists or adherents of the autopoietic theory of cognition (cf. Maturana and

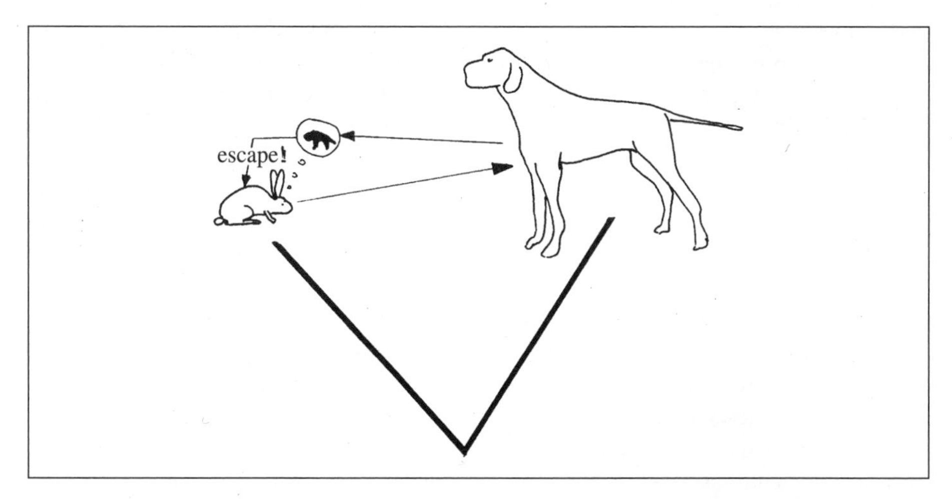

Simplified illustration showing a rabbit perceiving a dog. Rabbits and dogs are "connected" by their long evolutionary history, and billions of rabbits have made the experience that what looks like a "dog" is something dangerous, and that only running away helps. We of course do not know how a rabbit actually perceives a dog – for we do not know how, after all, it is to be a rabbit –, but even if the rabbit sees just a moving black spot it seems to "know" what to do. Perceiving a dog for a rabbit means that there is a feedback, a program in its brain, telling him to escape. Not much different is the situation in the case of humans when they are confronted with something dangerous.

Varela, 1980). It is so, simply because one can no longer ignore the importance of the organisms' activities, the organisms' capacities for calculating their own chances in the world. If we, as evolutionists, have really freed ourselves from the behaviourist doctrine and from the stimulus-response theory of cognition, then we have to abandon any view that suggests that animals are "impregnated" by the outer world and forced to learn reading the notes in the "musicbook of nature" – and to play according to the "notes given by their environment".

We can now put down postulates for evolutionary epistemology that might help developing a better approach to reality and the problem of realism.

First Postulate: We presuppose "reality" independent of any perceiving apparatus, but we have to be aware that any organism's perception reveals only parts of this "reality" – and that it reveals them in a specific manner, depending on the specific requirements of the respective organism. An important task for evolutionary epistemology is to reconstruct the evolutionary conditions for the development of different sense organs, brains and nervous systems. This is not a philosophical task, but a problem for empirical disciplines like physiology, ethology, evolutionary biology, and the neurosciences. Organisms can be regarded as problem solvers (Popper, 1963, 1972), however, "the kind of problems with which organisms are confronted... and their relative significance, varies from one species to another" (Hofman, 1988, p. 437). Evolutionary epistemologists should therefore be interested in how organisms solve their problems and, by doing so, how they act, which "world picture" they have developed.

Second Postulate: Physiological, ethological, and neurobiological data furnish ample evidence that different species act by means of their own constructs of (parts of) "reality" and process information according to what they need in order to survive. No organism has ever been forced to recognize reality in itself. Perception, one can argue, is initiated by the organism as an interaction with the environment (cf. Skarda, 1989). Only humans are attracted by the idea of a world in itself, but this idea can be abandoned for biological (evolutionary) reasons. Our everyday notion of "reality", even if it is just the expression of a naive realism, should be taken seriously by the evolutionary epistemologist. But evolutionary epistemologists should also pay attention and take seriously the fact that by means of science we can go far beyond our biologically constrained world view and transgress, as it were, our mesocosm.

The position of functional realism is a pragmatic one. It implies the idea that there is no Archimedian point, and it confirms Putnam's statement that "a use of 'exist' inherent in the world itself, from which the question 'How many objects *really* exist?' makes sense, is an illusion" (Putnam, 1987, p. 20). But doesn't this type of realism, if taken as an ontological position, necessarily lead to a kind of *relativism?* To a certain extent it does. Thus, for example, in our everyday life we may believe that the earth is a disk, and we will survive with such a belief as long as we move on the earth. (After all, for many millenia people *have* survived with this belief). But this does not mean that by this belief we have obtained the ultimate truth. Modern physics teaches us that the earth is a sphere. This may contradict our everyday experiences but nevertheless it is true. Therefore, relativism has also its (more or less clear) limits.

CONCLUDING REMARKS

I can imagine a variety of questions that arise from what I call functional realism, questions of epistemological and ontological relevance. But let me stress that, in the first instance, my notion of functional realism is basically a biological one rooted in the biologically trivial observation that all organisms have to "function" in a specific manner in order to survive. This functioning includes a coherent scheme of their environment(s) which is developed by sensory perception and information processing in the nervous system or similar organs or organelles (in the case of unicellular living beings). Yet it is clear that my position can challenge also the philosopher who is interested in "truth" or, better, in the question how we humans can – and whether we can at all - obtain the "truth about this world".

It seems important to concentrate on what organisms – including humans – actually are perceiving and how they are perceiving what appears to them as "reality". In this sense, David Hume's *Treatise of Human Nature* (1739 [1972]) can be considered as a forerunner of a modern version of evolutionary epistemology, for Hume had indeed troubles with the belief in a world beyond us (the world in itself). So had Williams James, the pragmatist, who accurately stated the following: "To 'agree' in the widest sense with a reality *can only mean to be guided either straight up to it or into its surroundings, or to be put into such working touch with it as to handle either it or something connected with it better than if we disagreed"* (James, 1908, p. 212-13). And with regard to truth, James (1908, p. 222) consequently stated: "'*The true'... is only the expedient in the way of our thinking, just as 'the right' is only the expedient in the way of our behaving*".

This, from the point of view of idealist philosophy, contemptuous position[3] includes negative statements about:

– the notion of idea in the sense of Plato (*essentialism* [cf. Popper, 1963]);
– notions of the "absolute" ("absolute" knowledge, "absolute" truth, etc.);
– the belief in the unknowable, and
– the belief in absolute values[4].

Therefore, as I already stated, one can be afraid that such a pragmatic position inevitably leads to relativism. This is certainly a serious point, for, after all, one might feel that "reality" will disappear, so to speak. It is a remarkable characteristic of humans that they are looking for reliability, for the unchangeable (not to say for the "absolute"), for certainty; that they like to regard a red rose as a *red* rose, a perceived tree as a *real* tree, and so on and so forth. This characteristic is a consequence of evolution: Our philogenetic ancestors and predecessors survived only because they could indeed rely on some regularities in their environment (not knowing, however, that they themselves were producing, by means of their perceiving apparatus, many of the regularities and harmonious features of the external world). Our everyday cognition and common-sense realism are not different, and therefore problems of "clearness" (*Anschaulichkeit*) arise whenever we are confronted with phenomena that do not match the expectations anchored in our perceiving apparatus (cf. Vollmer, 1982). So, what is left as "reality"?

To put it briefly: everything that makes any sense (for us or for other organisms)!

"Sense" can be found at the level of common-sense knowledge as well as at the level of scientific knowledge which goes far beyond our mesocosm. The position of functional realism does not plunge us into a thinking "without any constraints on knowledge" or into a world of subjectivism (or radical constructivism). For all those who are afraid of such a world where anything goes – everything and nothing is true - Ruse (1989, p. 220) found clear words: "We still have the real world, but it is the world as we interpret it. What is being rejected is not reality in any meaningful sense. No one is saying, for instance, that dinosaurs did not exist, or that if you see a fierce tiger, you can simply put your hand through it and wish it out of existence. It is simply to acknowledge that reality and thinking about it are inseparable and that the belief in something beyond this is meaningless and redundant".

This statement should provide some comfort to all those who have already found themselves in a world of cognitive chaos. I think that it does not require any further comment. Obviously, evolution has equipped us with "programs for perception" (Young, 1987) that allow us, in most cases, to see, hear, smell or taste what is of some significance for our survival, and that even permit us to make some necessary forecasts for living. And this is not nothing! In what aspect could the notion of things in themselves be of some additional help?!

Prof. Franz M. Wuketits
Institut für Wissenschaftstheorie
University of Vienna
Sesengasse 8/10
A-1090 Wien, Austria

NOTES

[1] This is what Karl Popper was stressing practically all the time in many of his books and papers (see, e. g., Popper, 1959, 1963). In his later work he developed a view of evolution and epistemology that closely resembles what I characterize here as the non-adaptationist view and what can be also characterized as "active Darwinism" (see also Perutz, 1986). This means that this view is still Darwinistic, but that it is a somehow modified version of the Darwinian (adaptationist) paradigm.

[2] Götschl discusses this topic in terms of a "hypercritical physical realism". His fundament is the paradigm of self-organization, his claim the categorial change in subject-object relations. Similarly, Carsetti (1989) discusses the relevance of complexity theory for models of cognition.

[3] I am aware that realism is – or can be – connected with idealism in different ways, and that idealism in philosophy and its history has many faces (see e. g. Ewing, 1961, for a critical survey).

[4] This is an issue in ethics, and therefore I separate it from "absolute" knowledge or truth, etc., although all these notions are interrelated. But "realism and morality" would be a special topic I cannot pursue further here.

REFERENCES

Carsetti, A., "Teoria della complessità e modelli della conoscenza", "La Nuova Critica" (Nuova Serie) 9-10, 3°, 1989, 61-103.

Clark A. J., "Evolutionary Epistemology and the Scientific Method", "Philosophica 37", 1986, 151-162.

Dennett, D., Consciousness Explained, Little, Brown and Company, Boston/Toronto/London 1991.

Diettrich, O., Kognitive, organische und gesellschaftliche Evolution, Parey, Berlin/Hamburg 1989.
–, "Heisenberg and Gödel in the Light of Constructive Evolutionary Epistemology", "Ludus Vitalis 2" (2), 1984, 119,131.
Eddington, A., The Nature of the Physical World, University of Michigan Press, Ann Arbor 1958.
Elitzur, A. C., "Let There Be Life: Thermodynamic Reflections on biogenesis and Evolution", "J. Theor. Biol." 168, 1994, 429-459.
Ewing, A. C., Idealism: A Critical Survey, Methuen, London 1961.
Falk, R., "Evolutionary Epistemology: What Phenotype is Selected and Which Genotype Evolves?", "Biol. & Philos." 8, 1993, 153-172.
Götschl, J., "Hypercritical Physical Realism and the Categorial Changes in the Subject-Object Relations", "La Nuova Critica (Nuova Serie)" 17-18, 2°, 1991, 5-19.
Hofman, M. A., "Brain, Mind and Reality: An Evolutionary Approach to Biological Intelligence", in: H. J. Jerison and I. Jerison (eds.) Intelligence and Evolutionary Biology, Springer, Berling/Heidelberg 1988, pp. 437-446.
Hume, D., A Treatise of Human Nature (1739), Collin, London 1972.
James, W., Pragmatism: A New Name for Some Old Ways of Thinking, Longmans, Green, and Co., New York/Bombay/Calcutta 1908.
Lewontin, R. C., "Organism and Environment", in: H. C. Plotkin (ed.), Learning, Development, and Culture: Essays in Evolutionary Epistemology, Wiley, Chichester/New York 1982, pp. 151-170.
Lorenz, K., "Kants Lehre vom Apriorischen im Lichte gegenwärtiger Biologie", "Blätter für Deutsche Philosophie 15", 1941, 94-125 (English translation in H. C. Plotkin, op. cit., pp. 121-143).
–, Behind the Mirror: A Search for a Natural History of Human Knowledge, Methuen & Co., London 1977.
Maturana, H. and Varela, F., Autopoiesis and Cognition: The Realization of the Living, Reidel, Dordrecht/Boston/London 1980.
Oeser, E., Psychozoikum: Evolution und Mechanismus der menschlichen Erkenntnisfähigkeit, Parey, Berlin/Hamburg 1987.
Perutz, M. "New View of Darwinism", "New Scientist" 1528, October 1986, 36-38.
Popper, K. R., The Logic of Scientific Discovery, Hutchinson, London 1959.
–, Objective Knowledge: An Evolutionary Approach, Clarendon Press, Oxford.
Putnam, H., The Many Faces of Realism, Open Court press, La Salle 1987.
Rensch, B., Biophilosophie auf erkenntnistheoretischer Grundlage, Fischer, Stuttghart 1968.
Rescher, N., "Reality and Realism", "Proceedings of the 10th International Wittgenstein Symposium", Hölder-Pichler-Tempsky, Vienna 1986, pp. 75-85.
Riedl, R., Biologie der Erkenntnis: Die stammesgeschichtlichen Grundlagen der Vernunft, Parey, Berlin/Hamburg 1980.
–, "Evolution and Evolutionary Knowledge: On the Correspondence Between Cognitive Order and Nature", in: F. M. Wuketits (ed.), Concepts and Approaches in Evolutionary Epistemology: Towards and Evolutionary Theory of Knowledge, Reidel, Dordrecht/Boston/Lancaster 1984, pp. 35-50.
Ruse, M., "The View from Somewhere: A Critical Defense of Evolutionary Epistemology", in: K. Hahlweg and C. A. Hooker (eds.), Issues in Evolutionary Epistemology, State University of New York Press, Albany NY 1989, pp. 185-228.
Simpson, G. G., This View of Life: The World of an Evolutionist, Harcourt, Brace and World, New York 1963.
Sjölander, S., "Some Cognitive Breakthroughs in the Evolution of Cognition and Consciousness and their Impact on the Biology of Language", "Evol. Cogn. (New Series)" 1, 1995, 3-11.
Skarda, CH. A., "Understanding Perception: Self-Organizing Neural Dynamics", "La Nuova Critica (Nuova Serie)" 9-10, 3°, 1989, 49-60.
Vollmer, G., "Probleme der Anschaulichkeit", "Philos. Nat." 19, 1982, 277-314.
–, "Mesocosm and Objective Knowledge – On Problems Solved by Evolutionary Epistemology", in: F. M. Wuketits, op. cit., pp. 69-121.
Weiss, P. A., "The Living System: Determinism Stratified", "Stud. Gen. 22", 1969, 361-400.
Wuketits, F. M., "Evolutionary Epistemology – A Challenge to Science and Philosophy", in: F. M. Wuketits, op. cit., pp. 1-33.
–, "Evolution as a Cognition Process: Towards and Evolutionary Epistemology", "Biol. & Philos." 1, 1986, 191-206.
–, "Evolution und Adaptation von Erkenntnisprozessen", "Biol. Rdsch. 25", 1987, 333-341.
–, "Cognition: A Non-Adaptationist View", "La Nuova Critica (Nuova Serie)" 9-10, 3°, 1989, 5-15.
–, Evolutionary Epistemology and Its Implications for Humankind, State University of New York Press, Albany NY 1990.

–, “Self-Organization, Constructivism, and Reality”, “La Nuova Critica (Nuova Serie)” 17-18, 2°, 1991, 21-36.
–, “Adattamento, Rappresentazione e Costruzione: Un Saggio di Epistemologia Evolutiva”, in: M. Ceruti (ed.), Evoluzione e Conoscenza, Pierluigi Lubrina Editore, Bergamo 1992, pp. 121-133.
–, “A Comment on Some Recent Arguments in Evolutionary Epistemology – and Some Counterarguments”, “Biol. & Philos”. 10, 1995, 357-363.
Young, J. Z., Philosophy and the Brain, Oxford University Press, Oxford/New York 1987.

DOMENICO PARISI

COMPLEXITY AND COGNITION

1. INTRODUCTION

How are complexity and cognition related? In this paper I will examine two aspects of their relationship. I will first ask in what sense (human) cognition can be considered as an example of a complex system. Then I will examine how human cognition reacts to complexity, how much we can understand systems that should be described as complex.

2. HUMAN COGNITION AS A COMPLEX SYSTEM

2.1 SIMPLE AND COMPLEX SYSTEMS

I will begin by clarifying the distinction between simple and complex systems. A system is simple if the system's behavior results from rules (laws, principles) that we are able to identify and formulate and that allow us to predict the system's behavior. An example of a simple system is a billiard ball and the rule: "Following the impact of another ball that moves with speed X, comes from direction Y, and has an impact angle W, a billiard ball will displace itself in direction Z with speed J". On the basis of rules of this type we are able to predict with good approximation the behavior of the ball and to explain or to identify the causes of the behavior of the ball. The billiard ball is a simple system.

A complex system is constituted by a very large number of elements. Each element interacts locally with a limited number of other elements on the basis of rules that we can identify and express. However, the system has global properties and behaviors that, although they result from the local interactions among the system's elements, we are unable to predict, to infer from the rules governing local interactions, and to explain by identifying in each particular instance their causes.

An example of a complex system is an artificial neural network (or a nervous system that is modeled by the neural network) (Hopfield, 1982; Rumelhart & McClelland, 1986; Parisi, 1989). A neural network is a (large) set of elements called units (neurons) linked together by connections (synaptic junctions between neurons). Each unit interacts with the other units with which it is connected in such a way that the unit's activation level is determined, via known rules, by the activation level of the other units and by the "weights" of the connections. The ensemble of the interactions among the different units determines the behavior of the network, i.e., how the network responds to external stimulation. The network's behavior is a global property of the system "neural network" which results from the many local interactions among the network's units. Given the high nonlinearity of the system, we are unable in general to infer or to predict the network's

A. Carsetti (ed.), Functional Models of Cognition, 39-49.

global behavior from a knowledge of the local interaction rules or to identify in each particular case a cause or a restricted number of causes that determine the network's response. The neural network is a complex system.

Various features distinguish complex from simple systems. One feature is how the two types of systems respond to external perturbations. A simple system responds to external perturbations in ways that are in general predictable and understandable. Another difference concerns sensitivity to initial conditions. If two simple systems have different initial conditions, their behavior in time will reflect linearly the difference in initial conditions in such a way that it will be possible to predict from the difference in initial conditions their different subsequent history. On the contrary, two complex systems with different initial conditions will be likely to have a history that reflects unpredictably their initial difference. For example, a small initial difference can translate into an increasingly large difference in subsequent history while a large initial difference can remain hidden for a long time and even for ever. Finally, complex systems are not only dynamical systems, that is, systems that change their state on the basis of fixed local interaction rules, but they can be evolutionary or adaptive systems, that is, systems that change their local interaction rules depending on the external environment - which cannot but increase the unpredictability of their future states.

All these differences between simple and complex systems are amplified if, as often is the case, a complex systems is hierarchically organized, i.e., it is constituted by a hierarchy of systems. A set of units of the first level system determines the behavior of a single unit of the second level system. In turn a set of units of the second level system determines the behavior of a single unit at the third level, and so on. Complex systems with this type of hierarchical organization are even less predictable, more sensitive to initial conditions, more capable of adaptation, with a more pronounced tendency to hide change at one level for a long time to make it emerge suddenly at a higher level at a later time (punctuated equilibria) (Miglino, Nolfi & Parisi, 1996).

2.2 THE GROWTH OF COMPLEX SYSTEMS

If we look at the phenomena studied by the various scientific disciplines, we observe that the proportion of phenomena that are complex systems becomes larger and larger as we go from physics to chemistry to biology to psychology to the social sciences. In physics there are many systems that are simple systems. In chemistry the proportion is smaller. In biology a large number of the phenomena studied are complex systems. In the behavioral sciences and especially in the social sciences practically everything is complex systems.

This progression of complex systems in the scientific disciplines has an interesting historical-temporal component. If we go backward in time (say, before the origin of life on Earth) the only phenomena that were in existence at that time were phenomena that could be studied by a physicist. There were no phenomena at that time that could have been of interest to a macromolecule chemist or to a biologist, let alone to a behavioral or social scientist. To find phenomena that could be studied by organic chemists or biologists it is necessary to move to more recent times, after the appearence of life on Earth. Behavioral scientists could have found phenomena of interest to them only after the much more recent appearence of organisms endowed with a nervous system. Social scientists (anthropologists, sociologists, economists, political scientists, historians) would have found a job only in the last tens of thousands of years or even more recently (Parisi, 1994).

It seems then that reality, as we know it through science, is subject to a general evolutionary trend that leads from an initial prevalence of simple systems to the progressive emergence and spread of complex systems. We are forced to recognize empirically that more and more parts of reality require us to study them as complex systems, and as complex systems that, since they are increasingly organized hierarchically, become more and more complex. Physical phenomena that are simple systems and that were already in existence before life continue to exist today and they can be studied as simple systems by today's physicists. However, new classes of phenomena have emerged in successive stages: macromolecules and then living systems and then behavioral systems and then specifically human social and technological systems.

The growth of complexity can be viewed as a growth in organization. The anti-entropic meaning of this process is of great interest. What must be asked is not what is the source of energy for this process (the answer is rather obviously the sun) but what are the detailed mechanisms that have been able to translate the energy of the sun into this process of complexification and organization.

Physics has made enormous progresses in the last 3-4 centuries by studying simple processes. It is only at the end of the last century, with thermodynamics, statistical mechanics, Poincaré's three-body problem, etc., that problems concerning complexity in physical phenomena have begun to emerge. But these problems have become an important and recognized area of research in physics only in the last decades. Even if experimental physicists are sometimes confronted with complex behaviors, for example in the physics of materials, many research areas in physics, including those that are recognized as paradigmatic of the discipline such as particle physics, are distant enough from problems of complexity. However, it must be recognized that physicists along with mathematicians have elaborated most of the concepts currently used in the study of complex systems (Hopfield, 1982; Mèzard, G. Parisi & Virasoro, 1987; G. Parisi, 1992). (But theoretical biologists such as Stuart Kauffman (Kauffman, 1993) and computer scientists such as John Holland (Holland, 1975) have also contributed in important ways to the field).

2.3 HUMAN COGNITION AS A COMPLEX SYSTEM

Human cognition emerges in a late or advanced stage of this long evolutionary process. We can properly speak of behavior and cognition only with the emergence of organisms with a nervous systems. Nervous systems can be modeled by using artificial neural networks that, as we have seen, are complex systems. Nervous systems actually are only one single level in a hierarchical organization that includes at least, at a lower level, the genotype and, at a higher level, the whole organism's interactions with the environment. The interactions among the parts of the genotype and among these parts and their products during the course of development determine the phenotypical nervous system, that is, the neural network's units and connections. The interactions among these units determine the behavior of the whole organism in the environment. The interactions of the organism with the environment, including other organisms, determine such global phenomena as the reproductive success of the individual, the properties of the population or social group of which the individual is a member, and the changes in these properties at various time scales.

Science is currently developing various tools to model and simulate on a computer the various levels of this hierarchical organization and the interactions between levels. Neural networks are used to model the nervous system and the resulting behavior (Amit, 1989; Rumelhart & McClelland, 1986). Genetic algorithms are used to model genotypes, the genotype-to-phenotype mapping (development), and the evolutionary processes at the population level that are caused by selective reproduction and the constant addition of variability through genetic mutations and sexual recombination (Holland, 1975; Mitchell & Forest, 1994; Nolfi & Parisi, 1995). Ecological neural networks are used to model the interactions between the organisms and the environment in which they live (Parisi, Cecconi & Nolfi, 1990).

The phenomena we have described concern all organisms or, more precisely, all animals endowed with a nervous system. In this sense, they are phenomena that belong to the biological or behavioral sciences. However, humans are an animal species exhibiting phenomena that can be called post-biological. When we say that humans are characterized by post-biological phenomena (more or less, the phenomena that are currently studied by the social sciences) what we mean is that (a) these post-biological phenomena have appeared at a later time than biological phenomena, and (b) they cannot be studied entirely within the limits of the biological sciences. (In this same sense one could say that biological phenomena are post-physical phenomena).

The specifically human, or post-biological, level of human cognition has various aspects that are very hard to study due to their intrinsic difficulty and to the limitations of the scientific disciplines that study them, i.e., (human) psychology and the social sciences. From a purely neural and behavioral point of view, human nervous systems have more complexity and organization because they are bigger and they have more complex internal circuits. The neural network of a simple organism can be viewed as a system that encodes the state of the local external environment with its input units, elaborates this input with its internal units, and responds by acting on the external environment with its output units. In human beings (and to some extent also in animals that are close to us such as the other primates) the internal elaborations become more complex (also due to the increase in brain size) and there is the emergence of self-feeding circuits: the network generates its own inputs.

But human cognition is complex not only neurally but because of some new developments that are specifically human. By acting on the external environment an organism to some extent determines its own input. (This is the most important phenomenon which is captured by the notion of an "ecological" neural network. See above). The organism can change its physical relationship with the external environment (by moving its body or body parts or its sensory organs) or it can change the external environment itself (by displacing objects, modifying their shape and other properties, creating new objects and structures). This has as its consequence that the inputs arriving from the environment to the organism are changed. Now, all animals do the first type of thing, that is, they change their physical relation to the environment, but only human beings have based their adaptation strategy on the modification of the external environment. All animals to some extent modify the external environment as a by-product of their behavior, for example they introduce carbon dioxide into the environment by breathing and they remove food from the environment by eating. But the modifications of the external environment caused by human beings are extremely more significant and, more to the point, they are adaptive modifications, that is, modifications that are caused because they increase the survival and reproductive chances in the modified environment. In other animals this occurs, marginally, for example in the construction of nests and spider webs (Collias & Collias, 1976).

This typically human strategy of adaptive modification of the external environment creates a new causal circuit. It is not only the external environment that has effects on the organism, but also the organism has effects on the external environment. This double circuit creates further complexity in that it causes the reciprocal interaction between two already complex systems: humans and the external environment constantly modified by them. The modifications caused by humans in the external environment are of various kinds. They are technological (creation of artifacts), social (creations of social institutions), and cultural (extra-corporeal transmission of behaviors). The various systems of artifacts and social institutions are themselves complex evolving systems (technological evolution, social evolution, economic evolution, cultural evolution) and human cognition is the cognition of organisms that live and interact with an environment containing these complex evolving systems. It is not surprising that the study of humans turns out to be rather difficult and that the sciences that are concerned with human-level phenomena are less advanced than the sciences of pre-human nature.

However, in this case too science is developing new tools that will allow it to deal with the complex systems created by humans more adequately. One sees the first models of the economy as an "evolving complex system" (Anderson, Arrow, & Pines, 1988), there are attempts to model cultural evolution (Hutchins & Hazelhurst, 1995; Denaro & Parisi, 1996) and to simulate social phenomena as emerging from the interactions among individuals (Castelfranchi & Werner, 1994; Gilbert & Conte, 1995; Conte & Chattoe, 1997; Parisi, 1997; Pedone & Parisi, 1997).

We can conclude this first part by saying that human cognition emerges as a complex system, or rather as a system made up of many reciprocally interacting complex systems. We often hear that the human brain is the most complex existing system and that it represents the next (ultimate) frontier of science. But human cognition is not entirely inside the brain. It is a larger system that includes many brains of organisms interacting with each other and with the products of their behavior in the external environment. Science is beginning to work out the theoretical and methodological tools (computer simulations) to deal with all this complexity. (To the role of the computer we will return in the second part). But the task appears to be formidable.

3. HUMAN COGNITION FACES COMPLEX SYSTEMS

3.1 LEARNING TO PREDICT THE CONSEQUENCES OF ONE'S OWN ACTIONS

In this second part I will discuss how human cognition reacts to complex systems. If reality is composed of both simple and complex systems, is human cognition as a knowing and understanding tool equally capable to understand simple as well as complex systems? The answer is No. The human mind appears to be tailored for the understanding of simple systems but it has enormous difficulty understanding (and, one should add, accept) complex systems. The computer, however, could play a critical role in making complex systems more understandable and more acceptable to human beings.

Let us come back for a moment to the adaptive strategy that characterizes rather uniquely human beings within the animal world. We can define this strategy as an active adaptation strategy. An adaptive strategy can be called passive when an animal species adapts to the environment as it is. The organisms modify themselves in order to increase their survival and reproductive chances in the environment as it is. An adaptive strategy can be called active, on the other hand, if it consists in modifying the external environment in order to increase one's survival and reproductive chances in the modified environment (Pedone & Parisi, 1997). All nonhuman animals adopt a passive adaptive strategy. Human beings are characterized by an active adaptation strategy. They radically and diffusely modify the external environment in which they live in order to make the environment more hospitable to them.

That this adaptive strategy is only found in a single animal species shows how special the pre-conditions necessary for this strategy to emerge and consolidate must be. We will consider only one of these pre-conditions, that is, the ability to predict the consequences of one's own actions.

As we have observed in the first part, when an organism that lives in an environment acts with its motor organs it causes some consequences that can be observed by the organism itself. For example, turning one's head or eyes or displacing oneself in space changes the visual input from the environment. In this case I do not modify the external environment itself, but I limit myself to changing the physical relationship between my body and the external environment. In other cases, however, I can modify with my motor actions the external environment itself. I can change the position of an object, or the object's properties, I can destroy or create objects. And of course I can observe the consequences of these actions of mine.

Now, a crucial capacity is the capacity to predict the consequences of my actions. The learning of this capacity can be simulated with a neural network. A neural network can receive as input a description of a planned motor act and a description of the current sensory input from the environment. The network's output is a prediction of what the next sensory input from the environment will be when the planned motor act will be actually (physically) executed. Neural networks can learn to make this type of predictions, both in the case that the organism's action is limited to changing the physical relation of the organism's body to the environment without actually modifying the external environment and in the case that the organism's action actually modifies the external environment. A neural network can learn this prediction ability by using the backpropagation procedure (Rumelhart, Hinton, & Williams, 1986). The network compares the predicted consequences of its motor act with the actual consequences that are observed when the motor act is physically executed, and it changes its connection weights in such a way that gradually predictions and observations tend to coincide and the network makes more and more correct predictions. What is interesting is that a neural network that has learned to predict the consequences of its own actions turns out to be more efficient in reaching its goals in the environment, i.e., in causing desired consequences with its actions. For example, a network that can predict the consequences of its actions with respect to an object in the environment is more able to approach the object or to reach the object with its "hand" or to throw the object to a desired location (Nolfi, Elman & Parisi, 1994; Cecconi & Parisi, 1990; Parisi & Cecconi, 1995).

This should not come as a surprise since a neural network that has learned to predict the consequences of its actions can be said to possess an internal model of the external environment. (To speak of a model or representation or theory of the external environment is only a metaphor. There are no explicit or symbolically expressed models or representations or theories in a network but only particular arrangements of the network's architecture and connection weights). A neural network that knows how to predict the consequences of its actions knows how the external environment, which includes the organism's body, reacts to these actions, and therefore it "knows" the external environment. This is where the roots of knowledge and cognition appear to lie.

As we know, human beings are specialized for doing actions that do not only change their relation to the external environment but change the external environment. This specialization includes an extremely sophisticated capacity to predict what modifications of the external environment will result from the various actions. In fact, it is likely that the ability to adaptively modify the external environment and the ability to predict the consequences of one's own actions on the external environment have co-evolved in the evolutionary history of *Homo sapiens.* It is because we are so good both at changing the external world and at predicting these changes that our knowledge of the world is so much better than that of the other animals.

The capacity to predict the consequences of our actions is at the root of our knowledge of the world and of science. What we call science is nothing but a systematization and amplification (by the use of instruments and quantification) of our prediction capacity. There are two types of predictions. We predict future states of the world on the basis of our knowledge of past and present states that are independent from our actions and we predict future states of the world that result from our actions. As is well known, science has made enormous progresses after the discovery of the experimental method. But the experimental method is nothing but the prediction of the consequences of our actions, plus the use of instruments and of quantification that make both our predictions and the actual observations that confirm or disconfirm our predictions more precise (cf. the "backpropagation" learning procedure above). In fact, our knowledge of the world is greatly extended if we learn not only to predict future states based on past states (this also can be simulated using neural networks; cf., e.g., Elman, 1990) but also to manipulate reality and learn to predict the changes resulting from our manipulations.

3.2 LEARNING TO PREDICT THE CONSEQUENCES OF OUR ACTIONS ON SIMPLE AND COMPLEX SYSTEMS

However, while simple systems can be brought into the experimental laboratory, complex systems cannot. More generally, if our knowledge of reality is based on our capacity to predict the consequences of our actions, it is not surprising that our knowledge of simple systems is much more advanced than our knowledge of complex systems, and we are more able to understand simple than complex systems. As we have seen, simple systems are predictable. Their behavior is the result of a single or a few causes. They respond to initial conditions and to external perturbations linearly and predictably. The behavior is the highly nonlinear result of a large number of causes, that is, of the many local interactions between elements. They respond to initial conditions and external perturbations nonlinearly and unpredictably.

When we manipulate a system, we change the system internally or we perturbate the system from outside. In the case of a simple system, for example our billiard ball, this can mean changing the physical matter of the ball or the angle of impact of another ball. In this case we are sufficiently able to predict how the ball will react. Complex systems also can be manipulated. We can change the system internally, for example we can modify the local interaction rules or we can change some of the system's elements, or we can perturbate the system from outside. In a neural network this can mean changing the rule that maps the algebraic sum of excitations and inhibitions arriving to a unit into the unit's activation level, or "lesioning" a unit (a neuron, as neurophysiologists do), or exposing the network to some particular input. But, unlike the billiard ball, these manipulations can have consequences that we are unable to predict at the global level of the network's behavior.

It is because the most important cognitive activity involved in laboratory experiments is predicting the consequences of our manipulations, that simple systems can be brought into the laboratory but complex systems cannot. In the laboratory we can observe phenomena in controlled conditions. The problem is that with simple systems we know what variables to control while with complex systems we don't. Experimental biologists (geneticists, molecular biologists, neuroscientists) cling to their experimental laboratory because the laboratory allows them to identify at least the local interaction rules of the complex systems they study. But to understand how the local interaction rules result in the global behaviors of these complex systems, the laboratory is not very helpful. Psychologists and social scientists are in greater trouble because in the laboratory they often cannot even identify the local interaction rules.

3.3 COMPUTER SIMULATION

If the experimental laboratory method is appropriate for simple systems but not for complex systems, what is appropriate for complex systems? The answer is: computer simulations. The traditional tools of science, the experimental method and theories expressed as systems of equations, do not work very well with complex systems. What does work is computer simulations. We can simulate a complex system and the rules that govern the local interactions between elements, and we can observe what happens at the global level. In a certain sense the observation of a simulated complex system replace the predictions. If we are able to construct a simulated system that essentially behaves as a real complex system, then we can say we have understood the real complex system even if we may be unable to predict the behavior of both the real system and the simulated system. In a computer we can simulate very complex systems, hierarchically organized complex systems, interactions of these systems with their environment, the changes in their local interaction rules with time, and so on. And we can manipulate in all possible ways the simulations we have constructed, certainly with many more degrees of freedom than is the case with real systems, in the laboratory or in the field.

Will this be sufficient to understand complex systems? It is too early to say. The analysis of phenomena of reality as complex systems is just beginning. We have still little experience, in many disciplines, with the method of computer simulation. Furthermore, there is a lot of resistance against a science of complex systems and the adoption of computer simulation as a research methodology. We have got used to simple systems and to traditional ways of doing science. And, perhaps, we are afraid to admit that so much of reality is complex systems. But even when these resistances will be conquered and we will have more experience with theories of complex systems and with computer simulations, it is an open question if we will understand complex systems as we like to think we understand simple systems. With complex systems it is not clear that we will be ever able to know and to tell other people "why things happen as they happen" (Frova, 1995).

But we will have to adapt and perhaps to change our vision of reality and our cognition of reality. What is even more important, we will have to change our vision of ourselves as agents that can obtain desired results by acting on reality. Cognition and science have practical roots. We make predictions about the consequences of our actions so that we become better able to decide, when it is needed, what action to choose to obtain some desired consequence. If complex systems are not very penetrable to our predictive abilities, they are also not very penetrable to our ability to obtain desired consequences by acting on them. If by acting in a certain way on a complex system, I am unable to predict what will happen, I am also unable to decide what to do with respect to the complex system to obtain some desired consequence. Things will probably improve when we will have better theories of complex systems and better simulations that will allow us to study complex systems in the computer. But it is very unlikely that our vision of the unlimited human power of knowledge and action will remain unchanged.

Prof. Domenico Parisi
Istituto di Psicologia del CNR
V. K. Marx
00137 Roma, Italy

NOTE

I thank Filippo Menczer with whom I have discussed the first part of this paper.

REFERENCES

Amit, D.J. Modeling brain functions. Cambridge, Cambridge University Press, 1989.

Anderson, P.W., Arrow, K. & Pines, D. The economy as an evolving complex system. Reading, Mass., Addison-Wesley, 1988.

Castelfranchi, C. & Werner, E. Artificial social systems. New York, Springer, 1994.

Cecconi, F. & Parisi, D. "Learning to predict the consequences of one's own actions". In E. Eckmiller, G. Hartmann, & G. Hauske (eds.) Parallel processing in neural systems and computers. Amsterdam, Elsevier, 1990.

Collias N.E. & Collias, E.C. (eds.) External construction by animals. Stroudsburg, Penn., Dowden Hutchinson and Ross, 1976.

Conte, R. & Chattoe, E. (eds.) Evolving societies. The computational study of societal complexity. London, UCL, in press.

Denaro, D. & Parisi, D. "Cultural evolution in a population of neural networks". Paper submitted to Annual Meeting of Cognitive Science Society, San Diego, 1996.

Elman, J.L. "Finding structure in time". "Cognitive Science", 1990, 14, 179-211.

Frova, A. Perché accade ciò che accade, (Why things happen as they happen). Milano, Rizzoli, 1995.

Gilbert, N. & Conte, R. Artificial societies. The computer simulation of social life. London, UCL, 1995.

Holland, J.H. Adaptation in natural and artificial systems. Ann Arbor, Mich., University of Michigan Press, 1975 (also Cambridge, Mass., Mit Press, 1992).

Hopfield, J.J. "Neural systems and physical systems with emergent collective computational abilities". "Proceedings of the National Academy of Sciences", Usa, 79, 2554-2558.

Hutchins, E. & Hazelhurst, B. "How to invent a lexicon. The development of shared symbols in interaction". In Gilbert, N. & Conte R. Artificial societies. The computer simulation of social life. London, Ucl, 1995.

Kauffman, S. Origins of order: self-selection and organization in evolution. Oxford, Oxford University Press, 1993.

Mezard, M., Parisi, G. & Virasoro, M. Spin glass theory and beyond. Singapore, World Scientific, 1987.

Miglino, O., Nolfi, S. & Parisi, D. "Discontinuity in evolution: how different levels of organization imply pre-adaptation". In R.K. Belew & M. Mitchell (eds.) Plastic individuals in evolving populations. Reading, Mass., Addison-Wesley, 1996.

Mitchell, M. & Forest, S. "Genetic algorithms and artificial life". "Artificial Life", 1994, 1, 167-290.

Nolfi, S. & Parisi, D. "Genotypes for neural networks". In M.A. Arbib (ed.) Handbook of brain theory and neural networks. Cambridge, Mass., Mit Press, 1995.

Nolfi, S., Elman, J.L. & Parisi, D. "Learning and evolution in neural networks". "Adaptive Behavior", 1994, 3, 5-28.

Parisi, D. “Science as history”. “Social Sciences Information/Information sur les Sciences Sociales”. 1994, 33, 621,647.

Parisi, D. “What to do with a surplus”. In R. Conte, R. Hegselmann & P. Terna (eds.) Simulating social phenomena. Berlin, Springer, 1997.

Parisi, D. & Cecconi, F. “Learning in the active mode”. In F. Moran, A Moreno, J.J. Merelo, & P. Chacon (eds.) Advances in artificial life. New York, Springer, 1995.

Parisi, D. Cecconi, F. & Nolfi, S. “Econets: neural networks that learn in an environment”. “Network”, 1990, 1, 149-168.

Parisi G. “La nuova fisica statistica e la biologia. (The new statistic physics and biology)”. “Sistemi Intelligenti”, 1992, 4, 247-262.

Pedone, R. & Parisi, D. “In what kinds of social grants can “altruistic” behaviors evolve?” In R. Conte, R. Hegselmann & P. Terna (eds.) Simulating social phenomena. Berlin, Springer, 1997.

Rumelhart, D.E. & McClelland, J.L. (eds.) Parallel distributed processing. Explorations in the microstructure of cognition. Volume 1; Foundations. Cambridge, Mass., Mit Press, 1986.

Rumelhart, D.E., Hinton, G.E. & Williams, R.J. “Learning internal representations through backpropagation”. In D.E. Rumelhart & J.L. McClelland (eds.) Parallel distributed processing. Explorations in the microstructure of cognition. Volume 1: Foundations. Cambridge, Mass., Mit Press, 1986.

HERMANN HAKEN

FROM VISUAL PERCEPTION TO DECISION MAKING: A SYNERGETIC APPROACH

1. WHAT IS SYNERGETICS ABOUT?

Since we do not require that the reader is familiar with synergetics, we shall give a brief outline of this field first. Synergetics is an interdisciplinary field of research [1]-[3], which deals with complex systems. Such systems are composed of many parts that interact with each other. We shall be concerned with those systems that may form spatial, temporal, or functional structures spontaneously, i.e. these structures are developed by the systems themselves without any specific interference from the outside. Such structures are, of course, formed when plants or animals are developing, but they may be even found in the inanimate world of physics and chemistry, for instance by the formation of specific structures in fluids. Synergetics asks whether there are general principles that govern the formation of structures. To this end, it has developed a specific strategy, namely to look at those situations in which qualitative changes occur. A typical and instructive example for such a qualitative change is the freezing of water. Though water and ice are composed of the same constituents, namely water molecules, at the macroscopic level water and ice have quite different mechanical and optical properties. On the other hand, this example does not cover all important aspects of systems treated in synergetics, because ice is, so-to-speak, a dead system, whose state is not maintained by any influx of energy. Rather synergetics deals with so-called open systems whose states are maintained by a continuous influx of energy and/or matter. In physics pattern formation occurs in fluids once they are heated from below. Such a fluid may form structures in the form of rolls, hexagons, or spirals, or still more complicated patterns (Fig. 1).

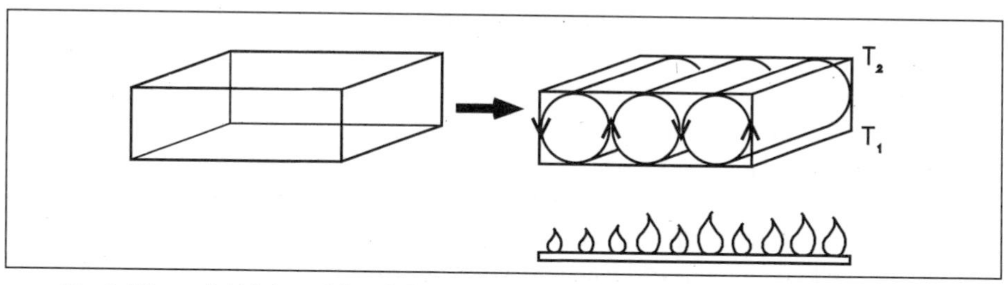

Fig. 1: When a fluid is heated from below, the fluid may start macroscopic motion in form of rolls

For what follows, we need to know a few basic concepts of synergetics. Though these concepts are quite general, we shall exemplify them by means of a fluid heated from below. The heating causes a temperature difference between the lower and the upper surface. This temperature difference is called a *control parameter*, because in this way we

A. Carsetti (ed.), Functional Models of Cognition, 51-66.

may control the fluid from the outside. For a small temperature difference, the fluid remains at rest, beyond a certain temperature difference the fluid starts a specific macroscopic motion for instance in the form of rolls. We say that the old state, i.e. in the present case the resting state, has become unstable.

Thus there is a specific control parameter value, at which the old state becomes unstable and the new state with its macroscopic motion forms. This new macroscopic motion, for instance in the form of rolls, is described and governed by one or several so-called order parameters. They may be visualized as some kind of puppet players, who let the puppets dance. In a fluid, the order parameter describes the motion of the individual molecules. In synergetics, the relationship between order parameter(s) and individual parts of the system is called the *slaving principle*. In contrast to the puppet player, the order parameters are, in turn, determined by the motion of the individual parts. Thus we have the relationship that the order parameters govern the motion of the individual parts by the slaving principle, but, in turn, the individual parts determine the behavior of the order parameters. This phenomenon is called *circular causality*. Because the order parameters describe the behavior of the individual parts, we need not describe the individual parts directly, which would require a high amount of information, but it is sufficient to know the order parameters. This implies an enormous *information compression*. Such information compression is, by the way, typical for any language. When we mention the word *dog*, this may imply different races, postures, and so on.

This article will be organized as follows: In section 2 we shall show how concepts of synergetics may be applied to visual perception. To this end we introduce a model, which can be implemented on the synergetic computer and allows one to recognize patterns. In section 3 we generalize this model, so to show that pattern recognition is not only unique but may show oscillations and hysteresis under specific situations. Finally, in section 4 show how decision making can be interpreted as some kind of pattern recognition.

2. VISUAL PERCEPTION - A MODEL OF PATTERN RECOGNITION

It is my deep conviction that by the study of visual perception or, more precisely speaking, of *pattern recognition* we can learn a good deal about *cognition*. Thus the considerations of this section will serve us later as a metaphor when we shall be dealing with decision making. In order to devise models for visual perception by humans, we may proceed at least in two ways: The by now more traditional way might be called *bottom up*. Here we start from model neurons and their connections and then derive the macroscopic properties of their network. This kind of approach that may be traced back to the seminal work by McCulloch and Pitts [4] runs under the headings of neurocomputers [5] or connectionism [6]. The other line of thought, which follows the spirit of synergetics, may be called a *top-down approach* [7]. Here we start with the tasks a macroscopic system has to fulfill and then only in a later step we look for realizations by means of networks of elements that we may again call *model neurons*.

So let us start with the task a macroscopic system for visual perception should fulfill, where the system may be either the human brain, an animal brain, or a highly developed computer. Visual perception is a vast field. For instance, when we - or an animal - see a bright spot, visual perception is at work. Our goal is, however, more specific and, at the same time, more ambitious. We want to study visual perception at the cognitive level, or in other words, we want to understand how a brain may recognize objects (or patterns) through vision. Therefore, in the following we shall speak of pattern recognition. First of all, what do we precisely understand by pattern recognition? When we see a face, we want to know the name of the person, or at least we wish to know whether we know this person from the past. Thus we wish to associate a name with a face. Or, in other words, pattern recognition may be interpreted as the action of an *associative memory* [8]. There are other examples of an associative memory, for instance, a telephone book. When we look up the name Alex Miller, the telephone book provides us with his telephone number. When we wish to express the action of an associative memory in abstract terms, we may say that is serves for the completion of a set of incomplete data to a complete set. But how do we realize such an associative memory?

In the following we wish to realize an associative memory by some dynamics. Quite often a dynamics can be visualized by means of the motion of a ball in a landscape. In such a landscape, we identify the recognized patterns with the bottoms of the valleys, whereas an incomplete pattern, i.e. a not yet recognized pattern, may be visualized as the position of a ball lying on a slope (Fig. 2). During the recognition process, the ball will

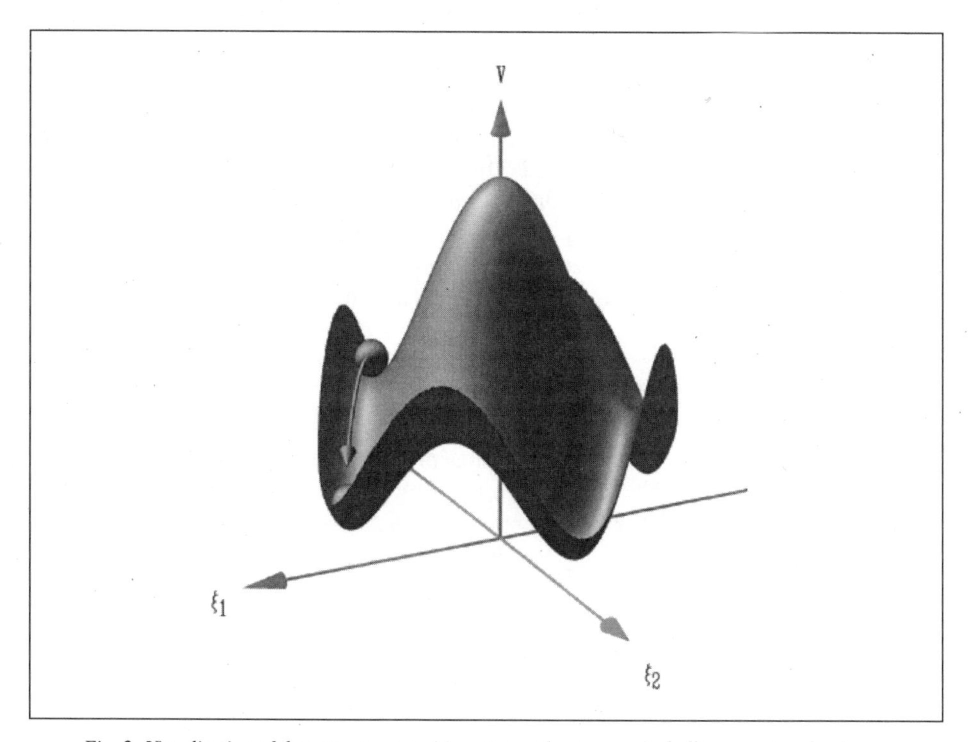

Fig. 2: Visualization of the pattern recognition process by means of a ball moving in a landscape

be pulled into its closest valley and thus the recognition task will be accomplished. The central question, of course, is in which way we can link the features of an individual pattern recognition and pattern formation. In other words, we claim that pattern recognition by a brain or a computer is nothing but pattern formation. To explain this idea in more detail, let us consider the example of a fluid heated from below. As we have mentioned in sect. 1, fluids heated from below may form roll patterns. Let us consider a circular vessel that is heated from below and where the critical temperature difference has been established so that, in principle, a roll pattern can evolve. Let us further assume that we prescribe one upwhelling roll in a specific direction (Fig. 3, left column). Then computer simulations show [9] that the liquid is able to complement this single roll to a whole pattern in the course of time. When we prescribe a different direction of the initial roll, a new roll pattern in that direction will develop (Fig. 3, middle column). Finally, we put the liquid under a conflict situation by prescribing two initial rolls in different directions, whereby one roll is somewhat stronger than the other one (Fig. 3, right column). As the computer calculation reveals, the originally stronger roll wins the competition and determines the evolving total pattern.

Let us interpret these results in terms of order parameters and the slaving principle. We first prepare an initial state that may be represented by means of a superposition of all possible roll patterns with their different orientations. Each of these roll patterns is governed by its specific order parameter, but one order parameter is strongest, namely the one which belongs to the initially prescribed roll. After the preparation of the initial state, a competition between the various order parameters sets in, which is won by the order parameter that belongs to the initially given strongest roll. This is also clearly exhibited by the right column of Fig. 3. After this order parameter has won the competition, it enslaves the whole system, i.e. it brings the whole fluid into its ordered roll-like state. In other words, a partially ordered system is brought into its fully ordered state via the order parameter competition.

What happens in pattern recognition? We claim exactly the same, namely originally some features of a face, like its nose and eyes, may be shown to the human brain or to a computer to define the initial state of that pattern. Then the corresponding order parameter is called upon that competes with all others, wins the competition and supplements the given features by new features that belong to the whole pattern. Thus a face may be complemented and also the family name may be added, if this is required (Fig. 4). In order to see how this procedure works, let us consider concrete examples, namely the recognition of faces. We decompose the pictures of faces into their individual pixels, say, hundred by hundred (Fig. 5). We denote a pixel by its index j. To each pixel we may attribute a grey value v_j. The set of all grey values v_j forms a vector $\mathbf{v} = (v_1, v_2 \ldots, v_N)$. When there are several faces given, we distinguish their corresponding vectors by an index k and write $\mathbf{v}_k$. We call the given test pattern vector $\mathbf{q}$. We then wish to devise a dynamics so that in the course of time the vector $\mathbf{q}$ develops and, eventually, reaches a vector $\mathbf{v}_{k_0}$, where $\mathbf{v}_{k_0}$ is one of the stored prototype pattern vectors. As some analysis shows, which we do not repeat here, however, we may construct the dynamics in a surprisingly close analogy to the fluid dynamics example. The pattern recognition dynamics is described by the eqs[1].

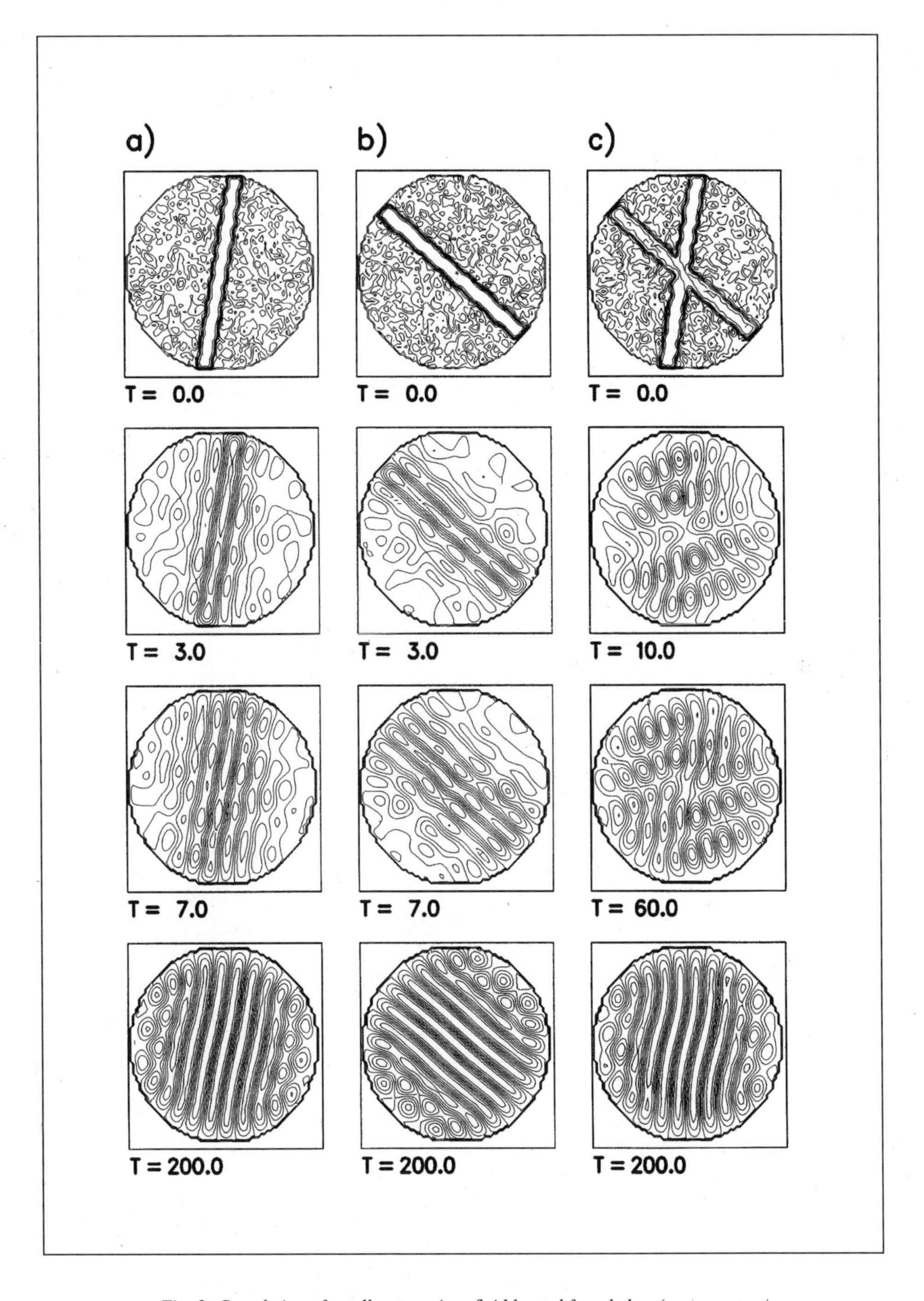

Fig. 3: Completion of a roll pattern in a fluid heated from below (compare text)

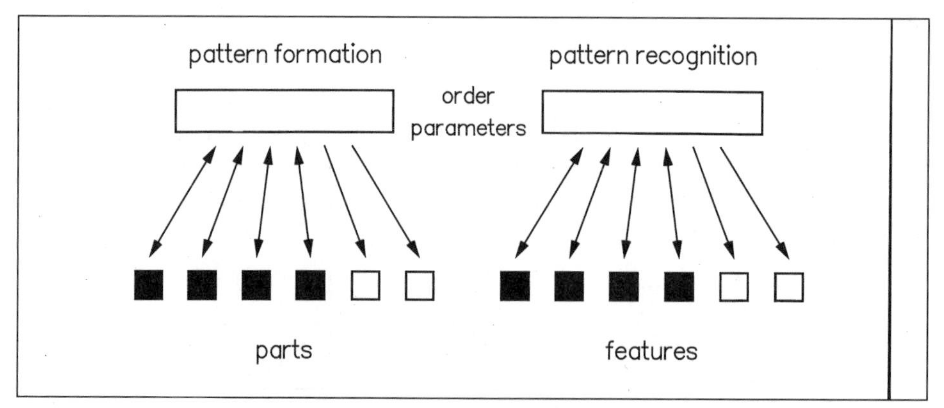

Fig. 4: Analogy between pattern recognition and pattern formation (compare text)

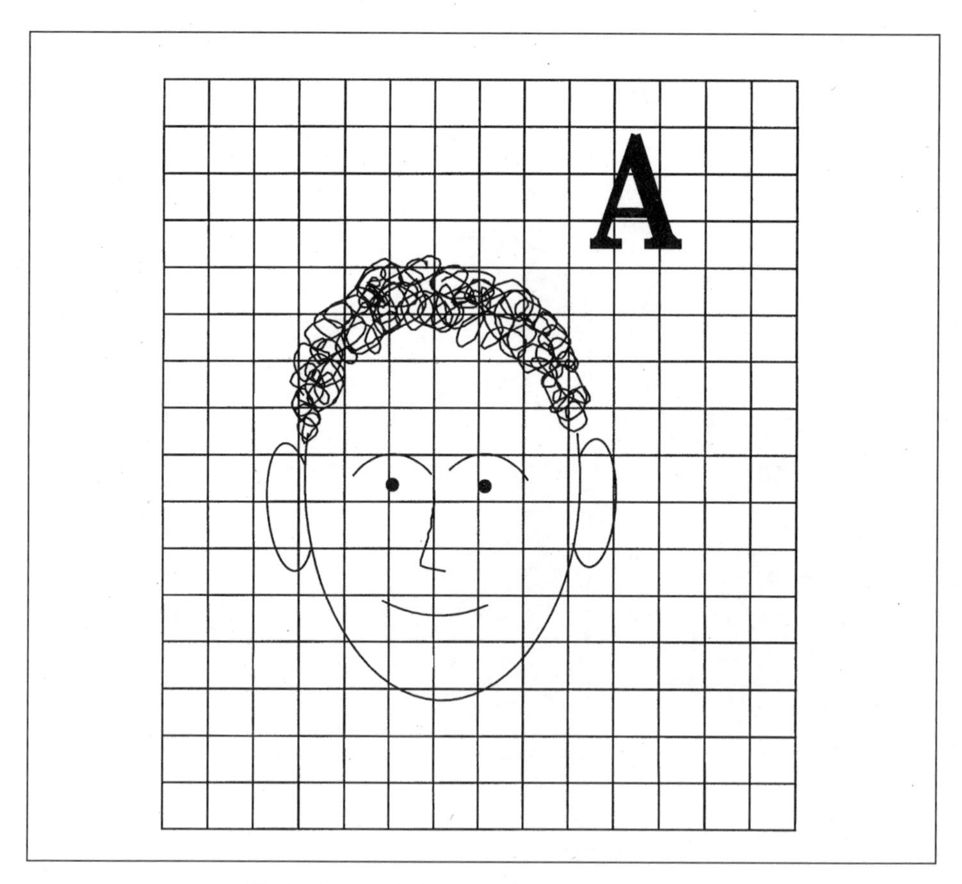

Fig. 5: A picture is decomposed into its individual pixels

$$\dot{\mathbf{q}} = \sum_{k=1}^{M} \lambda_k (\mathbf{v}_k\mathbf{q})\mathbf{v}_k - B \sum_{k,k'=1}^{M} (\mathbf{v}_k\mathbf{q})^2 (\mathbf{v}_{k'}\mathbf{q})\, \mathbf{v}_{k'} - C\,(\mathbf{q}\mathbf{q})\,\mathbf{q}. \tag{1}$$

where the individual terms on the right-hand side have the following meaning: The first term contains the *attention parameters* λ_k. We shall elucidate their role and their interpretation below in sect. 3. The term $\mathbf{v}^{(k)} \bullet \mathbf{v}^{(k')}$ represents the *learning matrix*. The third term serves for the discrimination between the patterns and the last term limits the growth of the grey values of the pixels. The dynamics of $\mathbf{q}$ can be visualized by the motion of a ball in a potential landscape. In general, this dynamics takes place in a high-dimensional space, but when we consider a two-dimensional example in which we have only two pixels and two prototype patterns, we may visualize the shape of the potential by means of Fig. 2.

A few explicit examples may illustrate the whole procedure. Fig. 6 represents a number of stored faces (or other patterns) jointly with their family names (encoded by letters). Fig. 7 shows the pattern recognition process, where part of a face is offered as a test pattern vector. Note that the recognition processes make use of all stored patterns simultaneously.

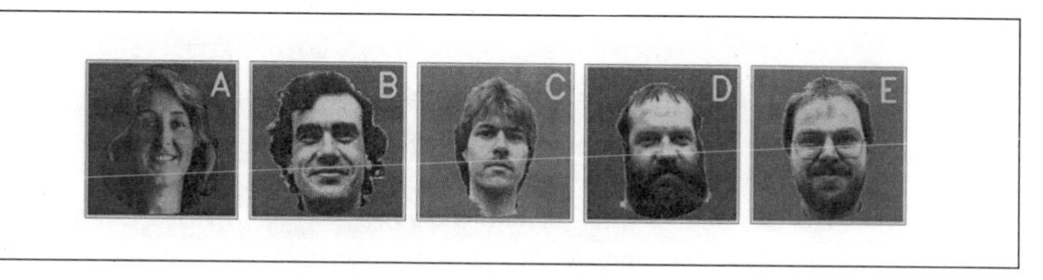

Fig. 6: A few examples of stored pictures of faces with their family names encoded by letters

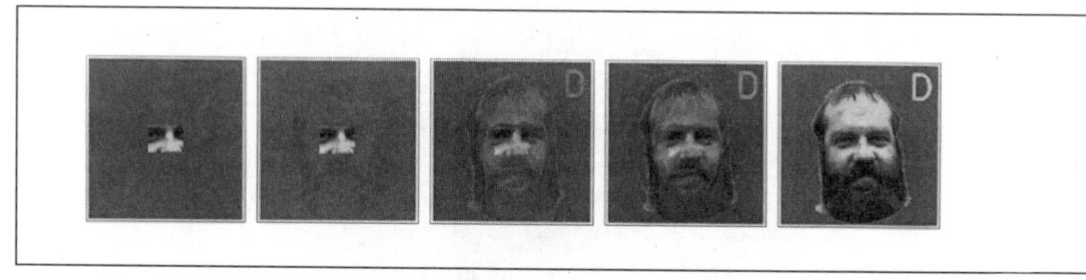

Fig. 7: From left to right: Example of the recognition process

In the context of this article it is important to note that equation (1) represents the action of a whole network of individual neurons and that our model allows us to study the connection between the individual activities and the origin of order parameters. The transition from the microscopic neuronal level to the order parameter level is achieved by the decomposition of $\mathbf{q}$ into the prototype vectors $\mathbf{v}_k$

$$\mathbf{q}(t) = \sum_{k=1}^{M} \xi_k(t)\, \mathbf{v}_k + \mathrm{w}(t). \tag{2}$$

ξ_k is the order parameter that belongs to the pattern $\mathbf{v}_k$. $\mathbf{w}$ represents a rest term that vanishes during the development of $\mathbf{q}$ in the course of time. Thus the evolving recognized patterns are determined by the order parameters ξ_k. When we insert (1) into (2), multiply this equation by $\mathbf{v}_k$ and form the scalar product, because of the orthogonality relation of the $\mathbf{v}_k$'s, we immediately obtain

$$\dot{\xi}_k = \lambda_k \xi_k - B \sum_{k' \neq k}^{M} \xi_{k'}^2 \xi_k - C \sum_{k'=1}^{M} \xi_{k'}^2 \xi_k. \tag{3}$$

This provides us also with a key to determine the initial values of ξ_k, namely by means of

$$\xi_k(0) = (\mathbf{v}_k \mathbf{q}(0)). \tag{4}$$

Eqs (2) - (4) establish the relation between the microscopic spatial patterns of activities of the model neurons and the order parameter level. As we see, each pattern k is governed by its order parameter ξ_k. In a way we may state that the order parameter ξ_k represents the "idea of the pattern k".

3. THE ROLE OF ATTENTION PARAMETERS. AMBIGUOUS FIGURES

Let us try to elucidate the role of the attention parameters λ_k.[7]. When we show a scene, such as that of Fig. 8, to the computer, it first recognizes the woman in the foreground. When we put the attention parameter belonging to the woman equal to zero and show the same scene again to the computer, it recognizes the man in the rear. In this way, the computer was able to consecutively recognize scenes composed of five partly hidden faces. I believe that this result may indicate how human brains deal with complex scenes. They

Fig. 8: Example of a scene that was recognized by the synergetic computer

first recognize a part of it, then the corresponding attention fades away, and attention is focussed on another part of the scene. This interpretation will be substantiated below.

In rare cases the computer failed to recognize the faces of a scene properly, but recognized, say, a third face instead. But as it turns out, humans can also be deceived by pictures. Consider to this end Fig. 9. Here most people first recognize Einstein's face, but when they look more carefully, they will recognize three bathing girls. This leads us to the field of ambiguous patterns. An example is shown in Fig. 10. Our brain deals with these ambiguous patterns in a very peculiar way. For instance, in the picture of vase/faces, we recognize a vase for a while, then we recognize two faces, then again the vase, a.s.o.. In other words, our perception oscillates back and forth between these two interpretations. In the beginning of the 20th century, the Gestalt psychologist *Köhler* suggested an explanation of this result [10]. According to his ideas once a pattern has been recognized the corresponding attention fades away and now new attention can be focussed on the other interpretation. He offered some physical explanation for these

Fig. 9: Einstein or ...?

Fig. 10: Vase or faces?

processes that are, however, nowadays considered as obsolete and no more shared by most neurophysiologists or psychophysicists. On the other hand, in the frame of our order parameter approach, it is quite simple to simulate these processes. To this end, we start from the order parameter equation for two order parameters ξ^1, ξ^2 representing two different percepts, such as vase and face. By specializing eqs. (3), we obtain

$$\dot{\xi}_1 = \xi_1 [\lambda_1 - C \xi_1^2 - (B + C)\xi_2^2], \tag{5}$$

$$\dot{\xi}_2 = \xi_2 [\lambda_2 - C \xi_2^2 - (B + C)\xi_1^2]. \tag{6}$$

Note the completely symmetric role played by ξ_1 and ξ_2.
In addition, we subject the attention parameters to a dynamics that takes into account saturation, i.e. an attention parameter decreases once the corresponding order parameter increases. The equations for the saturation of attention parameters then acquire the following form:

$$\dot{\lambda}_1 = \gamma (1 - \lambda_1 - c\, \xi_1^2), \tag{7}$$

$$\dot{\lambda}_2 = \gamma (1 - \lambda_2 - c\, \xi_2^2). \tag{8}$$

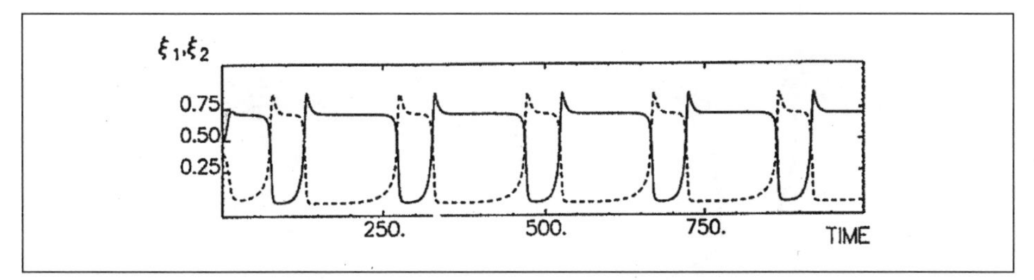

Fig. 11 Oscillations of perception. The size of the order parameters that correspond to vase or faces are plotted versus time

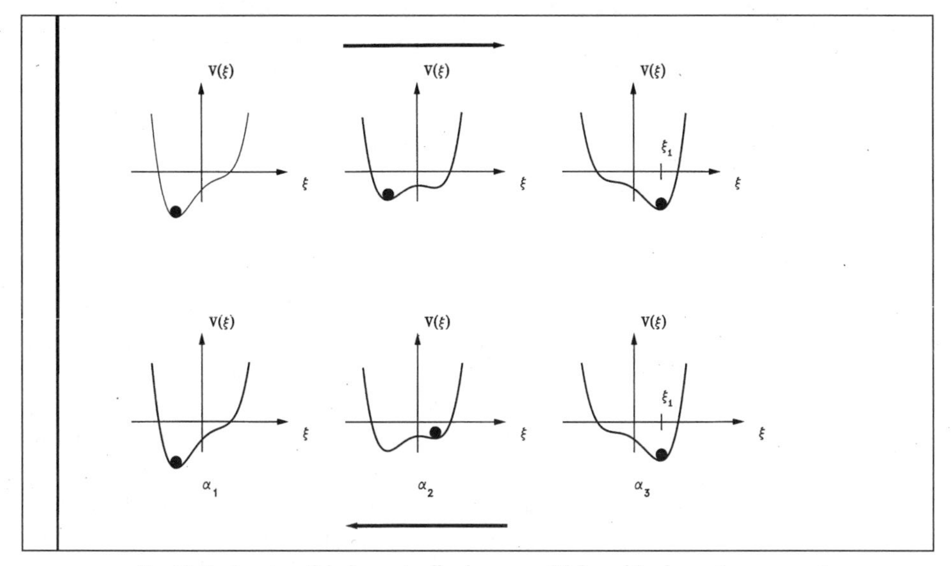

Fig. 12: Explanation of the hysteresis effect by means of deformed landscapes (compare text)

A solution of the coupled eqs. (5) - (8) is shown in Fig. 11, where we, indeed, find the observed oscillations of perception. An interesting phenomenon occurs, when a picture allows for three or more interpretations and when we include the impact of random fluctuations on the dynamics of the attention parameters λ_k. Then the sequence of perceived patterns changes in an irregular fashion.

Finally, we discuss the effect of hysteresis in visual perception. First we describe this effect by means of an order parameter moving in the landscapes shown in Fig. 12. As is shown in synergetics, a change of a control parameter may cause a change of the shape of the landscape. We first follow these changes in the upper part of Fig. 12 from left to right. The ball, whose position indicates the value of the order parameter, is in the middle part of this figure still in its original position. Only in the right part it has occupied a new position. Now look at the lower part of this figure, and start from the right side. When we change this landscape to that of the middle lower part of this figure, the ball remains at its

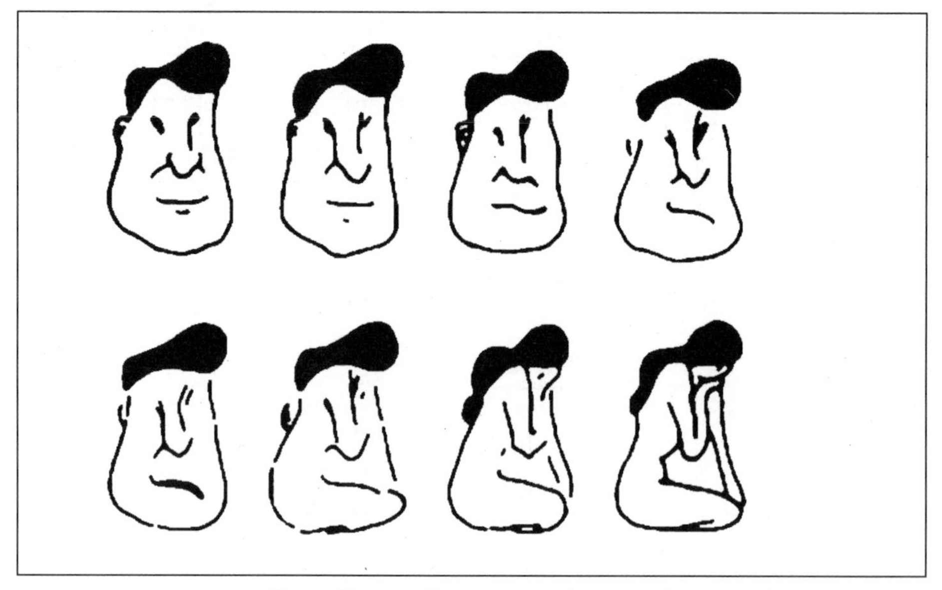

Fig. 13: Hysteresis effect in perception (compare text)

previous position. Thus we may draw the following important conclusion: In spite of the fact that in the middle of Fig. 12 in the lower and upper part the landscape is precisely the same, the order parameter (position of ball) is different, because it depends on its position in the foregoing landscape. Now consider fig. 13 in the first row from left to right, and then in the second row from left to right. You will recognize a jump of your perception in the second row. Now look at these pictures in the reverse direction. Your perception now switches in the first row! This is the effect of hysteresis in perception.

4. DECISION MAKING AS PATTERN RECOGNITION

In this section we wish to follow up our idea that visual pattern recognition may serve as a metaphor for the understanding of human cognitive abilities [11]. One typical problem humans are confronted with is decision making. This has to be done in our personal daily life, but also in economy and companies, especially by managers, and it is an important task in politics, a.s.o.. When we analyse the problem of decision making more closely, we rather quickly find that there are a number of intrinsic difficulties, and we mention a few of them. In general, the information we have about a problem on which we have to make a decision is incomplete. Quite often, in mathematical terms, the problem is ill-posed. Each specific decision bears its own risks. The problem of decision making implies that, in general, there are multiple choices and a repertoire of actions. In studying these problems, both quantitative and qualitative methods have been applied and there is, of course, a considerable literature on decision making.

In this section we wish to shed new light on this problem by invoking an analogy between decision making and pattern recognition. In general, there is a discrepancy between the known data and the required data for a decision to take a specific action. In the ideal case the known data coincide with the required data. In general, however, the known data are insufficient, i.e. there is a certain amount of unknown data. How do humans fill up the lacking gap of unknown data? This is what we want to analyse in the following. Consider to this end Fig. 14. It indicates that, at least in general, the known data can be complemented in a variety of ways to fill in the gap of the unknown data. Depending on how we fill in the unknown data, different decisions or actions can be taken.This figure may be oversimplified, because even if all the data are known there may still be several decisions possible and compatible with all the known data.

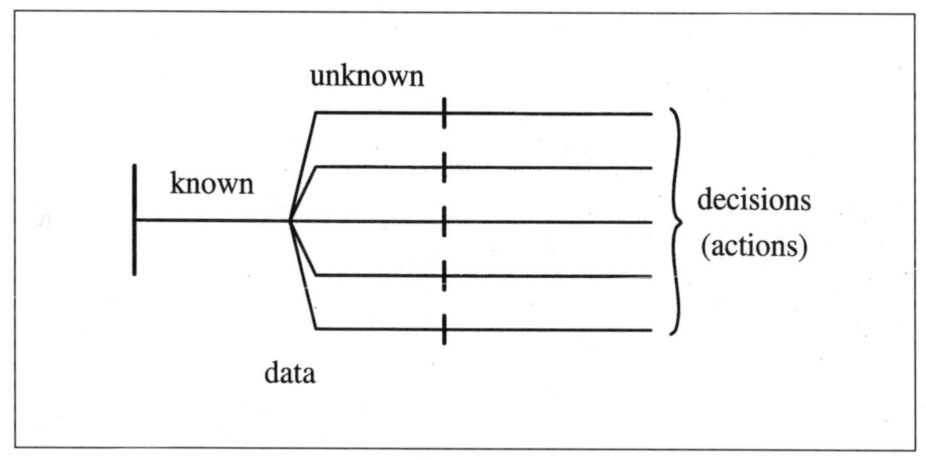

Fig. 14: The known data may be complemented in a variety of ways. Depending on these different ways, different decisions or actions can be taken

How do we fill in the unknown data? Our main theme will be that quite often we use a similarity between a given situation and a previous situation. When we want to cast this similarity into a mathematical frame, we have to look for similarity measures (see below). In more detail, we propose to draw the following analogies between pattern recognition and decision making (cf. also Table 1). In decision making the data correspond to patterns treated in pattern recognition. The data may be quantitative or they may consist of specific rules, laws or regulations. They may be in the form of algorithms or when we think of computers, in form of programs, or in the form of flow charts. Also diagrams may be considered as such data. In pattern recognition the patterns may consist of pictures or of the arrangements of objects. The patterns may be visual or acoustic signals. Quite often these patterns are encoded as vectors that are static or time-dependent. Of course, in decision making the data may be multi-dimensional.

So far we have been discussing the analogy between the objects decision making or pattern recognition are dealing with. In both cases the prototype patterns or the sets of known complete *data* may be learned or given. Incomplete data in decision making have their analogue in pattern recognition in the form of incomplete test patterns. How can we exploit this analogy to study decision making? In analogy to pattern recognition we may introduce a similarity measure, for instance, the overlap of prototype patterns $\mathbf{v}^k$ and and the test pattern $\mathbf{q}$, cf. (4). We then may establish a dynamics that is based on the similarity measure and may also include bias, attention parameters or awareness. So from a formal point of view the whole procedure, we came across in the previous sections on pattern recognition, may be transferred to a scheme on decision making.

What will be the consequences? They are listed up in Table 1. In both cases, pattern recognition and decision making, we may find a unique identification and a unique decision, respectively. But in a number of cases we may be confronted with oscillations between two or more percepts, to which oscillations between two or more decisions correspond. These oscillations are not unusual in our daily life as everybody knows. Here we can trace it back to a fundamental mechanism of the human cognitive abilities. A very

pattern recognition	decision making
patterns pictures arrangement of objects visual, acoustic signals movement patterns actions (often encoded as vectors)	data, quantitative, qualitative, yes/no rules, laws, regulations algorithms, programs flow charts, diagrams orders multi-dimensional in short: “data”
prototype patterns learned or given	sets of known complete “data” learned or given
test patterns	incomplete data in particular “action” lacking
similarity measure dynamics bias attention, awareness	→
unique identification or oscillations between two or more percepts hysteresis complex scenes saturation of attention	unique decision or oscillations between two or more decisions do what was done last time even under changed circumstances multiple choices failure, new attempt based on new decisions “heuristics”

Table 1

important analogy arises when we remember the hysteresis effect we came across in pattern recognition (cf. Fig. 12). Translating this effect into decision making means the following: A person does what he or she did last time even under changed circumstances. The analogy between pattern recognition and decision making can be carried further. In pattern recognition we dealt with complex scenes, where we saw that the computer and probably the human brain analyses such a scene by means of a saturation of attention. Once part of a scene has been recognized, we focus our attention on the next objects. In our analysis in decision making multiple choices correspond to complex scenes and the saturation of attention, we met in pattern recognition, can now be translated as follows: Based on our attention we make a first choice. When we encounter a *failure*, the *attention* parameter for that endeavor is put equal to zero. We then make a new attempt based again on our attention for a new kind of endeavor, a.s.o.. Depending on our previous experience there may be a hierarchy of attention parameters through which we work starting with the highest attention parameter. This interpretation is related to *Wagenaar's* notion of *heuristics* [12].

Summarizing these ideas we may state: The mechanisms we discussed in the case of pattern recognition can be translated into those of decision making. This can be done not only at a qualitative level but also quantitatively at the level of computer algorithms in analogy to the synergetic computer. Quite obviously, our analysis is by no means complete and other strategies may be of equal importance. Artificial intelligence and here especially the approach by expert systems must be mentioned. A problem encountered here is that of branching, where the various branches become extremely numerous and, eventually, decision making has become very difficult. We believe that this branching problem can be circumvented by the approach we outlined above, because as in pattern recognition the various possibilities are taken care of in a parallel fashion.

Prof. Hermann Haken
Institut für Theoretische Physik
University of Stuttgart
Pfaffenwaldring 57/IV
D - 70550 Stuttgart, Germany

NOTES

[1] We present a simplified version in which the pattern vectors are assumed to be orthogonal on each other.

REFERENCES

[1] H. Haken: Synergetics, An Introduction, 3rd ed., Springer, Berlin, New York (1983).
[2] H. Haken: Advanced Synergetics, 3rd, print., Springer, Berlin, New York (1993).
[3] H. Haken: Synergetics, The Science of Structure, Van Nostrand-Reinhold, New York (1984).

[4] W.S. McCulloch and W.H. Pitts: Bull. Math. Biophysics 5, 115-133 (1943).
[5] P.D. Wassermann: Neural Computing, Theory and Practice, Van Nostrand-Reinhold, New York (1989).
[6] J.L. McClelland and D.E. Rumelhart: Parallel Distributed Processing: Explorations in the Microstructure of Cognition, vol. 1: "Foundations". Cambridge, MA: Mit Press (1986).
[7] H. Haken: Synergetic Computers and Cognition, Springer, Berlin, New York (1991).
[8] T. Kohonen: Self-Organization and Associative Memory, Springer, Berlin, New York (1984).
[9] M. Bestehorn and H. Haken: Z. Phys. B - Condensed Matter 82, 305-308 (1991).
[10] W. Köhler: Die physischen Gestalten in Ruhe und im stationären Zustand, Vieweg, Braunschweig (1920).
[11] H. Haken: Principles of Brain Functioning, Springer, Berlin (1996).
[12] W.A. Wagenaar: "Heuristics: Simple Ways for Dealing with Complex Problems. Talk given at the Symposium": "Natural Sciences and Human Thought", Villa Vigoni, Italy, (29.3.-2.4.1993).

ELIANO PESSA

CONNECTIONIST PSYCHOLOGY AND SYNERGETICS OF COGNITION

1. INTRODUCTION

The birth of Connectionist Psychology (Feldman and Ballard 1982, McClelland & Rumelhart, 1986) has some resemblance with the one of modern Statistical Mechanics. In both cases we deal with the introduction of a new, two-level (microscopic and macroscopic) description of the world, aimed to avoid the inconsistencies of an explanation of the experimental facts within the usual theoretical framework. These inconsistencies, in the case of Statistical Mechanics, derive from the impossibility of reconciling the phenomenological description of physical events, like the ones considered by classical thermodynamics, with the traditional newtonian dynamics. But in the case of Connectionist Psychology, they derive from the impossibility of reconciling the phenomenological description of cognitive phenomena with the one given by information-processing Psychology. In the same way as Statistical Mechanics, which does not negate the relevance of macroscopic behaviors but tries only to explain them in a simpler way through a more fundamental level of description, also Connectionist Psychology does not negate the relevance of symbolic information processing in cognitive phenomena but tries only to explain macroscopic cognitive behaviors through the cooperative action of a large number of simple cognitive microscopic units.

These analogies between Connectionist Psychology and Statistical Mechanics seem to suggest that the former should be faced with the same problems as the latter. In this contribution we will try to show that, unfortunately, this is not the case: the problems that Connectionist Psychology must solve are more difficult than the ones considered by usual Statistical Mechanics. We feel that they can be dealt with only by using a suitable combination of the methods of Synergetics, the science of cooperative behavior in complex systems (Haken, 1978, 1983, 1988, 1989), with some methods of Theoretical Physics (for this see Pessa, 1998). In the following we will show, through examples, how this works. In this way the derivation of macroscopic cognitive behavior from the one of microscopic neural-like units appears to be the subject of a *Synergetics of Cognition*, able to generate a new class of cognitive models to be tested against experimental data.

2. NEURAL NETWORKS AS MODELS OF HUMAN COGNITIVE BEHAVIOR

Despite the ever growing number of models of cognitive processing based on neural networks, the problem of correspondence between network behavior and human

A. Carsetti (ed.), Functional Models of Cognition, 67-90.

behavior has not been yet solved in an entirely satisfactory way. What is lacking is an algorithmic rule which lets one to build, starting from some observational data on human subjects, the best "equivalent" neural network model for explaining them. To this regard, we remember that a similar problem arose when trying to build the so-called "user models" within the context of psychology of human-computer interaction (Card, Moran & Newell, 1983; Polson, 1987; Allen, 1990; Bonnie & Kieras, 1996). In that case, an interesting proposal for building the best model of human cognitive processing has been put forward by Rauterberg (1993) which introduced Amme, i.e., an automatic mental model evaluation to analyse user behavior. The implementation of Amme requires a complete knowledge of all possible states of the device to be manipulated, of all possible actions which induce transitions between these states, and a detailed transcription of a certain number of user's behavioral histories. Then a suitable algorithm, which is the kernel of Amme, lets one deduce, from these data, the best user mental model compatible with them, represented in the form of a Petri net. It is easy to see that such methodologies cannot be practically applied in the realm of neural network models of cognitive processing. Namely, in this case the device is nothing but the external world, with a virtually unlimited number of possible states and transitions, whereas the knowledge of cognitive behaviors of human subjects is very poor.

In the same way also the results obtained through an experimental investigation of human cognitive architectures by psychologists can offer to the modeler hints rather than detailed criteria for the design of suitable neural networks. What we actually know derives from the techniques employed in semantic memory research (such as reaction time measures, prototypicity ratings, etc.) or in experiments on concept learning. They can give some useful informations about the strength of connections between our internal representations of single features or concepts, but no more. Generally these informations have some value in the realm of symbolic cognitive processing but not in other contexts, such as, e. g., the one of visual perception.

It is to be added that in some cases it seems very difficult, if not impossible, to build neural network models of some forms of symbolic processing, at least through the network models most commonly used. A typical case is the one of the so-called "structural descriptions": whereas it is very easy to build a neural network able, as a consequence of suitable learning, to recognize particular attributes, such as "bottle" or "table", it appears nearly impossible to build a neural network which learns to bind the two previous attributes through the relational attribute "on the" in such a way as to represent the structural description of "the bottle is on the table". In recent times this Binding Problem has received many different solutions (see, e. g., Hummel & Biederman, 1992 for a review) which required a deep modification of the traditional network architectures and the use of the most advanced methods of Synergetics, at least as regards the network which performs a dynamic linking via phase coupling of oscillators (Von der Malsburg & Buhmann, 1992).

What has been said shows that the problem of the choice of the best neural network for explaining a given data set, relative to human cognitive behavior, is largely undeterminate. In such a situation it seems better to adopt an approach which could be rightly called "genetic". In it (see Pessa, 1994; Minati, Pessa & Pessa, 1998) the researcher uses a plurality, or a "population", of models, instead of a single model as happens in the traditional scientific paradigm. The modeling activity can be viewed as equivalent to the evolutionary dynamics of this "population" under the action of some "genetic" rules. These latter consist, essentially, of:

a) transformation rules wich let the modeler pass from an "ancestor" model to its "descendants" (such as when developing some consequences of a model generalizing or modifying it);

b) fitness criteria selecting the members of this "population".

The choice of the fitness criteria is, of course, strictly connected to the problems which the modeler tries to solve and, for this reason, these criteria can change with time following the changes of the modeler's goals. These latter, then, depend on the success obtained by the researcher by using the models selected in a particular situation. To give, therefore, some insight into the fitness criteria which actually appear as the most suitable for the "population" of models belonging to the Synergetics of Cognition, we need a list of the problems the researcher in this field should be prepared to meet, together with some considerations on the hopes we have of solving them with the methods actually available.

3. THE PROBLEMS OF A SYNERGETICS OF COGNITION

A rough subdivision of these problems could be centered around the following questions:

1) how to find the right form of dynamical laws at the microscopic level;

2) how to find the conditions granting the presence of given macroscopic structures;

3) how to derive the dynamics of macroscopic structures from the knowledge of microdynamics;

4) how to compare the findings eventually obtained in 1), 2), 3) with the experimental data on human subjects.

They can be dealt with by using two different approaches: bottom-up and top-down. In the former one starts from the existing phenomenology or from known experimental data, and then tries to find the best models able to explain them. In the latter one starts, on the contrary, from general, abstract, principles (such as the ones of symmetry, simplicity, etc.), and then deduces from them the model equations. If possible, these latter are solved and the results thus obtained are compared with experimental data. In the bottom-up approach the processes of induction and abduction play a prominent role, whereas in the top-down one deduction is the main tool. The two approaches, however, are not mutually exclusive. In most cases the researcher uses a mixture of both. In the following we will present, when speaking of the attempts to giving an answer to each question, the two approaches as separated only to evidence general trends.

4. THE MICROSCOPIC DYNAMICS

When we adopt a bottom-up approach, our goal would be that of finding a general algorithm for building the best neural network model able to explain a given, observed behaviour. As shown in the second paragraph, this is impossible. We are thus forced to conclude that every attempt to model the microscopic dynamics must start from a top-down approach. In it one chooses the dynamical equations starting from some general principles. As regards the net architectures, in most cases their choice is constrained by particular requirements imposed by the experimenter. These choices can be classified according to the type of constraint imposed.

Of course, the possibilities of choice of the dynamical equations are virtually unlimited. We will list briefly in the following only some choices effectively made, and others which appear as feasible on theoretical grounds but have not been introduced up to now.

a) Conventional abstract neural-like

This choice is based on the introduction of units whose transfer functions derive from a very abstract representation of some features of the activity of certain types of neurons. Generally these units are characterized by activation laws of the form:

$$x_i\,(t+1) = F\,[\Sigma_j w_{ij} x_j\,(t) + I_i\,(t)], \tag{1}$$

where $x_i(t)$ is the output of the i-th unit at time t, $I_i(t)$ is the external input to this unit, w_{ij} are the connection strenghts, and F is a suitable activation function which in some cases is linear but in most cases is threshold-like or sigmoidal, such as, e.g.:

$$F\,(x) = 1/[1+\exp\,(-x)]. \tag{2}$$

In other cases (1) is replaced by a differential equation of the form:

$$dx_i/dt = -x_i + F\,[\Sigma_j\, w_{ij} x_j + I_i\,(t)]. \tag{3}$$

In recent times many authors proposed models in which each unit behaves like an elementary oscillator. The simplest way to realize this situation is to model each unit as composed by two neural-like units, one excitatory of output u_i and the other inhibitory of output v_i, described by the coupled system of equations:

$$du_i/dt = -u_i + F\,(u_i - \beta v_i - \theta_u + I_i) \tag{4.a}$$

$$\tau dv_i/dt = -v_i + F\,(\alpha u_i - \gamma v_i - \theta_v), \tag{4.b}$$

where F is given by a function like (2), θ_u and θ_v are excitatory and inhibitory thresholds, respectively, I_i is the input coming from other units or from the outside, and τ, β, α, γ are suitable parameters (see, e.g., von der Malsburg & Buhmann, 1992; Yamaguchi & Shimizu, 1994).

As regards the connection strenghts w_{ij}, their dynamical evolution is driven by suitable learning laws, such as the well-known ones of error backpropagation, self-organizing maps, Hebbian learning, Grossberg's models and so on. These latter could be also written as special cases of a more general dynamical law of the form (Ramacher, 1993):

$$dw_{ij}/dt = G\ (x_i, x_j, w_{ij}) \tag{5}$$

where G is a suitable function. In some models also higher-order connections have been utilized (Lee & Maxwell, 1987), so that (1) is generalized as follows:

$$x_i\ (t + 1) = F\ [\Sigma_j w_{ij}\, x_j\ (t) + \Sigma_{ik}\, w_{ijk}\, x\ (t)\ x_k\ (t) + \ldots]. \tag{6}$$

b) More realistic neural-like

This choice is based on a more realistic representation of the activity of biological neurons, or of some parts of them, like the axonal membrane. Typical equations describing the neural-like units behavior are the ones of Hodgkin and Huxley, or their simplified form known as FitzHugh-Nagumo equations (see, e.g., Taylor & Mannion, 1991):

$$dx_i/dt = c\ [y_i + x_i - (x_i^3/3) + I_i] \tag{7.a}$$

$$dy_i/dt = (-1/c)\ (x_i + by_i - a) \tag{7.b}$$

where x_i is the unit output, y_i an auxiliary "internal" variable, I_i the external output and a, b, c suitable parameters. Until now, these and similar equations have found applications in connectionist network models of memory or learning by using architectures based, for istance, on arrays of coupled FitsHugh-Nagumo oscillators. Their mathematical complexity, of course, could be very high, owing to the richness of their possible behaviors.

c) Gauge-theoretical

There is another possibility of deriving microdynamical equations, widely used today in Theoretical Physics, which has not found cizitenship in Connectionist Psychology. We make reference, here, to the sort of arguments used in the so-called gauge-theoretical approach to field theories.

In this latter approach the starting point is constituted by the dynamical equations, describing the behavior of a "free", non-interacting, entity (such as an isolated particle). Generally these equations involve independent variables both of spatial and temporal nature, whereas their form is invariant with respect to suitable transformations of dependent state variables, which are of global type, i.e., independent from the space-time coordinates. The interaction forces between the free entities are introduced by letting the latter transformations become dependent on space-time coordinates. Namely, in this case the form invariance of the dynamical equations can be maintained only by adding suitable "compensating" terms which represent the equivalent of an "interaction" field. Reasonings, then, based on suitable symmetry properties let one derive from the nature of these compensating terms the form of the dynamical equations to be obeyed by the interaction field. Whereas this picture is mathematically equivalent to the usual one in which the interactions are introduced from the start in an explicit way, it has the advantage of giving rise to a unified representation, through a single field, of all complicated networks of interactions between the single entities. These latter would be very difficult to study; because to follow the dynamical effect of each single interaction along the entire system evolution would be a nearly impossible enterprise.

In order to sketch briefly how this approach could work within the context of connectionist models, let us start, as an example, with the equations (4.a), (4.b), describing an elementary neural oscillator. To simplify things, let us suppose that the input term I_i (t) has the form:

$$I_i (t) = \Sigma_k w_{ik} u_k (t) \tag{8}$$

where the connection weights depend on the distance d between the i-th and the k-th unit according to the "Mexican hat" function:

$$w_{ik} = (1/2) [2 - (d/\sigma)^2] \exp (-d^2/2\sigma^2) \tag{9}$$

where σ is a suitable parameter. It is well known that the convolution of a Mexican hat function with another function can be approximated by the Laplace operator applied to the convolution of a Gaussian with this latter function. If we deal with a "free" neural oscillator placed in a given spatial region, this input term becomes proportional to the Laplacian of the function u_i (t). Then, writing u = u (x,t) and v = v (x,t) instead of u_i (t) and v_i (t), the equation (4.a), (4.b) for a free oscillator will assume the form:

$$\partial u/\partial t = - u + F (u - \beta v - \theta_u + D\Delta_2 u) \qquad (10.a)$$
$$\tau\partial u/t = -v + F (\alpha u-\gamma v-\theta_v) \qquad 10.b)$$

where D is a suitable diffusion coefficient. We will suppose that (10.a), (10.b) have a unique stationary homogeneous equilibrium solution u_0, v_0. By imposing Dirichlet boundary conditions such that the value of u, v at the boundaries is given by u_0, v_0, it makes sense to study the small deviations of u, v from the homogeneous equilibrium solution u_0, v_0. By letting:

$$\xi = u-u_0, \eta = v-v_0 \tag{11}$$

and linearizing around u_0, v_0 it is possible to derive from (10.a), (10.b) the following system:

$$\partial\xi/\partial t = A\xi + B\eta + D_0\Delta_2\xi \qquad (12.a)$$
$$\partial\eta/\partial t = C\xi + E\eta, \qquad (12.b)$$

where the coefficients A, B, C, D_0, E can be obtained from (10.a), (10.b) through easy computations. If suitable conditions on these coefficients hold, the (12.a), (12.b), with null Dirichlet boundary conditions, have solutions of the form:

$$\xi = c_1 \exp [j(kx + \omega t)],\ \eta = c_2 \exp [j(kx + \omega t)],\ j = \sqrt{-1}, \tag{13}$$

which are stable for every value of k. Besides, (12.a) and (12.b) are invariant in form with respect to a global "phase transformation" of the form:

$$\xi' = \exp (\varphi)\xi,\ \eta' = \exp (\varphi)\eta. \tag{14}$$

At this point we can introduce an interaction between these free oscillators by letting φ be dependent from space-time coordinates, so that (14) becomes a local gauge transformation. If we utilize the unified notation $\nabla_\mu f$ for indicating both temporal and spatial derivatives of a generic function f (when $\mu = 0$, we have a time derivative, and when μ is nonzero, we have spatial derivatives), we see that, in order to maintain the form invariance of (12.a), (12.b) with respect to the local gauge transformation, $\nabla_\mu f$ must be substituted by the generalized derivative:

$$D_\mu f = \nabla_\mu f + (\nabla_\mu \varphi) f \tag{15}$$

To derive the dynamical equation for the new "compensating" field $W_\mu = \nabla_\mu \varphi$ which represents the effect of the interaction between free oscillators it is better to make a comparison with what happens in the most popular local gauge theory, i.e. the U(1) gauge theory of the electromagnetic field. In it the requirement of the form invariance of a scalar field dynamical equation with respect to local gauge transformations of the type:

$$\psi' = \psi \exp (j\varphi) \tag{16}$$

is satisfied only if the ordinary derivatives ∇_μ are substituted by the generalized derivatives:

$$D_\mu \psi = \nabla_\mu \psi + (j \nabla_\mu \varphi)\ \varphi = \nabla_\mu \psi + jeA_\mu \psi \tag{17}$$

where the "compensating" field A_μ is nothing but the usual electromagnetic potential (here e = electronic charge which will be put equal to 1 in the following considerations). If now one takes the commutator of two generalized derivatives, one obtains nothing but the usual definition of the fundamental physical quantity, i.e., the electromagnetic field tensor $F_{\mu\nu}$. Namely, we have:

$$[D_\mu, D_\nu]\psi = j\, F_{\mu\nu}\, \psi, \tag{18}$$

where:

$$F_{\mu\nu} = \nabla_\mu A_\nu - \nabla_\nu A_\mu. \tag{19}$$

At this point, it is enough to observe that every commutator must satisfy the Jacobi identity:

$$[D_\rho, [D_\mu, D_\nu]] + [D_\mu, [D_\nu, D_\rho]] + [D_\nu, [D_\rho, D_\mu]] = 0 \tag{20}$$

to derive, by substituting (17), (18) and (19) into (20), the "rotor" part of the electromagnetic field equations:

$$\text{Rot } F_{\mu\nu} = 0. \tag{21}$$

Besides, by putting:

$$[D_\rho, [D_\nu, D_\rho]]\, \psi = j\, J_\nu\, \psi \tag{22}$$

and substituting again (17), (18) and (19), we will obtain the "divergence" part of the electromagnetic field equations:

$$\text{Div } F_{\nu\rho} = J_\nu, \tag{23}$$

which connects the electromagnetic field tensor to the "current" J_ν deriving from external sources.

We can now repeat the same reasonings for our "compensating" field W_μ of neural interaction. By taking the commutator of the generalized derivatives (15), we will obtain:

$$[D_\mu, D_\nu]\, f = H_{\mu\nu}\, f, \tag{24}$$

where the field tensor $H_{\mu\nu}$ of neural interaction is given by:

$$H_{\mu\nu} = \nabla_\mu W_\nu - \nabla_\nu W_\mu. \tag{25}$$

The Jacobi identity (20) gives immediately the field equations:

$$\text{Rot } H_{\mu\nu} = 0. \tag{26}$$

By using the definition (22), then, we will obtain the connection between the neural interaction field and its sources:

$$\text{Div } H_{\mu\nu} = J_\mu. \tag{27}$$

In order to derive an explicit expression of J_μ in terms of the variables ξ and η, let us first multiply (12.a) by E and (12.b) by B, so that, by subtracting one from another the equations thus obtained, we will have:

$$\partial\rho/\partial t = (AE + BC)\, \xi + \text{div}\, (D_0 E\, \nabla_i \xi) \tag{28}$$

where:

$$\rho = E\xi - B\eta, \tag{29}$$

and the operator "div" refers only to spatial coordinates, as it is the case also for the derivative ∇_i Let us now introduce a vector χ satisfying the condition:

$$\text{div}\,\chi = (AE + BC)\xi. \tag{30}$$

By putting:

$$\chi = \nabla_i \sigma \tag{31}$$

where σ is a suitable scalar, we obtain that (30) will give rise to Poisson's equation:

$$\Delta_2 \sigma = (AE + BC)\xi. \tag{32}$$

As it is well-known (see, e. g., Novozhilov & Yappa, 1981), its solution, in an unlimited space volume, is given by:

$$\sigma = (AE + BC) \int (\xi / |\mathbf{r}\text{-}\mathbf{r}'|) dV'. \tag{33}$$

All these reasonings show that (28) can be written in the form of a continuity equation:

$$\partial\rho/\partial t = \text{div}\,\mathbf{j} \tag{34}$$

where:

$$\mathbf{j} = D_0 E \nabla_i \xi + (AE + BC) \nabla_i \int (\xi / |\mathbf{r}\text{-}\mathbf{r}'|) dV'. \tag{35}$$

The "current" $\mathbf{J}_\mu$ can be thus identified with:

$$\mathbf{J}_\mu = (\rho, \mathbf{j}). \tag{36}$$

If now we substitute in (12.a) and (12.b) the generalized derivatives (15) in the place of the ordinary derivatives, we will obtain the equations of neural oscillators in the presence of interaction:

$$(\partial\xi/\partial t) + W_0\xi = A\xi + B\eta + D_0\Delta_2\xi + D_0 S\xi + + 2D_0\Sigma_i W_i \nabla_i \xi + D_0 H\xi \tag{37.a}$$

$$(\partial\xi/\partial t) + W_0\eta = C\xi + E\eta, \tag{37.b}$$

where:

$$S = \Sigma_i \nabla_i W_i, \; H = \Sigma_i W_i W_i \tag{38}$$

and the indices represented by Latin letter correspond to spatial coordinates. As a conclusion, the equations (37.a), (37.b), together with (26), (27), and the definitions (25), (36), represent a system of coupled equations describing the interaction between the oscillator field and the "connection" field. Their solution can offer a representation of the dynamical evolution of a neural network very different from the usual one. We will not pursue further this topic, being satisfied for having shown how it is possible to translate the description of neural networks microdynamics into gauge-theoretical terms. This approach, however, has not been introduced so far into Connectionist Psychology, despite the fact that it makes possible to use the well-tested experimental tools of that part of Theoretical Physics which bears the name of Field Theory.

5. THE CONDITIONS OF EXISTENCE OF MACROSCOPIC STRUCTURES: THE BOTTOM-UP APPROACH

If we deal with this problem starting from a bottom-up approach, we must necessarily have some description of what we mean by a macroscopic cognitive structure. This description, then, must be compared with the results of some macroscopic observations of neural network dynamics, in order to investigate what are the conditions giving rise to the appearance of the structure we search for. The practical realization of this program can be done in three different ways:

a) entirely qualitative

Here the researcher uses subjective and qualitative evaluations of the features of the macroscopic cognitive structure, and subjective and qualitative evaluations of the features of neural network behaviour; this method is the most widely used in Connectionist Psychology research; thus, for example, the equivalence between the time variation of synaptic weights is entirely arbitrary and has nothing to do with a precise definition of what a learning process is;

b) semi-qualitative

Is this case we have a qualitative evaluation of the features of the macroscopic cognitive structure, to be compared with a quantitative evaluation of precise features of neural network behavior; this approach began to be taken into consideration only in very recent times; we will refer in the following to this topic in a more detailed way;

c) entirely quantitative

This is characterized by a quantitative evaluation of the features of the microscopic cognitive structure which are compared with a quantitative evaluation of the features of neural network behavior; this is the most favourable case, from the point of view of a solid scientific grounding of Connectionist Psychology; it is hard to see, however, that such a situation could arise at this stage of research.

As regards the approach b), we will quote here some examples of research on the comparison between the learning process in human subjects and the process of variation of synaptic weights in neural networks. The main point of departure is constituted by the observation that in many cases the learning curves (subject performance vs. number of trials) characterizing the learning processes by human beings show a sudden, nearly discontinuous, rise of performance in correspondence to a critical number of trials, very similar to that showed by the curves of some physical quantities, such as matter density, in correspondence to the critical temperature of a first-order phase transition (e.g., the solidification). Such an observation, added to other qualitative observations, such as the ones relative to "insight" or "restructuring" phenomena in learning and problem solving by humans and animals (see, e.g., Benjafield, 1992), led to the suggestion that the learning process is alike to a sort of phase transition process. This gives rise to some further questions: what is the type of phase transition involved? What is the relationship between the microdynamics of connectionist models of learning and the macroscopic features of the phase transitions taking place in them?

In order to give an answer to these questions, we first remark that, generally speaking, we can distinguish three broad classes of different processes to which, during the history of psychology, has been given the name of "learning", i.e.:

a) gradual learning

It is a process of acquisition of new knowledge or of new abilities which takes place in a gradual way, typically characterized by a power-law dependence of the learned knowledge or ability on the amount of training; once this process has attained a saturation level, the new structuration of knowledge thus acquired is very robust against spontaneous decay and external perturbations;

b) reversible learning

In this case the process of acquisition can be sometimes characterized by sudden changes, whereas in other situations it can be gradual; what matters is that, once this process has arrived at the end, e.g. when some criterion has been satisfied, the structures created by learning can be destroyed, i.e., we can return to the state existing before the beginning of the acquisition process; such a situation is typical of many laboratory experiments, such as, e.g., the ones on concept learning where the subject must learn items which have no meaning for his life, and which are rapidly forgotten after the end of the experiment or when he must learn new items;

c) irreversible learning

This process is characterized by sudden changes or restructuring of a subject's cognitive architecture; it is irreversible because the structural changes have a permanent character, and it is impossible to induce an apposite change which drives the cognitive architecture towards the state preceding the learning process; a typical example of this form of learning is given by the "insight" phenomena or by the phenomena which happen when a subject understands in a deep way some body of knowledge.

It is possible now to individuate some connections between these classes of learning processes and suitable forms of structural changes. In particular, gradual learning appears to behave like a process which has a tendency towards a statistical equilibrium state (corresponding to a maximum of the probability distribution) in a dynamical system with stochastic inputs. Namely, in this case the inputs given to the system act as stochastic perturbations and, when the law of dependence of the momentaneous ability acquired on the previous ability and the actual input has a nonlinear form, it is possible to show (e.g., by the methods of the theory of stochastic differential equations) that the maximum of the probability distribution of the dependent variable (i.e., of the ability) is determined essentially by the statistical properties of the distribution of inputs. This shows also that the position of this maximum is insensitive to the action of fluctuations. In this case, of course, it is inappropriate to speak of phase transitions.

As regards reversible learning, it appears to be comparable with the so-called equilibrium phase transitions. These latter appear to be characterized by a suitable control parameter which regulates the competition between "organizing" internal forces which tend to produce order and the "disorganizing" effects of thermal noise. When the value of this parameter exceeds a suitable critical point, this competition favours the "organizing" forces, and a phase transition takes place, with an increase of macroscopic order and the presence of discontinuities in the curves of some macroscopic quantities vs. the control parameter (such as the one of specific heat vs. temperature). However, when the control parameter is changed in the opposite direction, it is possible to have a phase transition from the more "organized" to the less "organized" state, letting the system destroy the effects of the restructuration which happened before.

Irreversible learning, on the other hand, seems alike to the so-called non-equilibrium transitions (Nicolis & Prigogine, 1977) which happen to be a system where the non-equilibrium conditions are maintained owing to the presence of suitable boundary conditions. In the case of physical systems when these transitions (which are associated to bifurcation phenomena dependent from a control parameter) give rise to a symmetry breaking, the systems themselves show, near the transition point, a very high sensitivity to external fields or (as in the case of superconductivity) to the presence of a macroscopic "condensate", which acts as a selection mechanism with regard to the choice of the newly organized system's state among the many states compatible with the boundary conditions.

We must remark, however, that the classification so far presented is useful only as a low-level schematization, for many reasons which we will list as follows:

1) in human subjects (and in animals) learning is never an isolated process but is a part of a complicated network of processes; besides, every learning process would be impossible without the presence of a pre-existing "structure", able to organize the input data relative to the process itself;

2) the learning of a new structuration of knowledge generally does not imply the forgetting of the previous one; even if a subject learns a new structural relationship between some elements of his cognitive field, he is nonetheless able to take into consideration, at the same time, also the old relations between the same elements;

3) in practice, none of the previously considered forms of learning happen in isolation; it is possible to have, in the same learning process, the coexistence of different conditions and different elementary types of learning, both in a diachronic and in a synchronic way.

All that implies some consequences:

a) it is better to concentrate the research on particular, very specialized forms of learning which can be easily controlled in laboratory experiments;

b) the characterization of learning can be done mostly by resorting to structural properties, such as the ones which, within the physics of phase transitions, are typical of the critical behavior near the transition point; the comparison between human learning and the dynamics of synaptic weights in neural networks, of which we spoke before, can be made essentially from this point of view.

In order to do such a comparison, we will consider an experimental situation like the one of concept learning by human subjects. In particular, we will refer to the experimental paradigm in which the experimenter presents to the subject, one at time, various items, asking, in correspondence to each one of them, if they are correct exemplars of the concept the subject must guess. After each subject's answer, the experimenter informs him whether his answer was right or wrong. The presentation of the items of the experimental set continues until the subjects reach a correct formulation of the concept to be guessed.

In Connectionist Psychology this form of learning can be modelled by resorting to a multilayered perceptron with a gradient descent learning rule for minimizing the total output error, such as error back-propagation (EBP). Taken for granted the possibility of finding the right network parameters in order to simulate an instance of human concept learning behavior (Kruschke, 1993), we will now concentrate on the classification of the observed neural network behavior, considered as representative of the human one, from the point of view of phase transition theory.

For this purpose, we remember that the EBP networks are plagued by the so-called catastrophic interference phenomenon (Ratcliff, 1990; Pessa & Penna, 1994a; see also Murre, 1996), consisting in the fact that the learning of a new item by the network destroys the performance on the items learned previously. This circumstance suggests that learning in EBP networks is reversible learning, equivalent to an equilibrium phase transition. As is well-known, the transitions of this type are described by a firmly grounded theory (Landau & Lifshitz, 1969; Ma, 1976; Rumer & Ryvkin, 1980; Amit, 1984), based on the idea that they can be viewed as a change of symmetry where the latter term means "invariance of some properties whith respect to suitable transformations". In this way a phase transition is considered as a passage from a symmetric to a non-symmetric phase (or vice versa). The latter is described through an order parameter, i.e., a suitable variable measuring the degree of departure from symmetry. For example, in the case of the second-order phase transition from the paramagnetic to the ferromagnetic phase, the order parameter is given by the specific magnetization, i.e., the magnetic moment per unit volume. Namely, it introduces a preferred direction which breaks the invariance of physical properties with respect to 3-dimensional rotations.

The behavior of a system undergoing an equilibrium second-order phase transition can be characterized through its response function when stimulated by an external field, expressed in terms of the *generalized susceptibility* χ. The latter is defined as the slope of the curve which gives the order parameter φ as a function of the energy E of the external field. In symbols:

$$\chi = \partial\phi/\partial E. \tag{39}$$

This quantity generalizes the ordinary susceptibility which appears in the linear relationship between the induced magnetic moment M and the external magnetic field H:

$$M = \chi H. \tag{40}$$

By applying the statistical theory of fluctuations it is possible to derive a formula which connects the generalized susceptibility to the order parameter:

$$\chi = (1/T)<\varphi^2> \tag{41}$$

where the symbol <...> denotes the average over the fluctuations and T is the absolute temperature.

Near the critical point T_c of a phase transition the behavior of specific heat c_s, of the order parameter, and of the generalized susceptibility can be described by the following universal power laws (we hypothesize the case in which the non-symmetric phase corresponds to $T>T_c$):

$$c_s = C'\tau^{-\alpha'} \quad \text{for } T < T_c \tag{42.a}$$
$$c_s = C(-\tau)^{-\alpha} \quad \text{for } T > T_c \tag{42.b}$$
$$\varphi = 0 \quad \text{for } T < T_c \tag{42.c}$$
$$\varphi = A\tau^{\beta} \quad \text{for } T > T_c \tag{42.d}$$
$$\chi = k'\tau^{-\gamma'} \quad \text{for } T < T_c \tag{42.e}$$
$$\chi = k(-\tau)^{-\gamma} \quad \text{for } T > T_c. \tag{42.f}$$

In these formulae τ corresponds to

$$\tau = (T_c - T)/T_c \tag{43}$$

It is possible to prove, by using the condition of thermodyinamical stability in equilibrium phase transitions, the following Rushbrooke-Coopersmith inequality to be satisfied by the critical exponents appearing in (42):

$$\alpha' + 2\beta + \gamma' \geq 2. \tag{44}$$

In order to adapt this theory to an EBP network we first observe that in the latter the learning process is nothing but a change of connection weights which, thus, are to be considered as the fundamental state variables. Besides, we can consider the learning process as a transition from an initial symmetric situation in which the connection weight values are uniformly distributed in a suitable domain of the weight space to a final non-symmetric situation where the values of the connection weights are distributed in a strongly structured way.

In order to characterize the symmetric phase, we remember that, from a theoretical point of view, the autocorrelation function of a perfectly uniform distribution is vanishing. From the experimental side it was possible to observe that the mean value, over all connection weights, of their autocorrelation function, defined as:

$$<m> = (1/N)\Sigma_{ij}\ m_{ij}, \tag{45}$$

where N is the total number of connection weights, and

$$m_{ij} = (1/N)\Sigma_{hk}\ w_{ij}\ w^{i}_{+h\ j+k}, \tag{46}$$

is very near to zero when we choose the weights at random, as initially happens in EBP learning procedure, this suggested to use <m> as order parameter.

In order to find a connection between the theory of phase transitions we presented before and the effective dynamics of an EBP neural network, we need a definition of the energy of the external field. Between the many possibilities we chose the one of identifying the input vector with the external field and of defining its energy as the sum of squares of its components:

$$E = \Sigma_i(x_i)^2. \tag{47}$$

Besides, we approximated (39) in such a way that the generalized susceptibility of an EBP network was defined by:

$$\chi = \Delta\phi/E = \Delta<m>/E. \tag{48}$$

Now, if in a rough approximation we identify φ with its average value, we can utilize (41) to find a formula which defines the generalized temperature T:

$$T = \varphi^2/\chi. \tag{49}$$

Once we know the value of T, it is possible to define the network specific heat c_s through:

$$c_s = E/T. \tag{50}$$

With this machinery we are able to test experimentally the hypothesis that learning processes in EBP networks are equivalent to an equilibrium phase transition. To this end we used (Pessa & Penna, 1995) a 3-layered feedforward network with 10 input units, 5 hidden units, and 3 output units. The training set was constituted by 10 different patterns, whereas we had only 4 different output patterns associated to them. We used EBP learning rules in "batch" modality, with zero momentum and learning parameter 0.8. The maximum admissible error needed to stop the learning process was chosen as 0.05. Each training pattern, both for input and for output, had components whose values were only 1 or 0, chosen initially at random. Also the initial weight values were chosen randomly, within the interval from -0.1 to 0.1.

Learning required 162 epochs. The analysis of the curves $c_s = c_s$ (T), $\varphi = \varphi$ (T), $\chi = \chi$ (T), drawn separately for each training pattern, gave evidence of phase transitions, with typical strong discontinuities, for all three quantities, in correspondence to particular critical values of T.

This allowed the calculation of critical temperatures. The results, relative to the patterns where the phase transitions occurred, are shown in the table below.

Pattern number	T_c	α'	β'	γ'	$\alpha'+2\beta'+\gamma'$
1	7.193095	0.34±0.05	0.15±0.04	0.21±0.05	0.9±0.1
4	3.293366	0.3±0.1	0.166±0.005	0.24±0.07	0.9±0.2
6	$3.621462\text{x}10^{-3}$	0.2±0.1	0.05±0.01	0.4±0,1	0.8±0.3
7	2.15015	0.19±0.08	0.16±0.05	0.15±0.06	0.7±0.2
7	30.47164	0.5±0.1	0.009±0.006	0.19±0.06	0.8±0.2
8	2.037515	0.28±0.05	0.22±0.04	0.20±0.08	0.9±0.2
9	9.635491	0.30±0.03	0.21±0.01	0.05±0.03	0.77±0.07

First of all, we remark that these data show that EBP learning has something to do with phase transitions. This circumstance appears to be connected to the fact that this form of learning is a supervised one. Namely, by applying the same method and the same input patterns to a Grossberg-Carpenter ART2 network (see Carpenter & Grossberg, 1987), characterized by unsupervised learning and by the right compromise between stability and plasticity, we obtained no indication of the presence of phase transitions. All experimental curves evidenced, without possibility of error, a gradual learning.

We must, however, take into account that EBP learning does not appear to be connected to equilibrium phase transitions, contrarily to what is expected on the basis of catastrophic interference phenomenon. First of all, the critical exponents have values very different from the ones predicted by classical theories of phase transitions. Besides, and this is the most important result, the Rushbrooke-Coopersmith inequality does no longer hold, contrarily to what was experimentally observed in all phase transitions so far known in physics. As the proof of this inequality is based on the hypothesis that we deal with equilibrium phase transitions, we are forced to conclude that our data imply that EBP learning can be considered as a phase transition, but not an equilibrium one.

This faces us with a dilemma, because the catastrophic interference phenomenon prevents us from considering EBP learning like an irreversible non-equilibrium transition. In order to solve this apparent contradiction, we start by observing that EBP learning dynamics, owing to its supervised character, is driven by suitable boundary conditions which maintain a situation very different from the equilibrium one. Namely, in the absence of inputs, outputs, and error signals, the connection weights would not change, and the learning process would not take place. However, differently from what happens in usual non-equilibrium thermodynamics, these boundary conditions are variable with time. This circumstance gives rise to a very strange situation which appears as different both from equilibrium phase transitions and non-equilibrium transitions.

A first step towards an understanding of what really is the state of affairs consists in the observation that, once the symmetry-breaking transition has happened, learning started and $\langle m \rangle$ became significantly different from zero, the system behavior appears as invariant with respect to continuous symmetry transformations. Namely, a given value of $\langle m \rangle$ can be realized by an infinite number of different choices of the w_{ij}, and every continuous transformation from w_{ij} to w'_{ij} which leaves $\langle m \rangle$ invariant is a symmetry transformation of the network. As it is well known, a similar situation characterizes many physical systems, such as the superconductors, the He^3 and the He^4 superfluids, the nematic liquid crystals. In such cases it is possible to show that this implies the existence of elementary excitations with zero mass which propagate within the system. They are known as *Goldstone bosons* (Goldstone, Salam & Weinberg, 1962). Their role is essentially that of transmitting to all system components information relative to the particular choice of the fundamental state made by the system among the infinite ones compatible with the boundary conditions presently holding. From a mathematical point of view the presence of Goldstone bosons is revealed by a pole in the Fourier transform of the response function of the system, connected in turn to the correlator of the transversal fluctuations.

Are there Goldstone bosons in our EBP network? The answer is yes, and we will sketch here briefly a proof of this assertion. To this end let us consider the simplest case of a 2-layered Perceptron (the extension of these reasonings to the 3-layered Perceptron with EBP learning rule is straightforward) with the Widrow-Hoff learning rule:

$$\Delta w_{ij} = -\eta \Sigma_k [y_i^{(k)} - t_i^{(k)}] \, [\partial F / \partial P_i^{(k)}] \; x_j^{(k)} \qquad (51)$$

where $y_i^{(k)}$ is the i-th output component corresponding to the presentation of the k-th pattern of the training set, $t_i^{(k)}$ is the corresponding desired output, $x_j^{(k)}$ is the j-th component of the k-th input pattern, F is the output unit activation function and:

$$P_i^{(k)} = \Sigma_r w_{ir} x_r^{(k)}. \qquad (52)$$

By supposing that every learning epoch has an infinitesimal duration dt, (51) becomes an ordinary differential equation for the dynamics of w_{ij}:

$$dw_{ij}/dt = G(w_{ij}, x_i^{(k)}, t_i^{(k)}), \tag{53}$$

where G denotes the second member of (51). At this point we can assume that $x_i^{(k)}$ and $t_i^{(k)}$ behave as stochastic variables, characterized by mean values $\bar{x}_i$ and $\bar{t}_i$. It is possible, then, to write:

$$x_i^{(k)} = \bar{x}_i + \xi_i^{(k)}, \; t_i^{(k)} = \bar{t}_i + \zeta_{i(k)}, \tag{54}$$

where $\xi_{i(k)}$ and $\zeta_{i(k)}$ are stochastic fluctuations which, if the number of patterns is great enough, can be considered like Gaussian white noises. If, now, we substitute the (54) into (53), we choose an explicit form for F (e.g., a sigmoid), and we develop G into a power series with respect to $\xi_{i(k)}$, $\zeta_{i(k)}$, having as starting point the values $\xi_{i(k)} = \zeta_{i(k)} = 0$, we will obtain, by retaining only the powers of $\xi_{i(k)}$ and $\zeta_{i(k)}$ not higher than the first, that (53) will transform itself this way:

$$dw_{ij}/dt = H(w_{ij}, \bar{x}_i, \bar{t}_i) + W(\xi_i^{(k)}, \zeta_i^{(k)}) \tag{55}$$

where H is a deterministic part and W is equivalent to a Gaussian white noise contribution. The explicit form of H and W can be determined through tedious but straightforward computations, but they don't matter here. But it is important to remember that Leroy (1992) showed how a system of this type, when invariant under a continuous transformation group, will lead to a pole in the Fourier transform of its response function. The existence of Goldstone bosons is thus proved, at least within the hypothesis we made.

What is the consequence of this result? We remember that some time ago Matsumoto, Sodano and Umezawa (1979) showed that Goldstone bosons can interact in a non-linear way, giving rise to macroscopic extended structures. In more recent times, Obukhov (1990) showed that the non-linear interactions of Goldstone bosons in a system underlying a phase transition can give rise to the appearance of a *self-organized critical state*. With this term we indicate a situation typical of very complex systems which are alway in metastable states far from equilibrium, such that a little external perturbation can generate a sequence of transitions from a state to another of arbitrary length. These latter can be viewed as "avalanches" of arbitrary extensions and duration in the space of the system's possible states. This critical state is called "self-organized" because its appearing is not due to the critical values of some control parameter (such as the temperature) but is intrinsic to systems dynamics and is an attractor for it, independently from the system's initial state.

The prototype of such systems is the so-called "sandpile model" (Bak, Tang & Wiesenfeld, 1987; Roskin & Feng, 1997). As shown, for example, by Kadanoff (1991) and Sornette (1992) the self-organized criticality (Soc) phenomenon is connected to a non-linear feedback mechanism of order parameter on control parameter. A close examination of curves $\varphi = \varphi$ (T) shows, indeed, a very complicated behavior. Also if this circumstance is not enough for granting the presence of Soc, it suggests, together with the presence of Goldstone bosons, that phase transitions observed in EBP learning are more similar to the "avalanches" which are typical of Soc systems. In order to test the validity of this hypotesis we did some numerical experiments on EBP networks to detect the eventual presence of Soc phenomena. In these experiments, however, we used parameters different from those used in the previous experiments on phase transitions in EBP learning. As a matter of fact, it is very difficult to individuate a particular critical value of $<m>$, defined by (45), corresponding to some equilibrium values of connection weights reached after a suitable learning phase. This value, to be useful, should be such that, when a new pattern to be learned is presented, we should be able to say that learning happens if $<m>$ is greater than this critical value and does not happen when $<m>$ is smaller.

For this reason we chose as control parameter, in our experiments on Soc, the instantaneous value of the global quadratic error of network output units, whereas the critical value of this parameter was identified with the maximum admissible error in the EBP learning process. These two quantities correspond, respectively, to the slope and to the critical slope in the sandpile model. As regards the sand "flux", it can be represented by the variation of connection weights in each learning epoch, determined by the difference between the actual slope and the critical slope, i.e., between the actual global quadratic error and the maximum admissible error. More precisely, we choose, as representative of the flux, the sum of the moduli of connection weight variations in each epoch.

In order to do a comparison between EBP learning and Soc behavior in the sandpile model, we remind that the latter is characterized by some scaling invariant power laws. Whith respect to these power laws we can mention the following (Tang & Bak, 1988):

$$D(s) \cong s^{-\tau+1} \quad (56.a)$$
$$D(t) \cong t^{-b} \quad (56.b)$$
$$t \cong \iota^{z} \quad (56.c)$$
$$s \cong \xi^{D} \quad (56.d)$$
$$\xi \cong (p_c-p)^{-\nu} \quad (56.e)$$
$$s \cong (p_c-p)^{-1/\sigma} \quad (56.f)$$
$$j \cong (p_c-p)^{\beta} \quad (56.g)$$

Here s is the cluster amplitude (i.e., the number of grains of sand triggering by an "avalanche"), D(s) its distribution, t the temporal duration of the avalanche, D(t) its distribution, ι the maximum linear amplitude of the clusters, ξ the spatial correlation length, j the flux, p the actual slope and p_c the critical slope. The critical exponents τ, b, z, D, ν, σ, and β must be connected, if we are in a Soc condition, by the relationships:

$$D = 1/\sigma\nu, \; \beta = (\tau-2)/\sigma. \quad (57)$$

In our experiments we used a 3-layered feedforward EBP network with 10 input units, 3 hidden units, and 3 output units. The training set contained 10 patterns, each one with

10 components having the values 0 or 1. The correct answer to each input pattern was determined by using a deterministic rule which required that the components of each pattern be subdivided into three groups: the first constituted by the first three, the second by the intermediate four, and the third by the last three. The activation of the i-th output unit was 1 if in the i-th group at least two values were equal to 1.

By using an EBP learning rule with learning rate = 0.4, zero momentum, maximum admissible error 0.01, initial connection weights randomly chosen between 0.1 and -0.1, the network learned correctly the training set after 3140 epochs. Starting from this situation we perturbed the network by performing a new learning procedure with a training set of 11 patterns, given by the 10 patterns previously learned, plus one additional pattern which played the role of avalanche trigger. We did 50 experiments, varying randomly this new additional pattern. As data we collected the learning time, the instantaneous global quadratic error, and the instantaneous "flux", defined through the variations of connection weights.

The main problem, however, was that of introducing some "spatial" features of the EBP learning process, in order to do a direct comparison with the corresponding "spatial" features of the sandpile model. To this end we remark that the initial learning, the one on the first 10 patterns, led us to introduce a "natural metric" in a "cognitive space" where we can represent each connection weight as a geometrical point. Namely, once this learning was completed, each hidden unit was characterized by a well-defined activation value in correspondence to each input pattern belonging to the training set. Thus, the behavior of this unit can be represented by a list containing the m activation values (m is the number of patterns of the training set, in this case 10) and the nm activation values of input units corresponding to training patterns (n is the number of components of each input pattern). This is equivalent to say that each hidden unit can be associated to a point in a (m+mn)-dimensional space. By adopting this point of view, it is easy to see that each connection weight relative to input-hidden connections can be associated with a point in a 2m-dimensional space, because its action can be characterized by a list of 2m numbers, i.e., the m possible activation values of the input unit from which the connection starts (one for each input pattern of the training set), and the corresponding m activation values of the hidden unit towards which the connection is directed.

Once we have obtained such a "spatial" representation of connection weights, it is possible to introduce a Euclidean metric on the 2m-dimensional space and to have a measure of the "spatial distance" between two connection weights. These reasonings can be repeated, *mutatis mutandis*, for the connection weights between hidden and output units. In order to do a direct comparison with the spatial features of the sandpile model we need only a criterion for deciding what variations of connection weight values, triggered by the learning of the new additional pattern, are to be considered as "activations" of the connection weights themselves. To this regard, we adopted the criterion of considering as "activated", in each epoch, the connections whose weights varied more than $1/(n^c\sqrt{n^c})$, where n^c is the total number of connections. This value represents the average value, for a single connection, of the mean global fluctuations expected in such a system.

Starting from these premises, all spatial quantities, such as the maximum linear dimension of the cluster of activated connections, the global cluster amplitude, the spatial correlation length, can be easily calculated as functions of learning time. It was, thus, possible, to compare directly the results of the numerical experiments on EBP learning with the power laws (56.a)-(56.g) of the sandpile model. In particular, the critical exponents were calculated through a linear regression on a log-log representation of data. The numerical results obtained, for the input-hidden connections were: $D = 3.22$, $\nu = 186.19$, $\sigma = 1.160 \times 10^{-2}$, $\beta = 88.20$, $z = -3.56$, $b = 1.01$, $\tau = 0.98$.

We remark that the second of the relationships (57) was satisfied almost exactly. This allows us to think of EBP learning as characterized by a Soc behavior.

As a conclusion, it seem that we can classify EBP learning, and therefore concept learning by human subjects, as something intermediate between equilibrium and non-equilibrium phase transitions. In other words, Soc represents a state in which the fluctuation amplitudes can grow to macroscopic dimensions but without a fixed boundary or inhibition mechanism able to fix the length scale of the macroscopic structures so formed. This state is a prerequisite for the appearance of non-equilibrium morphogenetic phase transitions; but some of the conditions which characterize the latter are missing. This suggests a search for finding which learning processes by human beings would be characterized by these conditions. The examples we have presented show the strong interplay between the Synergetics of Cognitive Processes and the experimental research within the domains of psychology and of neural network modelling.

6. THE CONDITIONS OF EXISTENCE OF MACROSCOPIC STRUCTURES: THE TOP-DOWN APPROACH

If we adopt a top-down approach, we must resort to the conventional methods of Synergetics (Haken, 1983; Mikhailov, 1990; Mikhailov & Loskutov, 1991). The latter can be better applied if we formulate the dynamics of connectionist models through macroscopic equations obtained by replacing the discretized neural network behavior with the one of a continuous neural field. For example, a locally connected 1-dimensional network described by:

$$x_i(t+1) = F[x_i-1(t)+x_i(t)+x_i+1(t)], \quad (58)$$

where F is a suitable activation function, can be described, within the hydrodynamical limit, by the following PDE:

$$\partial u/\partial t = -u+F[(\partial^2 u/\partial x^2)+3u] \quad (59)$$

where $u = u(x, t)$ is the neural field. This equation can be studied through the well-known methods of bifurcation theory (Sattinger, 1973). It is easy to see that, with Dirichlet boundary conditions in a segment of finite lenght L, such that $u(0) = u(L) = u_0$, where $u_0 = F(3u_0)$, the homogeneous stationary equilibrium solution of (59) is unstable with respect to great wavelength perturbations.

The analysis becomes more complicated if we generalize (58) to a 2-dimensional lattice, with different values of connection weights between neighbouring neurons. If we limit ourselves to the case of a uniform stochastic distribution of the connection weights and write the corresponding Pde, within the hydrodynamical limit, by using average weight values, bifurcation theory shows that there are suitable conditions of the parameters appearing in the Pde which grant the existence of stable stationary homogeneous solutions. The computer simulations, however, show the existence of a spatially nonhomogeneous stable equilibrium state (Pessa & Penna, 1994b), characterized by a very short spatial correlation length. This implies that the behaviour of models such as (58) and its 2-dimensional generalizations is determined mainly by local computations. These kinds of models, then, are suitable for doing "global" computations giving rise to the emergence of macroscopic structures.

With regards to this we must mention the solitary waves (localized structures) and quasi-particles (diffuse structures). As shown by Umezawa and his coworkers (cfr. Wadati, Matsumoto & Umezawa, 1978), quasi-particles can undergo a boson condensation, giving rise to extended objects with definite topological properties. In turn, these latter can be converted into solitary waves. We can thus assert that the different forms of macroscopic structures can be transformed one into the other. Much experimental and theoretical work remains to be done before obtaining a full understanding of the conditions granting the emergence of macroscopic structures from interacting microscopic neural activities.

7. CONCLUSION

The previous review of the experimental and theoretical work in the field of Synergetics of Cognition showed that, as a principle, it is impossible to have a model of cognitive processing which:

a) is able to explain all experimental data;
b) is completely solvable;
c) is unique.

In this situation we think that the right approach is that of using a plurality of models. In other words, the researcher should undertake the study of phenomena equipped with a "population" of models. From this point of view modelling activity could be considered as equivalent to the evolutionary dynamics of this "population" under the action of some "genetic" rules (cfr. Pessa, 1994). These latter can be classified as:

1) transformation rules which let one derive from an "ancestor" model its "descendants";

2) fitness criteria selecting the members of the population to be taken into consideration.

There has been a certain deal of work on fitness criteria (cfr. Holland et al., 1986; Sirgy, 1988). We feel, however, that one of the most important criteria is given by the degree of correspondence between the *structural relationships* within the model and the *structural relationships* characterizing cognitive behavior. These latter do not appear to be much studied in Cognitive Psychology. There are, anyway, some important developments such as:

a) the compatibility effects in memory;
b) the inverse duration effects in iconic memory;
c) the extralist cue effect in recall;
d) the attentional operating characteristics;
e) the mirror effect in recognition memory.

As a conclusion, we can assert that today we are very far from the building of a true Synergetics of Cognition. The only way for attaining this goal appears to be the comparative study, both from the experimental and the theoretical study, both from the experimental and the theoretical point of view, of the structural properties of both networks and human cognitive behaviours.

Prof. Eliano Pessa
Dipartimento di Psicologia
Università di Roma "La Sapienza"
V. dei Marsi, 78
00195 Roma, Italy

REFERENCES

Allen R.B. (1990), "User models: Theory, methods, and practice", "International Journal of Man-Machine Studies", 32: 511-543.

Amit D.J. (1984), Field theory, the renormalization group and critical phenomena, Singapore: World Scientific.

Bak P., Tang C., Wiesenfeld K. (1987), "Self-organized criticality: An explanation of 1/f noise", "Physical Review Letters", 59: 381-384.

Benjafield J.G. (1992), Cognition, Englewood Cliffs, NJ: Prentice-Hall.

Bonnie J.E., Kieras D. (1996), "Using GOMS for user interface design and evaluation: Which technique?", "ACM Transactions on Computer-Human Interaction", 3: 287-319.

Card S.K., Moran T.P., Newell A. (1983), The psychology of human-computer interaction, Hillsdale, NJ: Erlbaum.

Carpenter G.A., Grossberg S. (1987), "ART2: Self-organization of stable category recognition codes for analog input patterns", "Applied Optics", 26: 4919-4930.

Feldman J.A., Ballard D.H. (1982), "Connectionist models and their properties", "Cognitive Science", 6: 205-254.

Goldstone J., Salam A., Weinberg S. (1962), "Broken symmetries", "Physical Review", 127: 965-970.

Haken H. (1978), Synergetics. An introduction, Berlin-Heidelberg-New York: Springer.

Haken H. (1983), Advanced Synergetics, Berlin-Heidelberg-New York: Springer.

Haken H. (1988), Information and Self-organization, Berlin-Heidelberg-New York: Springer.

Haken H. (1989), Neural and Synergetic Computers, Berlin-Heidelberg-New York: Springer.

Holland J.H., Holyoak K.Y., Nisbett R.E., Thagard P.R. (1986), Induction, Cambridge, MA: MIT Press.

Hummel J.E., Biederman I. (1992), "Dynamic binding in a neural network for shape recognition", "Psychological Review", 99: 480-517.

Kadanoff L.P. (1991), "Complex structures from simple systems", "Physics Today", March: 9-13.

Kruschke J.K. (1993), "Human category learning: Implications for backpropagation models", "Connection Science", 5: 3-36.
Landau L.D., Lifshitz E.M. (1969), Statistical Physics, Elmsford, NY: Pergamon Press.
Lee Giles C., Maxwell T. (1987), "Learning, invariance, and generalization in high-order neural networks", "Applied Optics", 26: 4972-4978.
Leroy L. (1992), "On spontaneous symmetry breakdown in dynamical systems", "Journal of Physics A", 25: L897-L900.
Ma S.K. (1976), Modern theory of critical phenomena, Reading, MA: Benjamin.
Matsumoto H., Sodano P., Umezawa H. (1979), "Extended objects in quantum systems and soliton solutions", "Physical Review D", 19: 511-516.
McClelland J.L., Rumelhart D.E. (1986), Parallel distributed processing. Explorations in the microstructure of cognition, Cambridge, MA: MIT Press.
Mikhailov A.S. (1990), Foundations of Synergetics I, Berlin-Heidelberg-New York: Springer.
Mikhailov A.S., Loskutov A.YU. (1991), Foundations of Synergetics II, Berlin-Heidelberg-New York: Springer.
Minati G., Penna M.P., Pessa E. (1998) "Thermodycamical and Logical openness in General Systems", "Systems Res. and Behavioral Science", 15: 131-1-45.
Murre J.M.J. (1996), "Hypertransfer in neural networks", "Connection Science", 8: 249-258.
Nicolis G., Prigogine I. (1977), Self-organization in nonequilibrium systems, New York: Wiley.
Novozhilov YU.V., Yappa YU.A. (1981), Electrodynamics, Moscow: Mir.
Obukhov S.P. (1990), "Self-organized criticality: Goldstone modes and their interactions", "Physical Review Letters", 65: 1395-1398.
Pessa E. (1994), "Symbolic and subsymbolic models, and their use in systems research", "Systems Research", 11: 23-41.
Pessa E., Penna M.P. (1994a), "Catastrophic interference in learning processes by neural networks". In M.Marinaro, P.Morasso (Eds.), ICANN 94 (Vol. I, pp. 589-592), Berlin-Heidelberg-New York: Springer.
Pessa E., Penna M.P. (1994b), "Local and global connectivity in neuronic cellular automata". In Proceedings of the Third IEEE International Workshop on Cellular Neural Networks and their Applications (CNNA-94)". (pp. 97-102) Roma: IEEE.
Pessa E., Penna M.P. (1995), "Can learning process in neural networks be considered as a phase transition?". In M.Marinaro, R.Tagliaferri (Eds.), Neural Nets. WIRN Vietri-95. (pp. 123-129) Singapore: World Scientific.
Pessa E., (1998), "Emergence, Self-Organization and Quantum Theort" in G. Minati (ed.) First Italian Conference on Complex Systems, Milano, pp. 59-79.
Polson P.G. (1987), "A quantitative theory of human-computer interaction". In J.M.Carroll (Ed.), Interfacing thought. (pp. 184-235) Cambridge, MA: MIT Press.
Ramacher U. (1993), "Hamiltonian dynamics of neural networks", "Neural Networks", 6: 547-558.
Ratcliff R. (1990), "Connectionist models of recognition memory: Constraints imposed by learning and forgetting functions", "Psychological Review", 97: 285-308.
Rauterberg M. (1993), "AMME: An automatic model evaluation to analyse user behaviour in a finite, discrete state space", "Ergonomics", 36: 1369-1380.
Roskin H.J., Feng Y. (1997), "Self-organized criticality in some dissipative sandpile models", "Physica A", 245: 453-460.
Rumer YU.B., Ryvkin M.S. (1980), Thermodynamics, Statistical Physics, and Kinetics, Moscow: Mir.
Sattinger D. (1973), Topics in Stability and Bifurcation Theory, Berlin-Heidelberg-New York: Springer.
Sirgy M.J. (1988), "Strategies for developing general systems theories", "Behavioral Science", 33: 25-37.
Sornette D. (1992), "Critical phase transitions made self-organized: A dynamical system feedback mechanism for self-organized criticality", "Journal de Physique", 11: 2065-2073.
Tang C., Bak P. (1988), "Critical exponents and scaling relations for self-organized critical phenomena", "Physical Review Letters", 60: 2347-2350.
Von der Malsburg C., Buhmann J. (1992), "Sensory segmentation with coupled neural oscillators", "Biological Cybernetics", 67: 233-242.
Wadati M., Matsumoto H., Umezawa H. (1978), "Extended objects in crystals", "Physical Review B", 18: 4077-4095.
Yamaguchi Y., Shimizu H. (1994), "Pattern recognition with figure-ground separation by generation of coherent oscillations", "Neural Networks", 7:49-63.

RICCARDO LUCCIO

SELF-ORGANISATION IN PERCEPTION THE CASE OF MOTION

1. PRÄGNANZ

Prägnanz is definitely a cardinal concept in Gestalt theory, but, as Kanizsa and I have stressed in the last ten years (Kanizsa and Luccio, 1986, 1987, 1990, 1993, 1995), it has nevertheless given rise to a number of misunderstandings. The Gestaltists have often been criticised for having turned Prägnanz into a key to open all doors, without ever having given it a strict definition. The concept was introduced by Wertheimer (1912) in his essays on thought processes in primitive peoples in which he speaks of privileged, *ausgezeichneten* or "prägnant" zones in numerical series. However, Wertheimer spoke of a "law of Prägnanz" only two years later, affirming that, amongst many "Gestalt laws" of a general type, there is a "Tendenz zum Zustandekommen einfacher Gestaltung (Gesetz zur 'Prägnanz der Gestalt') (Wertheimer, 1914).

In Wertheimer's 1922/1923 essay, the very first systematisation of *Gestalttheorie*, traces can be found of the origins of some the ambiguities in the concept of Prägnanz (defined as *Ausgezeichnetneit*), which is a quality possessed by certain specific objects, forms or events belonging to our immediate perceptual experience which makes them "unique", "singular", "privileged". All the shapes which are phenomenally singular or "privileged" are "good Gestalten": it is the case of the equilateral triangle, of the circle, of the square, of the sinusoid, etc. In this sense, "prägnant" indicates phenomenal structures which are "regular"; they are endowed with internal coherence; all their parts go well together and can be said to "belong" to each other by mutual necessity.

But Wertheimer also gave a second sense of Prägnanz, that of the *lawfulness* of the process leading to the formation of visual objects. According to this second meaning, the term Prägnanz is used by Wertheimer to indicate the fact that it is rather a "meaningful" (*sinnvoll*) process. The principle of organisation acts as precise laws to which the process is forced to obey, overall in the sense of maximum economy and simplicity. Its result is a perfect balance of the forces at play and thus has also a maximum of stability and resistance to change.

According to Wertheimer, the process is such that any "almost good" Gestalt should end to be perceived as a prägnant one. For example, he says: "... that things are so is clearly demonstrated in experiments where the consistency of a tendency to a prägnant configuration is remarkable. *If an angle is tachistoscopically presented, even if its main difference from the right angle is noticeable, the viewer often simply sees a right angle, assimilating the shown angle to the prägnant one...*" (1923, p. 318).

After Wertheimer, the Gestalt psychologists used the concept always in the descriptive sense to indicate the "singularity" of a phenomenal outcome, or in explanatory way to indicate the conformity to rules of the perceptual process and its tendency towards a final state of stable equilibrium. The two concepts are not at all equivalent, in that a phenom-

A. Carsetti (ed.), Functional Models of Cognition, 91-100.

enal result can be completely stable but not necessarily at the same time *ausgezeichneten* in the sense of phenomenally "singular".

Very rare were the attempts to distinguish between the two meanings. Among them, A. Hüppe (1984) suggested such a distinction, calling phenomenal goodness *Primärprägnanz* and conformity of the process to rules and stability of the result *Sekundärprägnanz*. Prägnanz in the former sense, that is, "singularity" or figural "goodness", is then a given phenomenal *fact*, corresponding to a reliable description of visual experience, which was destined to play a leading role in later Gestalt theorising.

In any case, after Wertheimer the most important and interesting contributions to the development of the concept of Prägnanz in its first sense were made by Rausch (1966), who lists seven *Prägnanzaspekte* (bipolar dimensions): 1) *Gesetzmässigkeit - Zufälligkeit*; 2) *Eigenständigkeit - Abgeleitetheit*; 3) *Integrität - Privativität*; 4) *Einfachheit - Kompliziertheit*; 5) *Komplexität - Tenuität*; 6) *Asdrucksfülle - Ausdrucksarmut*; 7) *Bedeutungsfülle - Bedeutungsleere*: fullness of meaning as opposed to absence of meaning. He (1952) also distinguishes three zones around each point of Prägnanz: the zone of *formation (Verwirklichungsbereich)*, which is the exact point occupied by figures assimilated to the category of the prägnant one but which are experienced as badly made, "bad", and the *derivation* zone *(Ableitungsbereich)* to which belong figures which are categorically different from the prägnant ones whilst being connected to them in a relationship of derivation.

Opposite is the view put forward by E. Goldmeier (1937, 1982). Goldmeier's analysis differs from Rausch's in the degree of importance given to the two possible meanings, which may seem in a certain sense contradictory, of the concept of Prägnanz. Goldmeier emphasizes the fact that the zones of Prägnanz mark the points of discontinuity in a qualitative series. For Rausch, on the contrary, Prägnanz is above all a scalar property that can take on all the values of intensity lying between the two poles of the seven dimensions he distinguished.

For Goldmeier, the most salient characteristics of Prägnanz, which he significantly translates as "singularity", is the "uniqueness" possessed by some configurations in virtue of their having a quality that all others in a given series lack. As stressed in Goldmeier's view, one peculiar characteristic of our perceptual system that singularity highlights is that it has a high *sensitivity to change*. In the near singularity zone (which corresponds to Rausch's "approximation" zone) the slightest fluctuation of a singular value is noticed, whilst the threshold of discrimination rises considerably for those values which fall outside this area, where we are no longer able even to notice great differences between two adjacent elements in a series. But, note, this observation, that is very easy to check, is in full contradiction with the claimed "tendency to Prägnanz", in terms of the quotation from Wertheimer reported above. And it is in conctradiction with all other Gestalt theorists that claim that a tendency to Prägnanz exists, when Prägnanz is meant as singularity (Köhler, 1924, p. 531; Metzger, 1963, p. 207).

According to Stadler, Stegagno and Trombini (1974), a singular figure has less "tolerance of identity" *(Identitätstoleranz)* than a non-singular figure, or, in other words, it has a greater resistance to change. They used a stroboscopic transformation movement which consists in presenting, in rapid succession, two not too dissimilar forms, causing the perception of only one form, the first, which transform itself into the second while retaining its identity. With longer inter-stimulus intervals, differing according to the type of figure used, the succession of the two forms may be seen. Referring to Rausch's first four dimensions of Prägnanz (1996), they were able to show that the non-prägnant figures transform themselves into the prägnant one more easily than vice versa: in other words, they had a greater *Identitätstoleranz*. What we have here is a functional effect of singularity, like others (the already mentioned greater discriminative ability or greater processing speed as shown by Garner, 1962) which presumably demonstrate a greater ease in the encoding of singular structures.

So, Gestalt theorists use the term Prägnanz to mean *both* a tendency of the perceptual process to assume the most regular and economic course, given the constraints *(Randbedingungen)* present in each specific case, *and* a tendency towards the maximum *Ausgezeichnetheit* in the concrete phenomenal result of the process itself. It seems quite evident that for such theorists there is a close logical connection between these two facts. In general, scientists tend to take for granted that in nature processes governed by a minimum principle tend to produce regular, symmetrical results (Mach, 1885). The regularity is particularly apparent when we notice some kind of symmetry in the inanimate world (crystals, snowflakes, and so on), as well as in the natural kingdom (leaves, flowers, butterflies, and so on). Such instances are shown as conclusive evidence that natural phenomena have a character which is not casual, but stricly conforms to laws.

Even if one agrees to all this, a confusion arises when it is claimed that the tendency towards *Ausgezeichnetheit* is a natural consequence of the tendency towards *economy of the process*. Also in nature only few natural objects have a regular structure and most are amorphous or ill-shaped; so few phenomena objects and events have a "good" shape and are in this sense "better" than others, well done, *ausgezeichnet*.

It must be emphasied that such a process should be take place at the moment of the formation of the visual objects, a perceptual process in which it is convenient to distinguish two different levels, which, following Kanizsa (1979), I will call primary and secondary processes, respectively. The first process determines an immediate segmentation of the perceptual field, which therefore appears to awareness as being constituted by many phenomenal objects, distinguished from each other, before and irrespective of the attribution of a meaning to them, an attribution which is allowed by the secondary process.

There is a logical reason to distinguish these two processes. As Höffding (1887, pp. 195-202) emphasized, it is impossible to recognise an object if it is not already present. As a matter of fact, it is evident that the formation of a visual object as an entity distinct from other objects must take place *before* the objects can be recognised, and this is a logical requirement that cannot be refuted on the grounds that is is impossible to observe in a natural cognitive act a phase in which the visual data has not yet been identified.

The implications of Höffding's argument (or "step", as it is often said) were mainly developed in the Gestalt field (e.g., Köhler, 1924). The argument could be stated as follows: Let us take two associated mental contests, *a* and *b*. Let us now suppose that a new event *A* occurs, endowed with the same properties as *a*. Now, *A* leads to the evocation of *b*; and yet, *A* is not *a* and is not associated with it. The only way to explain the activation of *b*'s trace following *A*'s presentation is that *a* is activated because of its similarity links with *A*. In other words, Höffding's argument affirms that before an external event can be recognised and placed into the pertinent category, it must be constituted in such a way that it is endowed with characteristics which allow it to come into contact with the trace of a similar event. (For a discussion of Höffding's argument, see also Zuckerman and Rock, 1957; Rock, 1962; Kaniza & Luccio, 1987).

In my opinion, a consequence is that the tendency to Prägnanz is well recognisable *only* in the products of the secondary process, especially in transformations which are linked to the formation of memory traces, also in short term memory, and that a tendency to Prägnanz as singularity does not exist at all. In my view, the behavior of the visual system is not characterised by a tendency to singularity but by a *tendency to stability*. Though proximal stimulation undergoes a continous process of transformation, our phenomenal world is usually a *stable* world, constituted by objects that preserve a high degree of constancy as to size, shape, colour, identity. The stability is the result of a capacity of self-organisation displayed by the visual system. This has already been claimed by Gestalt theory using arguments that today many scholars have revisited, updated and refined, also in the light of progress made by neuroscience (Stadler & Kruse, 1986; Hatfield & Epstein, 1987; Haken & Stadler, 1990). The system regulates itself according to principles that are essentially the ones that Wertheimer specified (proximity, similarity, common fate, and so forth). The synergetic or conflicting action of such principles tends to a perceptual result that is better in the sense of maximal stability (i.e., less reversible, less ambiguous), and not to the better result in the sense of the esthetically agreeable, prototypical, or singular. Most cases that are referred to in the literature as evidence of a tendency to singularity are, according to Kanizsa and Luccio, casual results of these organisational principles. The possibility of a phenomenally "singular" appearance is only a by-product. The phenomenal solution preferred by the visual system does not show characteristics like regularity, symmetry, prototypicity, which are the peculiarities of Prägnanz, if understood as singularity.

Kanizsa & Luccio (1986) and Luccio (1966) have widely discussed this thesis, giving a number of examples that argue against this latter hypothesis. In this paper, I will confine myself to the case of motion.

2. THE DISTRUCTION OF THE ACTUAL SINGULARITY OF TRAJECTORIES

A strong counterdemonstration to the supposed tendency to Prägnanz as singularity is given by experiments on the perception of movement. It is possible to demonstrate that highly singular components of the perceptual field can be concealed, with a perceptual result which is all but prägnant.

It is known that often the perceived movement of an object doesn't correspond to its physical motion. This is true for *speed*, because it varies at the variation of the frame of reference inside which the movement occurs (Brown, 1931). It is true for *direction* also, as Ames's oscillating trapezoid (1955) and Johansson's analysis (1950) proved. The description of what one sees looking at a rotating wheel is very simple and univocal: the wheel accomplishes a movement of linear translation; and meanwhile all its parts accomplish circular movements around the axis of the wheel itself. Indeed, only one point of the wheel, its centre, goes on a "physical" path that corresponds to the phenomenal path. All other parts accomplish motions that are different from what one sees. No physical path is circular. To see the actual motion one needs to isolate a single part from all other parts, as Rubin (1927) and Duncker (1929) first did. One can accomplish this in a very simple way fixating a little lamp or a phosphorescent dot somewhere on a wheel. Then one lets the wheel roll in the dark along a plain. If the light is placed on the perimeter of the rolling wheel, one sees it running a path built up by a series of loops. This corresponds to the path physically followed by the lighted dot in the space. Mathematicians call this path a *cycloid*. In this case phenomenal path and physical path coincide. If one adds a second light to the periphery of the wheel, it is not so easy any more to see the two cycloids; phenomenally a rotatory movement of each point around the other prevails (Cutting & Proffitt, 1982). This phenomenal result stabilises itself and becomes coercive if we increase the number of lighted points on the perimeter. Although it is still true that all the lighted points actually trace cycloids, we are quite unable to see them. We see, on the contrary, the points that rotate around an invisible centre and that displace themselves all together along another invisible plane. This phenomenal decomposition of the actual cycloid motion in a rotatory and a translatory component has often been considered a particularly convincing proof of the existence of a *tendency to Prägnanz* in the perceptual system. Indeed, a circular movement is certainly "better" in the sense of regularity and fluidity than a discontinuous and jerking cycloidal motion.

One counterexample can be demonstrated for the perceived shape of the path of a movement. If three dots move along three circular paths which partially overlap (see Fig. 1), we don't succeed in seeing the actual paths. What we see is an elastic triangle rotating and twisting in space. If we increase the number of the dots moving on each path fom one to five, the patterns are still invisible. In the area in which the circular paths overlap, the dots form continuously transforming and disrupting groups. The overall impression is one of great disorder.

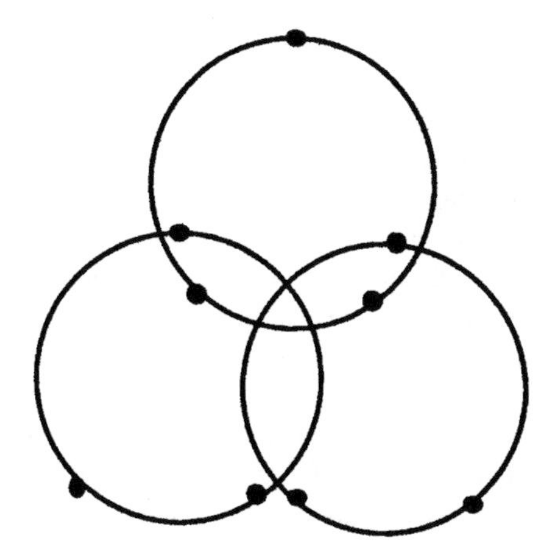

Figure 1 - Dots moving on three partially overlapping paths

The observer succeeds in detecting the circular motions only when there are more than six dots on each path. Obviousy, there is a problem of relative distance between dots. Note that the observer is quite aware of the existence of the three distinct circular paths; his or her task is precisely to succeed in detecting them. The phenomenal impression is one of confusion, of a brownian movement of dots surging from the middle of the configuration. This phenomenon was first seen informally by Kanizsa and Luccio in 1984. More precise conditions were established by Kanizsa, Kruse, Luccio and Stadler (1995) and by Luccio, Leonardi and Parovel (1966) in formal experiments.

In the first experiment by Kanizsa, Kruse, Luccio and Stadler, some dots appeared that were moving along three circular, partially overlapping paths (see Fig. 1). The number of the dots was equal in the three paths, and there were 8 conditions (15, 10, 8, 6, 5, 4, 3, 2 dots for each path). The positions of the dots were equidistant on each trajectory and occupied corresponding positions on the three trajectories. The task of the Ss was to assess the clearness with which they could detect the circular paths. The results were very clear-cut. From 6 dots up, the Ss could detect the paths with high confidence; but they were highly unable to do so from 4 dots down. From a qualitative point of view, we can say the following about the number of dots per path:

- one-two dots: Ss reported that they see the vertices of one (respectively two) virtual triangle that rotates (rotate together) on the screen.
- three dots: the prevailing perceptual result was a sort of pulsation, with dots moving alternatively inward and outward, with reference to the centre of the figure.
- 4-5 dots: any regularity disappeared: Ss saw something like a chaotic brownian motion in the centre of the figure; or dots springing up from the centre, in a process of continuous new generation. At the periphery of the figure the individual dots could trace fragments of circular paths, but these paths were completely lost towards the middle.
- 6-15 dots: the circular paths appear more and more clearly with the increasing of the number of dots.

In a paper on minimum principle and perceived movement, Cutting and Proffitt (1982), stressed the importance of the distinction between absolute, common, and relative motion. It is very clear what absolute motion is, mainly after the seminal work of Rubin, (1927) Duncker (1929), and more recently Johansson (1950, 1973). However, ideas are less clear about relationships between common and relative motion. The first is the apparent motion of the whole configuration relative to the observer, and the second is the apparent motion of each element relative to other configural ones. Cutting and Proffitt, however, have shown that there are two simultaneous processes that correspond to common and relative motion. In both, the minimum principle is involved. The prevalence of either is a matter of which process reaches a minimum first.

According to our results, in the situation with 1-2 dots there is a very clear prevalence of common motion: Ss perceive the whole figure as one or two triangles that move circularly all together. Once does not see the individual circular paths, because the three dots that are the vertices of the triangles belong each to a different path, and one sees the triangles rotating as a whole. With 6 or more dots, relative motion clearly prevails. Notice that in this case relative motion and absolute motion coincide. Here we see clearly the dots moving along circular paths. These dots, however, are seen *relative* to the other dots, forming three independent rotating circumferences. So, according to the minimum principle, one can consider the perceptual solutions for 1-2 and 6-15 dots to be "good" solutions. But what can we say about the 3-5 dots conditions? Hero no "good" solution is seen, neither for common motion (3-5 triangles rotating togeter), nor for relative motion (rotating circumferences). Moreover, why is there this dramatical change in the appearance of configurations, varying with the number of dots?

In this experiment, we noticed that when the average distance of each dot from the other dots of the same path (DS) is different, according to the class of the polygon, the average distance of each dot from the dots of the other paths (DO) is always the same, leaving constant the relative position of the three paths. If the centres of the paths are on the vertices of an equilater triangle, as in our case, DO is always the same, with 3, 4, 5… 15 dots per path. (We have discovered this geometric regularity with some surprise *a posteriori*). Now, in the figures that we used in the experiment, DO was always 4.96 cm. The DSs were respectively 3.96 cm (15 dots), 4.22 cm (10 dots), 4.27 cm (8 dots), 4.48 cm (6 dots), 4.65 cm (5 dots), 4.83 cm (4 dots), 5.20 cm (3 dots), 6 cm (2 dots). This suggested to us the hypothesis that the difference between DO and DS could be a crucial factor for the prevalence of either common or relative motion. When DO is clearly minor to DS, common motion prevails. When DS is clearly minor to DO, relative motion prevails. To test this hypothesis, a second experiment was performed, varying the distance between paths and leaving constant the number of dots per path. The task of the Ss was again to assess the clearness with which they could detect the circular paths.

In this second experiment, conditions were very similar to the ones of the first experiment. The number of dots was always three per path. The speed was 30 revolutions per minute. The centres of the paths were on the vertices of an equilateral triangle that in this case varied with regard to the length of the sides, in accordance with 5 conditions, according to different DOs. The DS was always the same.

Again the results are very clear-cut, replicating the ones of the first experiment and so supported clearly the hypothesis. When DOs are minor to DS, it is very difficult to detect the circular paths. A perceptual organisation in which the dots are moving along these paths is possible only if the average distance between dots on a path is less than the average distance of each of them to the dots of the other paths. Thus, the proximity of dots proves itself to be a crucial factor in the perceptual organisation of phenomenal motion.

This is no surprise. Several authors have stressed the importance of this factor in similar situations, for instance, Ishiguchi (1991). Ullman (1979, 1984) has demonstrated how important proximity is in the perceptual solution of the so called "correspondence problem", which arises when the apparent moving objects are more than 2 and when therefore there is a plurivocity of perceptual outcomes (exactly, for N objects, the possible solutions are N!) (see also Dawson & Phylyshyn, 1988; Marr, 1982, pp. 182 ff.).

But there is a difference between the situation of the described experiments and the other perceptutal situations previously investigated. The perceptual result that one gets when no circular paths are seen is far from being simple, regular, prägnant. Indeed, one can think that the condition to have such perceptual outcomes are *present* in the stimulus. The dots actually move along circular paths, and circles are highly prägnant figures.

And when common motion prevails, as in conditions 1 and 2 of the first experiment, we have equilateral triangles moving together, again highly prägnant figures. Yet, in conditions A and B, as in conditions 3-5 of the first experiment, neither circles nor triangles can be seen.

We can conclude that with the increasing of the number of dots in the first experiment the visual system undergoes a phase transition from common motion to relative motion passing an instable situation with no clear perceptual organisation. The relevant control parameter (in terms of "Synergetics"; Haken & Stadler, 1990) or the higher order variable (in terms of "ecological perception") is the difference between the average distance from the dots of other paths (DO) and the average distance from the dots of the same path (DS). If DO is clearly minor to DS, the order parameter of common motion emerges, and the system is in a stable attractor state. If DS is clearly minor to DO, the order parameter of relative motion emerges and the systems is in a totally different stable attractor state. In this theoretical view of the experimental data it can be predicted that one sould be able to demonstrate a hysteresis effect by gradually approaching the phase transition from the situation of stable common motion and from the situation of stable relative motion.

This interpretation is strongly supported by the results obtained by Luccio, Leonardi and Parovel (1996). In these experiments, we had three dots per path, and four paths that were moving approaching alternatively the center or the vertices of a rectangle. The Ss had to individuate the two transition points between common motion and perceptual chaos, and between perceptual chaos and relative motion. We were able to demonstrate a strong perceptual hysteresis, according to the direction of the movement of the paths. Similar phenomena were individuated in other perceptual domains (e.g. speech categorization), and are assumed to be a strong support for nonlinear dynamic models of perception.

CONCLUSION

It seems to me that, for all I have said and shown, a *tendency* to Prägnanz in the perceptual field *does not exist* when Prägnanz is seen as a tendency to singularity. However, it does not mean at all that nature is not lawful. The world which surrounds us is in fact normally perceived as a highly *stable* world. Therefore, if one wants to speak about tendency, one must say that there is an autonomous tendency of the field to *stability*. In my opinion, till now the more convincing interpretation of the tendency to stability is the one in terms of economy and simplicity, that is in terms of a "minimum" principle (cfr. Hatfield & Epstein, 1985; Zimmer, 1986).

Notice that, apart from the very special cases of multistability, nearly any stimulus situation, although it is in principle plurivoque, and can therefore give rise to many phenomenal outcomes, tends *to come perceptually to a unique outcome*: not towards the most singular solution, but in general towards the most stable one. This probably occurs because the structural factors which in any stimulus situation are usually numerous are often in antagonism to each other (proximity *vs* closure *vs* continuity of direction, etc.); therefore, the more stable situation is the one with the maximum equilibrium between the tensions generated by the counteracting factors. Such tensions, however, find a point of balance in configurational structures, that only by accident have also the property of the figural "goodness". Only in special cases, particularly those in which only one factor acts, can one presume that the tendency to stability coincides with the tendency to Prägnanz. But the more numerous the interacting factors are, and consequently the more complex the occurring configurations, the more rarely stable solutions coincide with the prägnant ones.

The case of motion appears to me specifically apt to support this point. But the case of motion (and similar considerations could be made in other domains of perception, particularly in the case of multistable displays: see Kanizsa and Luccio, 1995; Haken, 1995; Stadler and Kruse, 1995) is particularly apt to show the heuristic value of nondynamic linear systems in this area. In particular, Synergetics can be shown to be a very powerful tool for the interpretation of perceptual phenomena (Luccio, 1991, 1993, 1994).

Prof. Riccardo Luccio
Dipartimento di Psicologia
Università di Firenze
V. S. Nicolò, 93
50125 Firenze, Italy

REFERENCES

Ames, A. Jr (1955). An Interpretative Manual. The Nature of Our Perceptions, Prehensions, and Behavior. Princeton: Princeton Un. Press.

Brown, J.F. (1931) "The visual perception of velocity", "Psychologische Forschung", 14, 199-232.

Cutting, J.E. & Proffitt, D.R. (1982) "The minimum principle and the perception of absolute, common, and relative motion", "Cognitive Psychology", 14, 211-246.

Dawson, M., & Pylyshyn, Z.W. (1988). "Natural constraints in apparent motion". in: Z.W. Pylyshyn (ed.), Computational Processes in Human Vision. Norwood, NJ: Ablex, 99-120.

Duncker, K. (1929). "Über induzierte Bewegung". "Psychologische Forschung", 12, 180-259.

Garner, W.R. (1962). Uncertainty and Structure as Psychological Concepts, New York, NY: Wiley.

Goldmeier, E. (1937). "Über Ähnlichkeit bei gesehenen Figuren". "Phychologische Forschung", 15, 146-208.

Goldmeier, E. (1982). The Memory Trace: Its Formation and its Fate, Hillsdale, NJ: Lea.

Haken, H. (1990). "Synergetics as a tool for the conceptualisation and mathematisation of cognition and behaviour - How far can we go?". in: H. Haken, H. & Stadler, M. (1990). Synergetics of Cognition. Berlin: Springer.

Hatifield, G. & Epstein, W. (1985). "The status of minimum principle in the theoretical analysis of visual perception", "Psychological Bulletin", 97, 155-176.

Höffding, H. (1887). Psychologie in Umrissen auf Grundlage der Erhahrung. Leipzig: Fus's Verlag.

Hüppe, A. (1984). Prägnanz. Ein gestalttheoretitischer Grundbegriff. München: Profil-Verlag.

Ishiguchi, A. (1991) "Effects of change of spatial distance between moving dots on their apparent linkage", "Japanese Psychological Research", 33, 86-96.

Johansson, G. (1950). Configurations in Event Perception, Uppsala: Almkvist Wilsell.

Johansson, G. (1973). "Visual perception of biological motion and a model for this analysis", "Perception & Psychophysics", 14, 201-211.

Kanizsa, G. (1979). Organization of Vision, New York, NY: Praeger.

Kanizsa, G. Kruse P., Luccio, R., & Stadler M. (1994). "Conditions of visibility of actual paths". "Japanese Psychological Research", 36, 113-120.

Kanizsa G. & Luccio R. (1987). "Die Doppeldeutigkeiten der Prägnanz". "Gestalt Theory", 8, 99-135.
Kanizsa G. & Luccio R. (1990). "The phenomenology of autonomous order formation in perception". in: H. Haken & M. Stadler (eds), Synergetics of Cognition, Frankfurt: Springer, 186-200.
Kanizsa G. & Luccio R. (1993) "Visual cognition". "Nuovi Argomenti", 1-2 (Nuova Serie), 210-225.
Kanizsa G. & Luccio R. (1995). "Multistability as a research tool in experimental phenomenology". in: M. Stadler & P. Kruse (eds): Ambiguity in Mind and Nature, Berlin: Springer, 47-68.
Köhler, W. (1920). Die physischen Gestalten in Ruhe und im stationären Zustand, Braunschweig: Vieweg.
Köhler, W. (1924). "Gestaltprobleme und Anfänge einer Gestalttheorie". "Jahresberichte über die gesamte Physiologie", 3, 512-539.
Köhler, W. (1940) Dynamics in Psychology, New York, NY: Livertight.
Luccio R. (1991). "Complessità e auto-organizzazione nella percezione". "Atque", 4, 91-109.
Luccio R. (1993). "Gestalt problems in cognitive science". in: V. Roberto (ed): Intelligent Perceptual Systems. New Directions in Computational Perception, Berlin: Springer, 2-19.
Luccio R. (1994). "Visual Thinking. Stability and Self-Organization". in: V. Cantoni (ed): Human and Machine Vision. Analogies and Divergencies, New York: Plenum Press, 277-287.
Luccio R. (1986). "An essay on Prägnanz". In: L. Albertazzi (ed), The Shapes of Forms. Klüwer.
Luccio R. Leonardi G. & Parovel E. (1996). "Hysteresis in detecting viasual paths". Trieste Symposium on Perception and Cognition.
Mach, E. (1885). Die Analyse der Empfindungen und das Verhaeltnis des Physischen zum Psychischen, Jena, Fischer.
Marr, D. (1982). Vision. S. Francisco, CA: Freeman.
Metzger W. (1963) Psychologie, Darmstadt: Steinkopf.
Rausch E. (1952) Struktur und Metrik figural-optischer Wahrnehmung. Frankfurt: Kramer.
Rausch E. (1966) "Das Eigenschaftsprobem in der Gestalttheorie der Wahrnehmung". in W. Metzger & H. Erke (Hsg.), Handbuch der Psychologie. Bd 1/1. Wahrnehmung und Bewußstein. Göttingen: Hogrefe.
Rok I. (1962). "A neglected aspect of the problem of recall: the Höffding function". In: M. Scher (ed): Theories of the mind. New York: Free Press of Glencoe.
Rubin E. (1927). "Visuell wahrgenommene wirkliche Bewegungen". "Zeitschrift für Psychologie", 103, 384-392.
Stadler M. & Kruse P. (1986). "Gestaltheorie und Theorie der Selbstorganisation". "Gestalt Theory", 8, 75-98.
Stadler M., Stegagno L., and Trombini G. (1987), "Über evidente und funktionale Phänomene der Prägnanchedenz: Ein Diskussionsbeitrag zu Kanizsa und Luccio". "Gestalt Theory", 8, 136-140.
Ullman, S. (1984). "Maximizing rigidity: The incremental recovery of 3-D strucutre from rigid and non rigid motion". "Perception", 13, 255-274.
Wallach, H. (1949). "Some considerations concerning the relation between perception and cognition". "Journal of Personality", 18, 6-13.
Wertheimer, M. (1912). "Über das Denken der Naturvölker", "Zeitschrift für Psychologie", 60, 321-368.
Wertheimer, M. (1922) "Untersuchungen zur Lehre von der Gestalt I", "Psychologische Forschung", 1, 47-58.
Wertheimer, M. (1923). "Untersuchungen zur Lehre von der Gestalt II", "Psychologische Forschung", 4, 301-350.
Zuckermann, C.B. & Rok I. (1957). "A reappraisal of the role of past experience and innate organizing processes in visual perception". "Psychological Bulletin", 54, 269,296.

PART II

SELF-ORGANIZATION, COMPLEXITY AND TRUTH

GIORGIO AUSIELLO - LUCA CABIBBO

EXPRESSIVENESS AND COMPLEXITY OF FORMAL SYSTEMS*

ABSTRACT

Formalization has a central role in computer science. Communication between man and computers as well as information processing within computer systems require that concepts and objects of the application domain are formally represented through some artificial language. In the last fifty years, the development of formal models has concerned practically all aspects of human life, from computing to management, from work to playing games and sex. Such development is one of the important contributions of computer science to human epistemological and intellectual development which expands the powerful role that mathematical formalization has played through the centuries. In building a formal model of reality, though, an important issue arises, the complexity of the formal system that is adopted; in fact, the stronger the expressive power of the system, the higher the complexity of using the model. An "optimal" balance between expressive power and complexity is the key that explains the success of some formal systems and languages that are widely used in computer science and applications. After discussing the contrasting needs of expressiveness and computational tractability in the construction of formal models, in the paper two case studies are considered: formal Chomsky grammars and relational database languages, and the balance between expressive power and complexity for such systems is analyzed in greater detail. Finally, the concept of succinctness of a formal system is taken into consideration, and it is shown that the role of succinctness in affecting the complexity of formal systems is crucial.

1. REPRESENTATION, MODELING, FORMALIZATION

Communication between man and computers as well as information processing within computer systems require that concepts and objects of the application domain are formally represented through some artificial language. For this reason computer science has emphasized the importance of formal modeling and of analyzing formal properties ("expressiveness", "complexity" etc.) of modeling languages.

Before discussing formal modeling in computer science, let us briefly address the more familiar issue of mathematical modeling. As it will become evident, no clearcut line can separate a discussion on formalization in computer science from the issue of mathematical modeling or, even more, from the issue of representation of reality throughout mankind history.

The need for man to represent and understand physical phenomena is at the base of the use of mathematical models. We might say that mathematics itself was born to satisfy

A. Carsetti (ed.), Functional Models of Cognition, 103-120.

such need and, also, to provide formal tools for answering questions raised by the early engineering problems. It is well known, for example, that land measurement and astronomical observations led to the development of geometry in Egypt and in Greece, in ancient times.

Indeed, looking back in the history of mankind we may realize how profound the need for representing reality was for man, way before the beginning of mathematics. In fact, a long time before starting to "formalize" concepts of the real world in mathematical terms, man started to feel the need to "represent" reality. Since the beginning of man's social life, the representation of reality had a magic and an epistemologic role at the same time. By drawing horses and wild oxen on a cave's walls the hunters of the neolithic age could "capture" the animals with their minds before chasing them with their primitive weapons.

In the same way, "capturing" concepts has been for millennia the aim of the intellectual activity by which man, through an abstraction process that from the observation and representation of reality leads to the classification of "objects" and to the formalization of their "interrelationships", have expanded their knowledge.

Clearly, mathematics has played a major role in such a process, extending the ability of man to model reality in a formal way, and this role has continued through the centuries, first with the applications to commerce and finance, in the Middle Ages, and then, in the Modern Age, with the great theoretical developments that accompanied the physical discoveries (mechanics, electromagnetism etc.) and the industrial revolution.

In the last fifty years, due to the use of computers, the need for formal modeling has expanded in all directions of human activity. Indeed, it has been in connection with the first military applications of computers in organization and logistics that new mathematical modeling and problem solving methodologies (namely, operations research) were developed. Subsequently, new formal tools were introduced for modeling the various aspects of productive life where computers were gradually introduced, such as industrial production, information management, office automation and workflow management, etc. More recently, the large increase of computer applications in everyday life has led to the development of formal systems for representing and automating a wide variety of human activities, ranging from natural language processing to common sense reasoning, from image understanding to interactive graphics.

Beside enhancing the ability of modeling reality, the use of computers has brought the need for formalizing and understanding aspects of computing itself: semantics and efficiency of programs, computation models, concurrency, man-machine interaction, etc. Within the field of computer science a substantial body of theoretical knowledge has been created around these subjects during the last three decades [18]. In particular, computer scientists have developed a research area which is not only of practical interest (in connection with software engineering methodologies) but also of great epistemological interest: the study of formal methods and languages for building computer applications [17].

In fact, the development of formal models in computer applications often proceeded in parallel with the evolution of programming methodologies. While new domains were attacked, new formal programming concepts and new information structures were needed for representing reality. While imperative programming was suitable for (and, indeed, born together with) scientific numerical applications, information management eventually led to database models, declarative languages, and to structured design methodologies. Recently, more advanced interactive applications have led to the widespread use of the object-oriented approach in which the real world is represented in terms of objects, classes, and interaction methods [19]; then, the essence of modeling is captured by offering the programmer the paradigms of abstraction, generalization, inheritance, etc.

The study of modeling and formalization processes, hence, has become one of the major areas of interest in computer science. Its meta-theoretical character directly relates this field with the logical and meta-mathematical studies that, between the Thirties and the Fifties, investigated the power of computation processes and computability [20].

In this paper two particular aspects of formal models are considered, expressiveness and complexity, and their conflicting characters are discussed by drawing examples from some well established fields of theoretical computer science. In the next section the concept of "life cycle" of a formal model is introduced, and it is observed how the search for more expressive power is in contrast with the need for computational efficiency in the use of the model. In Section 3 and Section 4 the issue of complexity of formal models is discussed with reference to two fundamental areas of computer science: the theory of formal languages and the theory of database languages, respectively. In Section 5, the source of complexity in a formal model is analyzed and identified in the concept of succinctness of representation. Finally, in Section 6 some conclusions are outlined.

2. LIFE CYCLES OF FORMAL MODELS

Formalization in computer science can be seen as the process of reducing a fragment of reality to a simplified system of symbols through abstraction with the aim of computing solutions of problems, proving properties, running simulations, etc.

Typically, the construction of a formal model goes through various stages. Without referring to any specific design methodology, we may roughly characterize the various stages as follows.

First, the real world domain that we wish to model is analyzed; the entities (objects) that characterize it are identified and classified through an abstraction process and so are their mutual relationships and their relationships with the outside world.

Then, a representation system (the "syntax"), consisting of formal symbols, is introduced. These symbols, that may be characters of a given alphabet, pictures, icons, are then mapped onto the objects and the relationships of the real system.

Finally, the behaviour of the real system is taken into consideration, and the behaviour of the symbolic system is defined in such a way as to mimic it (for example, by means of state transition rules).

An interesting example, in this respect, is provided by Petri nets [15], a formal model that has been introduced in 1962 for describing systems organization and, since then, has been widely used for modeling and analyzing systems of concurrent processes and, more generally, systems in which regulated "flows" play a role, such as parts' flow in a production line, jobs' flow in a computer system, work flow in an office.

In their simplest version [16] Petri nets are used to model systems (called "condition/event systems") in which objects flowing in the net determine "conditions". In suitable cases conditions determine "events" that, in turn, determine the new flow of objects in the net. A situation of this kind arises, for example, in a library. A request is submitted to the loan clerk (first condition) and, if the book is available (second condition), the loan event is determined. As a consequence, the following two conditions are created: the book is given to the customer while the request is stored in the borrowed books file. When a book is returned and the corresponding request is in the borrowed book file, the return event is determined, and the book is placed in the stacks again.

From the syntactic point of view, a Petri net of this type (called Condition/Event Net, C/E Net) consists of a triple <P, T, F>, where P, the set of "places", represents the conditions, T, the set of "transitions", represents the events, and F, the "flow" relationship, relates places and transitions. Graphically, places are usually drawn as circles, transitions as bars, and the flow relationship is represented by lines and arrows (see Figure 1).

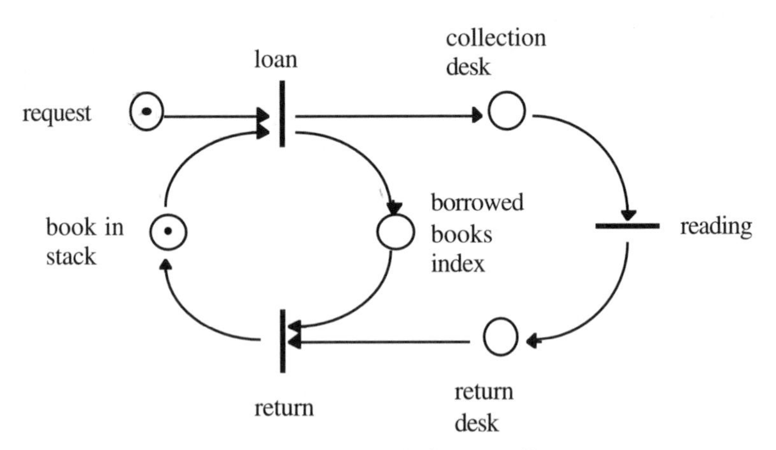

Figure 1. Petri net model of loans in a library.

In a C/E Net the objects flowing in the system are represented by means of "tokens". The behaviour of the net is then characterized in terms of transition rules: if all places preceding a transition contain a token, the transition is enabled and, as a consequence, the tokens in such places are "consumed" while a token is "produced" in any place following the transition.

The use of a Petri net model allows to study the properties of the domain of interest, in this case a concurrent system. Typical properties that we wish to analyze are the absence of deadlock, the reachability of particular condition configurations, the fact that a given transition will eventually be enabled (liveness), etc.

Figure 1 contains a simple example of C/E Net modeling the loans in a library.

Once a formal model is established we have to check its validity by matching its features against the system that we want to represent. This is achieved by defining the semantics of the model, that is, by establishing an interpretation which maps syntactic structures of the model back onto the system, and by checking whether the model satisfies two fundamental characteristics: soundness (may the properties of the model be suitably interpreted in the system?) and completeness (does the model capture all the properties of the system?).

As a consequence of the semantic analysis we may realize that the formal model does not represent all aspects of reality that we want to simulate and process. In other words, the expressive power of the formal model is insufficient and an enrichment is necessary. If we go back to the Petri nets example, we notice that the Condition/Event model is actually rather simplistic. Suppose that books are classified (some can be borrowed for one week, some other for one month). This case cannot be modeled by a C/E Net. In fact, tokens representing requests and books should be labeled and only corresponding labels can activate a loan. More complex types of Petri nets have been introduced for dealing with situations of this kind (for example, Colored Petri Nets [8]).

An immediate consequence of the increase in expressive power of the model is a corresponding increase in complexity. In particular the flow relationship, which relates places and transitions, becomes more complex. Running the model on a computer may become heavy, proving its properties may become computationally intractable or even undecidable.

When constructing a formal model we are therefore faced with two contrasting needs: on the one hand, the need to make the model more expressive, more powerful, in order to capture more and more aspects of reality; on the other hand, the need to keep the complexity of the model low, in order to preserve the tractability of its properties. Usually formal models "stabilize" in an equilibrium point that constitutes an "optimal" compromise between the two needs. This position is maintained until the advances of technology push toward the adoption of a new model which will also reach, after a while, through different stages, its optimal equilibrium point.

Thus, the practical success of various formal systems in computer science is due to the fact that for such systems the optimal compromise between expressiveness and complexity has been reached. A typical example is represented by finite state automata, introduced in the Fifties by E.F. Moore and used for specifying sequential machines, for which most properties can easily be proved and whose computational power can be characterized under various points of view: in terms of classes of input sequences (called regular events in [11]), by means of equivalence with respect to the neural net model of McCulloch and Pitts (see again [11]), in terms of closure under algebraic operators (see [7]), etc. Other examples of formal systems in which the expressive power is conveniently balanced by processing efficiency are deterministic context free languages (whose syntactic properties provide the paradigm for the structure of most programming languages and at the same time allow the efficient compiling of programs) and relational database languages (whose strong expressive power, characterized in terms of both relational algebra and calculus, is nevertheless compatible with efficient implementation of query processing in commercial database systems).

In the next two sections we will see some specific aspects of this correlation between expressive power and complexity, occurring in the case of such particular, well-studied families of formal systems.

3. CHOMSKY GRAMMARS

Chomsky grammars were introduced by Noam Chomsky in the Fifties [2] with the aim of providing a formal basis for the understanding of the structure of natural languages. Chomsky grammars are a generative formal system, based on "production rules" (a particular kind of rewriting rules, related to earlier work of the logicians Axel Thue and Emil Post, see [6]) and allow the characterization of classes of languages, in relation to the structural properties of the corresponding production rules. Formally, a *Chomsky grammar* consists of a 4-tuple <T, V, S, P> where T is the alphabet of *terminal symbols*, V is the alphabet of *non-terminal* (*variable*) *symbols*, S is a particular symbol of V, called *axiom*, the seed of the generation process, and P is the set of *production rules*. A production rule is of the form:

$$u \rightarrow v$$

where u and v are strings of, possibly, terminal and non-terminal symbols with the only constraint that u contains at least one non-terminal symbol. Such a rule specifies that, if the substring u occurs in the string that we are in the process of generating, then we can replace u by v. The *language generated by the grammar* is the set of all strings consisting of only terminal symbols that can be generated starting from the axiom and repeatedly applying the production rules. An example is the following. Consider the grammar $<\{a, b\}, \{A\}, A, P>$, where P is the set of production rules:

$$A \rightarrow aAb$$
$$A \rightarrow ab$$

The grammar generates the language consisting of all strings of the form $a^n b^n$ for all $n \geq 1$, that is, all strings consisting of a sequence of n a's followed by a sequence of an equal number of b's.

The example is indeed rather simple and corresponds to a particular case of grammar that is called *context free grammar*. In this type of grammar, in each production the left hand string consists of just one symbol (say, A), which can be replaced by the right hand string, no matter the context in which it occurs. Other important particular classes of grammars are *context sensitive grammars* where the right hand string and the left hand string are of general type but the right hand string cannot be shorter than the left hand string, and *regular grammars*, a particular case of context free grammars in which the right hand string consists of a single terminal symbol or of a terminal symbol followed by a non-terminal symbol.

Chomsky grammars (especially context free grammars) played an important role in the study and understanding of natural language concepts; however, after a while it became clear that their structure was too simple for such aim and other formal generative devices were introduced [4]. Nevertheless, despite being inadequate to formalize natural languages, Chomsky grammars had a remarkable effect on the syntactic development of artificial languages used for computer programming. A few years after the seminal work of Noam Chomsky, John W. Backus [1] and Peter Naur [14] introduced the so-called Backus Normal Form (or Backus Naur Form, BNF, for short) for the description of the programming language ALGOL, which was based on a sort of production system and indeed could be seen as a reformulation of context free grammars. In BNF, for example, the structure of an if-statement in a programming language could be expressed in the following way:

<if-statement> :: = **if** <condition> **then** <statement> **else** <statement>

which is clearly structured as a context free production with **if**, **then**, **else**, as terminal symbols and <if-statement>, <condition> and <statement> as non-terminal symbols.

Since the early Sixties, therefore, the study of syntactic aspects of programming languages as well as compiler design techniques have been based on the structure and properties of Chomsky grammars. The main problem we have to solve in this context is the following. We are given a text consisting of several thousands of lines of source code, and we want to translate it efficiently into machine code. This means that by scanning the text (possibly only once) we want to perform at the same time the following operations:

- check for syntactic correctness and, in case, report errors;
- interprete the meaning of the code;
- translate into machine language.

Moreover, we don't want to waste too much space in doing this. Clearly the cost of the said operations changes according to the syntactical structure of the text, that is, according to the type of grammar that defines the language. Table 1 reports the cost of verifying syntactic correctness with different types of grammars. More precisely, the table indicates time and space needed for solving the so called *recognition problem*, that is, the problem of deciding, given a grammar of a certain type, whether a string has been derived according to the rules of such grammar. Note that, for the general class of unrestricted Chomsky grammars, the problem is undecidable. Actually, Chomsky proved that the computational power of such grammars coincides with the computational power of Turing machines [3].

Type of grammar	**Time**	**Space**
Regular	$O(n)$	$O(1)$
Deterministic Context Free (LR(k))	$O(n)$	$O(\log n)$
Context Free	$O(n^3)$	$O(\log^2 n)$
Context Sensitive	$O(2^{n^2})$	$O(n^2)$
Unrestricted	undecidable	

Table 1. Complexity of the recognition problem for various classes of formal languages.

From the table the following facts can be seen. Languages defined by regular grammars can indeed be recognized very efficiently, in linear time with a constant amount of memory; unfortunately, the syntactic structure of regular grammars is too simple for real programming languages. For example, it allows to express sequences of simple commands (such as the ones that are used in an electronic mail system) but does not allow to define algebraic expressions in infix notation. For representing all types of algebraic expressions that are needed in computer programming as well as the parenthetical structure of nested programming constructs (if statements, while statements, for statements etc.) context free grammars are needed, but then, in general, the syntactic analysis costs become prohibitively high.

In order to overcome this difficulty several new classes of languages have been introduced, properly included in the general class of context free languages (actually all of them included or coinciding with the class of the so called *deterministic context free languages*) but strictly more powerful than regular languages, for which both time and space-efficient parsing algorithms could be defined. In particular, the *LR(k) languages* [12] are a class of languages that can be efficiently parsed from left to right if we allow a *k* symbols look-ahead. These languages are expressive enough to allow the definition of the syntactic constructs required by most programming languages while achieving, at the same time, very good efficiency in syntactic recognition, parsing, and translation. Besides, for deterministic context free grammars the *equivalence problem* has been recently proved decidable [21] while it is undecidable for the general class of context free grammars. These languages represent a lucky example, in this context, of the optimal balancing between the needs for expressiveness and the needs for efficiency of a formal system that was discussed above.

4. DATABASE QUERY LANGUAGES

A *database* is a collection of structured data to represent some aspects of the real world for a specific purpose. Typically, databases are large, persistent, and shared by several users. Database technology offers the software tools for an efficient and effective management of databases. The theoretical foundation of current database management systems is provided by the *relational model of data*, a formal model proposed by E.F. Codd in the early Seventies. (See [10] for a survey of relational database theory.)

Intuitively, data in a relational database consist of sets of rows, each row representing a relationship among a set of values; rows with uniform structure and intended meaning are grouped into tables. More precisely, a *relational database* is a collection of *relations*. Each relation has a structural part (called the *scheme*) and an extensional part (called the *instance*). A relation scheme consists of a name (unique in the database) together with a tuple of distinct names, called the *attributes* of the relation. A relation instance is a finite set of tuples over the attributes specified in the scheme.

Flights (*Company, Flight-No, From-Airport, To-Airport*)
Airports (*Airport, Full-Name, City*)
Trains (*Train-No, From-City, To-City*)

Flights

Company	*Flight-No*	*From-Airport*	*To-Airport*
Alitalia	AZ2010	FCO	LIN
Alitalia	AZ2011	LIN	FCO
Alitalia	AZ638	FCO	CHI
KLM	KL264	FCO	AMS
TWA	TW312	JFK	CHI
UsAIR	US913	AMS	JFK

Airports

Airport	*Full-name*	*City*
AMS	Schiphol	Amsterdam
CHI	O'Hare Intl.	Chicago
FCO	Leonardo da Vinci	Rome
JFK	J.F. Kennedy Intl.	New York
LIN	Linate	Milano

Trains

Train-No	*From-City*	*To-City*
E654	Rome	Florence
EC699	Paris	Frankfurt
IC412	Florence	Milan
EC511	Milan	Paris

Figure 2. An example transportation database.

As an example, consider the transportation database (shown in Figure 2) representing information about flight and train connections. Data are split into three relations, named **Flights**, **Airports**, and **Trains**, each of them representing facts about a specific kind of information. Relation **Flights** has attributes *Company* (the name of the company offering a flight), *Flight-No* (the flight number), *From-Airport* and *To-Airport* (the codes of the departure and arrival airports). Each tuple in this relation contains a value for each attribute, establishing a relationship among the values; for instance, the tuple (*Alitalia*, *AZ638*, *FCO*, *CHI*) states that Alitalia offers a flight, coded AZ638, from airport FCO to airport CHI. The information about the cities connected by the flight can be found by means of the relation **Airports**: since FCO and CHI correspond to airports in Rome and Chicago, respectively, this means that AZ638 is a flight connecting Rome to Chicago. The fact that different information are spread across different relations is for the sake of succinctness, to avoid redundancy. However, information can be extracted by correlating tuples in multiple relations, mainly relying upon equality of values. This activity is referred to as *querying* the database, the main topic in database management together with *updating*. Conversely, an update consists of a set of additions, removals, and/or modifications of tuples within relations. (We will not consider updates, since a discussion of this topic is beyond the scope of the paper.)

Formally, a *query* is a function from databases to databases, specified by means of an expression of some *query language*. (With a little abuse of terminology, the term "query" is often used for both the function, the expression, and a natural language description of the function.) Possible queries over the transportation database are the following.

Q1 Is there any direct flight connection from Rome to New York?
Q2 What are the cities for which there is a direct connection (either by flight or by train) from Rome?
Q3 What are the pairs of cities for which there is a direct flight connection but not a direct train connection?
Q4 Is there any flight connection (possibly involving intermediate stops) from Rome to New York?
Q5 What are the pairs of cities having an airport but for which there is no flight connection?

The relational model adopts the so-called *closed world assumption*, according to which the facts stored in a database are the only ones to be true, and the ones that are not present in the database are assumed to be false. (For example, our transportation database assumes that there is no flight connection from Rome to Paris, since there is no explicit fact stating it.)

The relational model is provided with two basic query languages, stemming from different paradigms. The *relational algebra* is a "procedural" language, based on a few algebraic operators. In order to allow for composition, all the operators produce relations as results. On the other hand, the *relational calculus* is a "declarative" language, based on the first order predicate calculus. By "procedural" we mean that a query specifies the actions that must be performed to compute the answer to a query. Conversely, "declarative" means that a query is specified in a high-level manner, essentially by stating the properties that the result should satisfy; in this case, an efficient execution of the query has to be worked out by the interpreter of the language. Thus, "declarative" concerns "what" and "operational" concerns "how."

In spite of the differences in the two languages, they are equivalent to each other, that is, they can express exactly the same queries. This equivalence is known as Codd's Theorem and can be specialized to suitable restrictions and extensions of the relational calculus and algebra. Codd's Theorem has great practical significance; the translation of calculus into algebra reveals a procedural evaluation for a query defined declaratively by a calculus expression.

Instead of introducing the relational algebra or calculus, we will discuss some examples referring to a declarative query language, stemming from logic programming, called *da-*

talog. Intuitively, queries are specified in *datalog* by means of sets of rules, called *programs*. The left-hand-side of a rule is a conjunction of literals, referring to relation names and variables: if there are values for the variables such that all the literals in the left-hand-side are known to be true in the database, then we can infer the truth of the literal in the right-hand-side of the rule. Figure 3 shows the programs implementing the queries **Q1** to **Q5** over the transportation database.

Q1	Ans 1	←	Flights (C, N, F, T), Airports (F, FN, "Rome"), Airports (T,TN, "New York")
Q2	Ans 2 (X)	←	Flights (C, N, F, T), Airports (F, FN, "Rome"), Airports(T,TN,X)
	Ans 2 (X)	←	Trains (N, "Rome", X)
Q3	Conn-by-train (X, Y)	←	Trains (N, X, Y)
	Ans 3 (X, Y)	←	Flights (C, N, F, T), Airports (F, FN, X), Airports (T, TN, Y), **not** Conn-by-train (X, Y)
Q4	Flight-conn (X, Y)	←	Flights (C, N, X, Y)
	Flight-conn (X, Y)	←	Flight-conn (X, Z), Flight-conn (Z, Y)
	Ans 4	←	Flight-conn (F, T), Airports (F, FN, "Rome"),
		←	Airports (T, TN, "New York")
Q5	Flight-conn(X,Y)	←	Flights (C, N, X, Y)
	Flight-conn(X,Y)	←	Flight-conn (X, Z), Flight-conn (Z, Y)
	Ans5(X,Y)	←	Airports (SX, NX, X), Airports (SY, NY, Y), **not** Flight-conn(SX,SY)

Figure 3. Examples of Datalog queries.

Datalog has many variants, some of which are described as follows.

- *Conjunctive* programs are made of a single rule, not involving negated literals.
- *Positive existential* programs are made of multiple rules, with no negated literals and no recursive definitions.
- *First order* programs are made of multiple rules, with no recursive definitions.
- *Datalog* programs are made of multiple rules, with no negated literals.
- *Stratified* programs are made of multiple rules; a technical condition is imposed to avoid that recursive definition involves negated literals.
- *Datalog with 1-sets* are stratified programs, allowing for the management of sets of values.
- *Datalog with k-nested sets* are stratified programs with sets, allowing for *k* levels of nesting.
- *Datalog with sets* are stratified programs with sets, allowing for unbounded levels of nesting.

The *expressiveness* of a query language is related to the class of queries that the language can express. Specifically, we say that a language L expresses a query q if there is an expression E of L whose semantics coincides with the function q. For example, the query **Q1** can be expressed by means of all the above cited languages, whereas query **Q3** can be expressed by a first order program, but not by a conjunctive program (since negation is required). We say that a language L is *more expressive than* another language L' if any query expressible in L' is expressible in L as well. For example, it turns out that *first-order datalog* is more expressive than *conjunctive datalog*. Note that the relation "more expressive than" is a partial order and not a total one, since some languages are not related according to it. For example, *datalog* and *first-order datalog* are unrelated, since there are queries that are expressed only by one of the two languages.

It should be noted that *first-order datalog* is another language having the same expressiveness of the relational algebra and calculus, and that *conjunctive datalog* and *positive existential datalog* have equivalent counterparts in suitable restrictions of both the relational algebra and calculus. Thus, we can consider classes of languages having the same expressiveness. The relative expressiveness of the various classes is summarized in Figure 4. (The arrows denote greater expressive power, in the strict sense.)

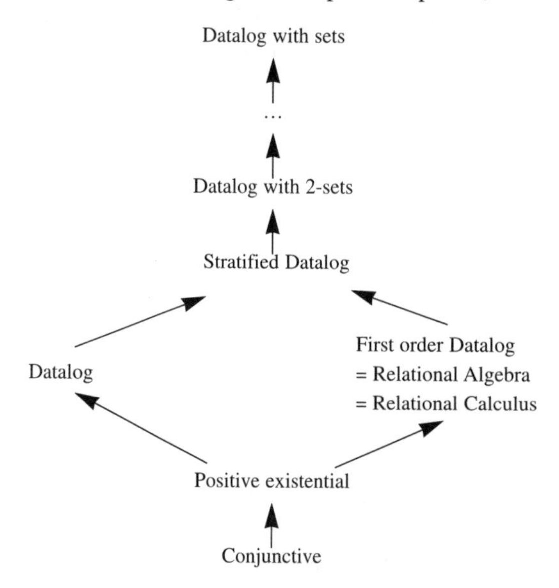

Figure 4. Hierarchy of query languages based on their expressiveness.

Database theoreticians consider different complexity measures, differing in the parameters with respect to which the complexity is measured. Intuitively, the complexity of evaluating a query with respect to a database is related to both the size of the query expression and that of the database. In practice, the size of the database typically dominates, by many orders of magnitude, the size of the query, and is therefore the parameter of interest. The *data complexity* of a query is defined as the computational complexity of testing whether a given tuple belongs to the result of the query with respect to a database *d*.

It turns out that the relational algebra and calculus express only queries whose data complexity is in LOGSPACE. Since the class LOGSPACE is contained in the class NC^1, this implies that the two languages have a lot of potential parallelism. The other side of the coin is that the two languages cannot express queries whose data complexity is higher than LOGSPACE. Of course, the trade-off is between expressiveness of a language and the possibility of efficient execution.

There is another point to ponder, namely, query optimization. Usually, queries are written in a high-level language (such as the relational calculus), and automatically translated into an equivalent expression of a procedural language (such as the relational algebra), using Codd's Theorem. Then the database management system tries to optimize the expression by rewriting it into another equivalent expression that is better under some measure of complexity. The rationale for optimization consists in trying to minimize the computational cost of evaluating the query which is mainly related to the size of intermediate results. An optimizer has a module for estimating the evaluation cost of an expression. The goal is to find an equivalent rewriting of the query with minimal esteemed cost. To do so, the optimizer generates a set of rewritings, by means of suitable heuristics, and verifies the equivalence with the initial expression. Of course, this is convenient only if the optimization process does not consume more resources than the evaluation of the initial expression. Specifically, the computational cost of optimization is mainly related to the complexity of testing for the equivalence of the rewritten expression with the initial one. Unfortunately, testing for equivalence of relational algebra expressions is undecidable, but decidable for more restricted languages. Again, there is a trade-off between expressibility of a language and possibility of efficient optimization.

Table 2 summarizes, for each class of queries, the data complexity and the complexity of testing expression equivalence. (See [9] for a definition of the complexity classes mentioned in the table.)

Class of queries	Data complexity	Complexity of the equivalence problem
conjunctive	LOGSPACE	NP-complete
positive existential	LOGSPACE	P_2-complete
first order	LOGSPACE	undecidable
datalog	PTIME	undecidable
datalog with stratified negation	PTIME	undecidable
datalog with 1-sets	EXPTIME	undecidable
datalog with k-sets	k-EXPTIME	undecidable
datalog with sets	unbounded	undecidable

Table 2. Data complexity and complexity of the equivalence problem for various query languages.

We now briefly discuss the compromise reached between the needs of expressiveness and complexity in the case of relational query languages. In practice, that is, in real database management systems, the query language often used is SQL (for Standard Query Language). SQL is a declarative language based on a variant of the relational calculus. With respect to the expressive power, it is essentially equivalent to both the relational calculus and algebra, and thus its queries can be evaluated rather efficiently. On the other side, Table 2 shows that finding the optimum rewriting of an SQL query is, in general, an unsolvable problem. In principle, the optimization could be still carried on for some SQL queries (the ones that syntactically are conjunctive or positive existential). In practice, however, the systems apply some set of predefined heuristics, usually finding "good" rewritings, but without a guarantee to find the best ones.

5. EXPRESSIVENESS, SUCCINCTNESS AND COMPLEXITY

In the preceding sections, by means of various examples, we have seen how an increase in the expressive power of a formal model entails a corresponding increase in the cost of the computations that we want to perform on the model (such as testing for membership in formal languages or answering queries in databases).

In this section we will see that a major role in the expressiveness-complexity trade-off is essentially played by the succinctness of the encodings that the formal system allows. In particular, in strongly expressive formal systems we may encode computational processes (such as for example Turing machine computations), and what makes the properties of certain classes of formulae intrinsically hard to decide is the fact that "short" formulae may describe "long" computations.

The first evidence of this property is provided by Cook's Theorem [5]. This result, proven by Steve Cook in 1971 and, independently, by L. Levin in 1973 [13], establishes the intrinsic complexity of deciding the satisfiability of propositional formulae by showing how such formulae can encode Turing machine computations. More precisely, the result shows that, given a nondeterministic Turing machine M operating in time bounded by a polynomial p, and given a string x, of length n, we can build a propositional formula w (of length at most $p^4(n)$) that is satisfiable if and only if x is accepted by M in time p(n). As a consequence, the problem of deciding the satisfiability of propositional formulae is proven to be at the highest level of complexity in the class NP (the class of all problems that can be solved in polynomial time by means of nondeterministic Turing machines) or, as it is customary to say, "NP-complete". It is well known that still, 25 years after Cook's result, we do not know any polynomial-time algorithm for solving either the satisfiability problem or any of the thousands of NP-complete problems that have been discovered since then. At the same time, thanks to Cook's result, we know that if any polynomial-time algorithm would be discovered for the satisfiability problem (an unlikely possibility, though), all other NP-complete problems could be solved in polynomial time and the class NP would be proven to coincide with the class P (the class of problems solvable by means of deterministic Turing machines in polynomial time), giving positive answer to the well known question "P = NP?".

The formula built in Cook's Theorem's proof, somehow, encodes the rules that all Turing machine computations have to satisfy (e.g., at any given instant t the head of the machine can read exactly one tape cell), encodes the initial and final configurations (at time 0 tape cells 1 through n contain string x, all other tape cells are blank; at time p(n) the machine state should be the final state), and, finally, encodes the relationship between instantaneous configurations of the tape that are implied by the transition function of the machine M.

What is particularly important is that the construction is logarithmically succinct because a nondeterministic computation of polynomial depth p(n) is indeed a tree of, possibly, $2^{p(n)}$ instantaneous configurations while in the proof the same information is compacted in a polynomially long formula. This gives the proof of Cook's Theorem all its remarkable power.

The fundamental idea to represent "long" computations with "short" formulae is at the base of practically all proofs of complexity hardness of problems (see [9]), and it is interesting to observe that the shorter the formulae, the more complex the properties of the formalism. Since in general, in this context, computations are modeled in terms of Turing machines, we may conclude that the intrinsic complexity of a formalism can eventually be related to the power of the formalism in describing Turing machine computations. In order to clarify this claim, let us consider the formalism of regular expressions.

Regular expressions have been introduced by S.C. Kleene (in connection with the definition of the already mentioned regular events [11]) and are used for describing sets of words on a given alphabet (sets of input sequences to an automaton, paths on a labeled graph, etc.).

In their basic formulation, given an alphabet $\{a,b\}$, regular expressions are inductively defined as follows:

- a,b are regular expressions;
- if e_1 and e_2 are regular expressions, then $e_1\ e_2$, $e_1 + e^2$, and $e_1{}^*$ are regular expressions.

The meaning of a regular expression is a language defined as follows:

- a and b represent the languages $L_a=\{a\}$ and $L_b=\{b\}$, respectively;
- if Le_1 and Le_2 are the languages represented by the regular expressions e_1 and e_2, then:
- $e_1\ e_2$ represents the concatenation of languages, $Le_1 \circ Le_2$;
- $e_1 + e_2$ represents the union of languages, $Le_1 \cup Le_2$;
- $e_1{}^*$ represents the iteration (Kleene's star operator) of a language, $(Le_1)^*$.

For example, the regular expression $ab^*(aa+bb)^*$ represents the language consisting of strings that begin with a, and are followed by a (possibly empty) sequence of b's, and by a (possibly empty) sequence of strings of the type aa or bb.

The formulation of regular expressions may be enriched with the complement operator (that is, $e_1{}^c$, whose meaning is $\{x \mid x \text{ is not a string in } L_{e1}\}$) and with the squaring operator (that is, $e_1{}^2 = e_1\ e_1$, whose meaning is $\{x \mid x = uv,\ u,v \in L_{e1}\}$), without changing the expressive power. In other words, if we write a regular expression e^1 with the complement and/or the squaring, we know that there is another regular expression e_2 which does not make use of either the complement or the squaring and such that L_{e1} equals L_{e2}.

It is important to notice that the complement and squaring operators allow to express in a more succinct form the same languages that we can represent with expressions that do not make use of such operators. Table 3 provides the complexity of the problem of deciding whether a regular expression is not equivalent to {a,b}* for various classes of regular expressions.

Regular expressions with operators	Complexity	Time complexity
°, +	NP-complete	$O(n^2)$
°, *, +	PSPACE-complete	$O(2^{n^k})$
°, *, +, squaring	EXPSPACE-complete	$O(2^{2^n})$
°, *, +, complement	non elementary	$O(2^{2^{\cdot^{\cdot^{\cdot 2}}}}$}n times)

Table 3. Complexity of the equivalence problem for various classes of regular expressions.

In all cases, except the first one, the expressions represent the same class of languages (regular languages), but the increase of problem complexity that accompanies them derives from the succinctness of the formulae. For example, it is easy to see that the squaring operator allows a logarithmic compression of strings: in fact $(\ldots((a^2)^2)^{2\ldots})^2$}n times, corresponds to a sequence of the type *a…a* of length 2^n. Even stronger is the compression that we may achieve by making use of the complement operator and, as a consequence, the complexity of deciding whether a regular expression with complement is equivalent to {*a*,*b*}* becomes *non elementary*. In other words, its complexity cannot be expressed by any elementary function obtained by composition of sums, products, and exponential functions, because it grows as $2^{2\ldots^2}$}n times.

6. CONCLUSIONS

The complexity of deciding properties of a formal model is strictly related to the expressive power of the model. By referring to the well-known models of Chomsky formal grammars and of relational databases, we have seen that the success of a formal model somehow derives from the fact that in such a model an optimal balance is established between two factors: (i) the need of achieving a strong expressive power in order to represent and study the fragment of reality of interest (for example, in those cases, the syntactic structure of programming languages or the structure of queries allowed for a database); and (ii) the need of preserving the efficiency in the use of the model (syntax analysis and query processing in the mentioned examples). Finally, by examining the growing difficulty of deciding equivalence in various families of regular expressions, we pointed out that, when formalisms with the same expressive power are considered, the computational complexity of the properties of the formal expressions used in a formalism strictly depends on their succinctness and in particular on the fact that very short formulae may be used for representing long computations.

Prof. Giorgio Ausiello
Dipartimento di Informatica e Sistemistica
Università di Roma "La Sapienza"
V. Salaria, 113
00198 Roma, Italy

Prof. Luca Cabibbo
Dipartimento di Informatica e Automazione
Università di Roma Tre
Via delle Vasca Navale, 79
00146 Roma, Italy

NOTES

* The first author gratefully acknowledges the support of the Italian MURST National Project "Efficienza di algoritmi e progetto di strutture informative." The second author has been partially supported by MURST and CNR.

[1] NC (Nick's Class, from the name of Nick Pippenger, who defined this class) is the class of problems that can be efficiently parallelized. (See [9] for more details).

REFERENCES

[1] Backus, J.W. "The syntax and semantics of the proposed international algebraic language of the Zürich ACM-GAMM conference". "Proc. Intl. Conf. on Information Processing", UNESCO, 125-132, 1959.

[2] Chomsky, N. "Three models for the description of language". "IRE Trans. on Information Theory", 2:3, 113-124, 1956.

[3] Chomsky, N. "On certain formal properties of grammars". "Information and Control", 2:2, 137-167, 1959.

[4] Chomsky, N. Knowledge of Language. Its Nature, Origin and Use. New York, Praeger, 1986.

[5] Cook, S.A. "The complexity of theorem-proving procedures". "Proc. of the Third Annual Symp. on the Theory of Computing", 151-158, 1991.

[6] Davis, M. (ed.) Solvability, Provability, Definability: The Collected Works of Emil L. Post. Birkhäuser, Boston, 1994.

[7] Hopcroft, J.E., and J.D. Ullman. Introduction to Automata Theory, Languages, and Computation. Addison-Wesley, 1979.

[8] Jensen, K. Coloured Petri Nets. Basic Concepts, Analysis Methods and Practical Use. Volume 1. EATCS Monographs on Theoretical Computer Science, Springer-Verlag, 1992.

[9] Johnson, D.S. "A catalog of complexity classes". In J. van Leeuwen, editor, Handbook of Theoretical Computer Science, volume A, pages 67-161. Elsevier Science Publishers (North-Holland), Amsterdam, 1990.

[10] Kanellakis, P.C. "Elements of relational database theory". In J. van Leeuwen, editor, Handbook of Theoretical Computer Science, volume B, pages 1073-1156. Elsevier Science Publishers (North-Holland), Amsterdam, 1990.

[11] Kleene, S.C. "Representation of events in nerve nets and finite automata". In C.E. Shannon, J. McCarthy, editors, Automata Studies, pages 3-42. Princeton University Press, 1956.

[12] Knuth, D.E. "On the translation of languages from left to right". "Information and Control", 8:6, 607-639, 1965.

[13] Levin, L.A. "Universal sorting problems". "Problemy Peredaci Informacii", 9, 115-116, 1973 (in Russian); English translation in: "Problems of Information Transmission", 9, 265-266, 1973.

[14] Naur, P. et al. "Report on the algorithmic language ALGOL 60". "Comm. ACM", 3:5, 299-314, 1960.

[15] Petri, C.A. "Kommunikation mit Automaten". Schriften des Institutes für Instrumentelle Mathematik, Bonn, 1962.

[16] Reisig, W. Petri Nets. An Introduction. EATCS Monographs on Theoretical Computer Science, Springer-Verlag, 1985.

[17] Clarke, E.M., Wing J.M. et al. "Formal Methods: State of the Art and Future Directions", in Strategic Directions in Comparty Research, ACM Computer Surveys, 28, 4, 1996.

[18] van Leeuwen, J. Ed., Handbook of Theoretical Computer Science, Elsevier Science Pub. (North Holland), Amsterdam 1990.

[19] Heileman, G.L., Data Structures, Algorithms, and Object Oriented Programming, Mac Graw Hill, New York, 1996.

[20] Kleene, S.C., Introduction to Metamathematics, Van Nostrand, New York, 1952.

[21] Sénizergues G., "The equivalence problem for Deterministic Pushdown Automata is decidable", in Automata, Languages and Programming, LNCS 1256, Springer, 1997.

CORRADO BÖHM

IMPOSING POLYNOMIAL TIME COMPLEXITY IN FUNCTIONAL PROGRAMMING

1. INTRODUCING TIME COMPLEXITY OF FUNCTION EVALUATION

Recursive functions are functions whose values can be effectively computed, hence they are the most interesting part, from a pragmatical point of view, of the definable functions on natural integers.

At the beginning of computer science the computable functions were essentially numerical functions. The contemporaneous trend toward multimedia information processing suggested the creation and the use of a variety of different data structures, like lists, trees, graphs, text strings, sounds and images. It is true that all this polymorphic information is always coded, inside a computer, by sequences of binary strings with alphabet { 0,1}; but the human user prefers to describe data and procedures modifying data in a less uniform way.

As a consequence data are still evolving toward algebraic data types and programming languages toward functional programming languages. At the same time, as a consequence of program diagnostics assisting today's compilers, loops occur only sporadically during runs.

If you are running a program and the result keeps you waiting too long, probably one of the following two things has happened:

- ◦ The function whose value you are computing is of moderate complexity, but your definition is unnecessarily inefficient.
- ◦ The function is inherently intractable; the better thing to do is to retrench your target.

Let me give an example of the first case. Let the function to be evaluated be the integral part of the quotient of two natural numbers. Addition and multiplication are supposed to consume, as internal computer operations, a fixed amount of time since the maximum length of their operandi is fixed. We consider here the integer division, defined recursively on the unary successor operation and based on the simple fact that, if you increase the dividend by one maintaining the divisor unaltered, the quotient remains the same or increases by one.

DIV (0, m) = 0
DIV (n + 1,m) = **if** n+1 < (DIV(n,m)+1)*m **then** DIV(n,m) **else** DIV(n,m)+1

The units of time you must wait to obtain respectively 1,2,3 and 4 are

5\ 5	10\ 5	15\ 5	20\ 5
156	5 116	163 836	5 242 876

The second definition is again recursive, the **ifthenelse** function and the > predicate have been eliminated, but negative integers are introduced by decreasing 0 by 1,-1 by

A. Carsetti (ed.), Functional Models of Cognition, 121-126.

1,.., etcetera. The definition is based on the fact that, if you increase the dividend by the divisor, the quotient is increased by 1.

div (0, m) = 0
div(n + 1,m) = 1 + div(n + 1 - m,m)
div(n - 1,m) = -1

The waiting time is much more satisfactory:

5\ 5	10\ 5	15\ 5	20\ 5
3	5	7	9

An example of a function inherently intractable could be the following super-exponential function

sexp(0)=1 sexp (x+1)= exp(x+1,sexp(x))

where exp(a,b) means a to the power b. The succession of values of sexp corresponding to the succession of natural numbers is

$$1,\quad 1^{1},\quad 2^{1^{1}},\quad 3^{2^{1^{1}}},\quad 4^{3^{2^{1^{1}}}} = 4^{9}, \ldots$$

that is

1, 1, 2, 9, 262144,...

Now, sexp(5) is still practically computable, but the computation of sexp(6), measured on the usual human scale, would exceed the space and the duration of our universe.

2. HISTORY

About thirty years ago Manuel Blum introduced the theory of complexity of computation of the functions over natural numbers as a selected chapter of the recursive function theory, introducing explicit reference to Turing machines and to some measure for the time and space used during the evaluation of functions.

Very soon all researchers interested in complexity agreed on some decisions:

- To choose the binary number system to represent arguments and values of functions.
- To define that a function has polynomial time-complexity if all its values are computable within a time which is the value of a polynomial of one variable, representing the (total) length (number of binary digits) of the argument(s) of the function.
- The degree and the coefficients of the polynomial will depend only on the function to be computed.

The notion of a practically computable function may then be identified with its belonging to the class of poly-time functions. This complexity class shall not depend on the choice of computing model (Turing machine or Von Neumann machine) nor on a programming language (imperative or functional). In this paper, I would like to draw your attention to the use of some functional programming languages where functions or even higher order functions are defined equationally or by rewriting systems. Since the same function can be defined, more or less efficiently, in different ways as we did for the division function, we will say that a function is poly-time if it possesses at least one definition logically implying a poly-time computation.We have just arrived at the first seminal question of this paper:

◦ Looking at an equation system defining a function recursively, is there some way to decide whether this function has poly-time complexity?

Due to a lack of information on the function to be defined, the answer was negative before 1993.

The poly-time family of functions was characterized by only adding to the usual machinery "starting from a finite functional basis and closing with composition and primitive recursive operators" the clause that every function defined by primitive recursion must be bounded by a poly-time function.

3. DESCRIPTION OF SOME RECENT RESULTS

At this point it may be convenient to introduce useful names in connexion with the composition operators. A function f of n arguments is obtained by composition of a function h of m arguments (the compositor) with m functions $g_1,..,g_m$ (the components) each of n arguments $x_1,...,x_n$ if the following equality holds:

$$f(x_1,...,x_n) = h(g_1(x_1,...,x_n),...,g_m(x_1,...,x_n)).$$

We add the possibility that the components g_j do not need to depend from *all* its n arguments, but only from a *proper subset* of them. For example in the case n=4 and m=2, we will consider as legal the following composition

$$f(x_1,x_2,x_3) = h(g_1(x_1, x_2),g_2(x_2,x_3))$$

since we have followed the convention to write only the arguments of each function from which it really depends.

In the PhD thesis (1993) of Bellantoni, Cook's student, a positive answer to the main question is given, in the sense that no more subsidiary information is needed, and the question can be settled by just verifying some circumstances which can be read off the equation system.

Another positive solution of the problem of how to characterize poly-time complexity of functions can be found in a paper by Leivant (1993) about poly-time characterization of functions represented by some second-order typed lambda-terms. By the way, he uses some device described in 1985 in a paper by Berarducci and myself. The Leivant method is very elegant from a theoretical point of view, but the checking of the poly-time property has been reduced to a particular type checking of terms which is algorithmically describable but not so immediately translatable into an answer to the previous question.

Here is the second question, motivating the title of this paper:

◦ Can the Bellantoni-Cook method be generalized to equation or rewriting systems defining functions on any algebraic data type?

In order to answer this question, which I asked a PhD student of mine, Vuokko-Helena Caseiro, it is convenient to describe the new ideas developed by Bellantoni-Cook in a way that can be easily generalized to meet our aims.

First, the recursive function scheme over natural numbers is essentially translated into a scheme of functions over binary strings (this is mandatory if you want to talk about complexity).

Secondly, in the composition scheme and in the primitive recursive scheme the variable arguments of all mentioned functions are partitioned into two sets: Normal and Safe.

Then, the class of poly-time functions is characterized by the usual machinery with the following explicit prohibitions (this is a rather personal interpretation, logically oriented, of Bellantoni-Cook's thoughts):

◦ to replace normal arguments by values of the function to be defined

◦ to recur to safe arguments

◦ to replace normal arguments of the composer by components possessing safe arguments.

We will show two examples of poly-time functions, conc and mult, and a supposed counter-example, exp.

Conc is the function concatenating two binary strings. Mult is the function concatenating its second argument with itself n times, where n is the length of the first argument. Exp is the binary equivalent of the exponentiation of natural integers with base 2.

The algebra of binary strings is a homogeneous algebra with 3 constructors:

the empty string # of arity zero and 2 constructors of arity one: s_0, s_1 (concatenating a 0,1 to the right of the string). All the constructors are poly-time functions.

A definition of conc is:

$$\mathrm{conc}(\#,x) = x$$
$$\mathrm{conc}(s_i(y),x) = s_i(\mathrm{conc}(y,x)) \qquad (i= 0,1)$$

In order to prove that conc is poly-time it is sufficient to choose the first argument of conc as normal and both the second argument and the unique argument of s_i as safe.

Since

conc(y,x) replaces a safe argument, the
recursion is done on a normal argument,
the argument of the composer is safe,
no forbidden action has ocurred.

Notice that an alternative proof would be choosing both arguments of conc as normal.

A definition of mult is:

$$\mathrm{mult}(\#,y) = \#$$
$$\mathrm{mult}(s_i(x),y) = \mathrm{conc}(y,\mathrm{mult}(x,y)) \quad (i= 0,1)$$

mult is poly-time, since we may confirm the choice of normal, safe for the arguments of conc (first proof) and we may choose both arguments of mult as normal. Notice that the second proof for conc would introduce a forbidden composition.

The simplest example of a definition of a function which cannot be proved poly-time is the following:

$$\exp(\#) = s_1(\#)$$

$$\exp(s_i(x)) = \mathrm{conc}(\exp(x),\exp(x))$$

From the previous proofs it follows that at least the first argument of conc must be normal. But it is replaced by exp(x), the value of the function to be defined, which is a forbidden action. Hence exp so defined cannot be evaluated in polynomial time.

4. LAST YEAR RESULTS

Since this is an improved version of the author's talk given more than a year ago, it may be useful to summarize the results obtained by Vuokko-Helena Caseiro.

First, she defined normal and safe variable positions in a constructive way, depending only on the structure of the system recursive equations normalized as in the Appendix.

Second, she proved that, in the case of the recursive function definition on integers, her definitions characterizes, as in Bellantoni and Cook paper, the class of poly-time functions.

Third, she showed that, in the case of a more general data type, her previous definitions were not sufficient to cover all cases of poly-time functions. Hence she defined some classes of functions structurally, that is, as only depending from the structure of the equation systems warranted to be poly-time. The characterization of poly-time recursive functions with normalized shape still remains an open problem, but some step has been done in that direction.

Prof. Corrado Böhm
Università di Roma "La Sapienza"
P. A. Moro, 5
00185 Roma, Italy

REFERENCES

C. Böhm and A. Berarducci "Automatic Synthesis of Typed Lambda-programs on Term Algebras", "Theoretical Computer Science 39", p. 135-154, 1985.

D. Leivant and J-Y. Marion "Lambda characterization of poly-time", in J. Tyurin, ed. "Fundamenta Informaticae", Special issue in Typed Lambda Calculi, 1993.

S. Bellantoni and S. Cook "A new recursion-theoretic characterization of the polytime functions", in "24th Annual ACM STOC" (1992), 283-293.

M. Hoffman, "A mixed modal/linear Lambda calculus with application to Bellantoni-Cook safe recursion", in Proceedings of CSL '97, Aarhus, Springer, LNCS, 1998.

APPENDIX A pure functional language

Before concluding I would like to satisfy the possible curiosity of some readers about how it was possible to produce the table

	5\ 5	10\ 5	15\ 5	20\ 5
Number of operations by				
Div	156	5 116	163 836	5 242 876
div	3	5	7	9

Let us sketch now the structure of a pure functional programming language. In order to stick to reality. I will refer to the programming language CuCh (acronym derived from Cu-rry and Ch-urch) of which we possess, in our department, an interpreter first written in C by Stefano Guerrini (1991) and later improved by many other students of my course.

A program describes an input-output transformation. Inputs and outputs belong to some union of term algebras. It follows that (h.o.) functions are, in a natural way, polymorph, that is, they are applicable to more than one algebra. A term algebra is roughly an algebra in which each element can be written in a unique way.

Every term algebra is either built-in or defined by the programmer inside the program itself.

Let us consider built-in algebras : Boole, Integer and List algebras. Every algebra is inductively constructed by nullary, unary, binary,... functions called constructors.

Here is the definition of the built-in algebras in a Backus-like style:

```
Boole ::={ boolean: true | false }
Int   ::={ int: succ int | zero | succ- int }
List  ::={ list: cons P list | nil }
```

where P is a name for an element of a not yet specified algebra, or union of algebras, which may remain unspecified if we do not require type checking.

We can write elements of built-in algebras like

succ(succ(succ zero)), succ-(succ- zero) cons 3(cons 2 nil) etc.

which we may abbreviate in input-output into

3,-2 [3,2] etc.

Obviously, built-in algebras possess also built-in operations directly executed in machine code like + , - , * and \ the four operations on integers (the last one means the integer part of the quotient). Notice that all operations are written in prefix notation with the exception of a synonym of cons which may sometimes be written as [a | b] instead of cons a b.

Algebras may be defined by the programmer in a similar way as built-in algebras.

The algebra name, the constructors names, their arities, and the type of the arguments are freely chosen by the programmer.

Example: the algebra of binary strings

Bstrings::={ bstr: # | s_1 bstr | s_0 bstr }

Let us return now to pure functional programs. A program is a set (not a sequence) of definitional rules. A computation is initiated by an assignment statement having the shape:

identifier := expression

where 'identifier' is a name for a variable and 'expression' means any expression formed by means of defined functions, built-in functions and specific elements of defined algebras. Any definitional rule has the following shape:

f.identifier(c x_1 ..x_m)y_1...y_n := expression

where f.identifier is a name of a (possibly h.o.) function, and c is the name of a constructor of an algebra to which the first argument of the function belongs. The important point is that x_1,..,x_m,y_1,...y_n are *bound* variables in the sense that their values will be assigned automatically, by a pattern matching process, only at run time.

The expression at the r.h.s. is arbitrary; it may contain defined functions, constants and also variables. Such variables either occur at the l.h.s. of the *same* line or may be also *free* variables whose values are to be assigned before any computation of that particular function. Summarizing: every function is defined by cases determined by the head constructor of its first argument. The mentioned restrictions on the way of writing definitional equations do no weak the definitional power of the language. Mutual or multiple recursive functions on any data types may be defined, possibly adding new functions and consequently equations.

Let use give an example. The first is a program defining the well known map functional. Given a list [x_1,..,x_n] and a unary function f, map constructs the list [f x_1,.., f x_n]. This means that the value of

map [1,2,3] square is [1,4,9].

The program is

```
@env;
map(cons x y )f := [f x | map y f];
map  nil  f     := nil;
```

The running is induced by

```
@lambda;
square := lmb:x. * x x,
r := map[1,2,3]square;
```

HENRI ATLAN

SELF-ORGANIZING NETWORKS: WEAK, STRONG AND INTENTIONAL, THE ROLE OF THEIR UNDERDETERMINATION

SELF-ORGANIZING NETWORKS AND EMERGENCE OF FUNCTION

Since A. Turing did his pioneering work on morphogenesis by coupling diffusion and chemical reactions, many models of cooperative phenomena, making use either of sets of continuous differential equations or discrete networks of cellular automata, have continually produced the same kind of phenomenon: starting from an homogeneous initial state (which may be set by a random distribution of states over the elements of a network), the system evolves "spontaneously" towards a stable state where a macroscopic structure in space and/or in time can be recognized. This evolution is the result of local laws of interaction between the coupled phenomena or the connected automata. This phenomenon is very general and is observed for a wide range of local laws of interaction (continuous, boolean, threshold functions, etc.) and connections.

The microscopic structure can be set up randomly, at least partially, as in random boolean networks, where different boolean laws are distributed randomly on the elements of the network (Atlan 1987; Kauffman 1984).

Even for relatively small networks of a few coupled processes, as studied in immunology (Atlan 1989; Cohen & Atlan 1989) and cell biology, and all the more in larger networks which aim at simulating the cognitive capacities of the brain, it is very difficult (and, in most cases, practically impossible) to predict the emerging macroscopic structure from a mere inspection of the microscopic structure and initial state without actually running the computer simulation.

In this sense, these models show how a mixture of local determinations and randomness can easily produce apparently "self-organizing" structures which were not explicitly programmed. Even systems completely deterministic at the level of *local* interactions can produce this kind of phenomenon providing that:

1. their dynamics are rich enough so that their integration leads to several different stable solutions;
2. they are submitted to random fluctuations which drive them towards one or another of these solutions. The extreme case is that of so-called "deterministic chaos" or "strange attractors" where the number of possible solutions is as large as the number of possible initial states (i.e., infinite for continuous variables) so that the smallest possible fluctuation may drive the system to a completely different, unpredictable solution.

Self-organization has become popular in artificial intellingence as well, with the use of neuron-like automata networks to perform tasks of pattern recognition by non-directed learning and distributed associative memory ("neural net computation") (for a minireview see Weisbuch 1986). This method, which makes use of several competing algorithms, is a set of more extended, more sophisticated, versions of the perceptrons devel-

A. Carsetti (ed.), Functional Models of Cognition, 127-142.

oped in the 1960's. For this reason, this approach has been termed neo-connectionism and has become well-known in cognitive sciences under that name. Then the prefix "neo-" has been dropped.

However, even at this level of computer simulations of physicochemical processes where no intentional, conscious or unconscious definition of the self is assumed it is appropriate to make a clear distinction between two kinds of self-organization. I suggest that self-organizing systems in a strong sense should be distinguished from those in a weak sense.

Examples of *self-organization in the weak sense* are given by most Artificial Intelligence applications of neural network computation to the design of learning machines and associative distributed memory (Hopfield 1982; Kohonen 1984). A typical performance of such networks consists of recognizing an infinite number or variations of a given pattern (such as a handwritten letter of an alphabet) after exposure to a limited, relatively small sampling of such variations. Contrary to classical pattern recognition procedures, the class of patterns to be recognized is not explicitly defined by specific features. Rather, very general learning rules are set up, so that exposure to samples of the class member will trigger the modifications of the networks's connection structure. Then, further exposure to different pattern members of the same class will lead to recognition with a reasonably high rate of success. It is in this sense that the network organizes itself, since the learning rules are not specific for recognition of any particular pattern. They are the same for all possible classes of patterns to be defined by the experience which constitutes the exposure itself, during the learning period. Specific connections are automatically established, but in a non-explicity programmed manner, by application of the rules, under exposition of the network to external stimuli i.e. to the samples of the pattern which, implicitly, define the class. This self-organization of the connection structure is real in the sense that it is not explicitly programmed. However, one can speak of self-organization in a weak sense only, because the task to be achieved is defined, although implicitly, from the outside: the goal of this learning is to recognize a given pattern and, more generally, to solve a given problem, set up from the beginning by the designer of the network. The learning rules are considered to be efficient (or inefficient) according to an *a priori* established criterion, namely, the extent to which the specific problems are solved and the proposed tasks performed, i.e., to the extent that they achieve the goal set by the user. In other words, the meaning of the functioning of the network is still given *a priori* by the human designer.

Self-organization in a strong sense means that even the goal, i.e., what constitutes the meaning of the structure and function of a machine, emerges from the evolution of the machine itself. The work of our group in Jerusalem, in collaboration with G. Weisbuch and F. Fogelman in Paris, on random boolean networks mentioned above, can give some primitive, although demonstrative, example of such a phenomenon (Atlan 1987; Atlan *et al.* 1986).

Fig. la shows an instance of such a network after it has reached its attractor in the form of a macroscopically structured state. Its initial state was macroscopically homogeneous, being a random distribution of 0's and l's on the elements. One can see that it is now divided into subnets, characterized by their different temporal behavior. The elements in the network indicated as S are stable, i.e., their state, either 0 or 1, does not change in time. Those indicated as P oscillate periodically, i.e., their state changes in time according to a given relatively short sequence of 0 and 1, cycling indefinitely on itself. This is an example of a well-documented phenomenon of macroscopic spatiotemporal structures emerging from microscopic laws (in this case boolean functions) randomly distributed and certain kinds of connections. However, we have also found that this kind of network, after it has organized itself into such a spatiotemporal structure, has some striking properties of pattern recognition. If binary sequences are imposed from the outside on one of the elements in order to perturb the network after it has reached its attractor, the perturbation is transmitted to the rest of the network in two possible ways (see Fig. lb). While some stable units (S) are destablized, others which oscillated periodically in the unperturbed network become, surprisingly enough, stabilized by the perturbating sequence. The mechanism of stabilization is a complex resonance between the incoming perturbing sequence and the periodic cycle of some of the oscillating elements. However, this resonance is not restricted to one single periodic incoming sequence. It is also observed for partly random sequences, the periodic structure of which is limited to a few bits, while the others are indifferent. Thus the whole class of sequences (potentially infinite if their length is not limited), which share this minimal periodic structure, will have the property of stabilizing a given element of the network.

When this happens, i.e., when a sequence imposed on a given (input) element has the property of stabilizing another (output) element, we say that the network functions as a pattern recognition device and that this sequence has been "recognized". Thus, for a given network in its steady state and a couple of elements functioning as input-output of the pattern recognition device, a class of sequences are recognized and those sequences which do not belong to that class are not recognized.

An example of sequences belonging to such a class is given in Fig. 2. The class is defined by the displayed partly random sequence where the stars stand for 0 or 1 indifferently, at random.

The important point in this phenomenon is that the definition of the recognition class, i.e., the criterion for recognition, has not been programmed and is the result of the overall spatiotemporal structure of the network in its attractor, and more particularly of the states of the input-output element, together with the elements which join them, and their neighbors.

As a matter of fact, when the input and output elements for a particular recognition channel are far apart in the network, it is very difficult to figure out the details of the mechanism whereby the output element is stabilized when a member of a given class of sequences is imposed on the input. It is not enough that the criterion for recognition has not been programmed. It can be discovered only a posteriori, by experiments, as in a true natural recognition device. For instance, one could think of a region in the brain of an animal which responds to some stimuli and not to others when it is explored by a couple of electrodes, one stimulating and the other recording the response. The task of the neurophysiologist is to discover the criterion for this discrimination and its mechanism.

Thus, this recognition property is an emerging one which accompanies the emergence of the macroscopic structures, as a function accompanies a structure.

This kind of phenomenon exemplifies self-organization in the strong sense since, as in a natural (non-man-made) self-organizing system, e.g., a biological one, the goal has not been set from the outside. What is self-organizing is the function itself with its meaning and not only a structure adapted to achieve a predetermined goal. In this example, where the function is pattern recognition, the criterion for recognition is the goal and the origin of meaning for the recognizing system. The demarcation between recognized and unrecognized sequences according to this criterion transforms the whole undifferentiated set of possible binary sequences into those which are meaningful and those which are not, or alternatively, into those which have one meaning and those which have another. Since the criterion for this demarcation is itself the result of self-organization, one can say, in some sense, that the origin of meaning in the organization of the system is also an emergent, self-organizing property.

Our distinction between self-organization in a weak and a strong sense may seem too schematical in the light of recent work in A.I. on computer simulations of cognitive performances exhibiting properties of self-organization in the strong sense. The distinction seems to be blurred because the real biological systems that these models simulate are self-organizing in the strong sense, although the computer models themselves, whether they use classical programming or neural net computation, are self-organizing in the weak sense, as mentioned above. That is the case for example of performances of our visual system (and that of animals) (Grossberg 1990a,b). Similarly, that is the case of some aspects of our analogical reasoning (Hofstadter 1985a,b; Mitchell & Hofstadter 1990), and of our capacity to state and solve new problems and to find ways out of unexpected difficulties in dealing with new situations which seems to be reproduced by the "Soar" program of A. Newell, J. Laird and P. Rosenbloom (Waldrof 1988, Newell 1990). But that is also the case for much less sophisticated biological systems such as the immune network, some self-organizing properties of which we are trying to simulate (Atlan & Snyder 1989; Cohen & Atlan 1989; Weisbuch & Atlan 1988).

In any case, indeed, the question of the internal or external origin of meaning still remains of importance as an indicator of a qualitative change in the logic of self-organization.

We shall argue in Section III that even self-organization in a strong sense is not sufficient to account for a true emergence of meaning and intentional goals, since the emerging behavior of the boolean network in our example has the meaning of a pattern recognition only to our eyes as human observers. Thus, if the origin of meaning is to be found in the operation of observation, one could argue that it should be looked for in the observer's brain. The latter, as a neural network, would thus exhibit self-organizing properties in an even "stronger" sense which we shall describe as intentional self-organization.

UNDERDETERMINATION OF THE THEORIES BY THE FACTS

Before coming back to this critical question, it is appropriate to underline a general property of neural network computation, trivial in a sense, but with far-reaching consequences. Moreover, the consequences are different according to whether neural network computation is used as a mathematical tool for modelization or is considered to represent the actual functioning of our brain. In the first instance, underdetermination of the theories is obviously a weakness of the tool, irreducible if no other more powerful tool can be used. However, it is a property of robustness when it pertains to our networks of actual neurons. Thus, the same phenomenon will also help us to understand how different complex cognitive systems, with different histories and different structures, can nevertheless share some features and be able to communicate in a meaningful fashion with one another.

Neural network models of natural phenomena exhibit in a conspicuous way a property of underdetermination previously described in different, philosophical contexts. Today we come across this phenomenon in the analysis of automata networks even of small size, i.e., in much more operational and simpler situations.

By underdetermination of theories by facts I mean the following: many different, non-redundant theories predict the same facts, and there is no way one can decide between them, based on the observable facts alone.

Let us consider any network of interconnected units, even relatively simple ones in the sense that it is composed of a small number of such units.

Such a network may be used to represent the behavior of a system composed of several elements or processes working in a coupled fashion; for example, different biochemical reactions in cell physiology, or different cell populations in immunology or different neurons or groups of neurons in the nervous system. Then, in general, the observable facts are represented by the states of the network and more precisely by its stable states. Now, the stable states are determined by the structure of the network; they can be computed on the basis of the connection structure between the units. That is why every connection structure of a given network represents a theory which allows one to explain and predict the stable states of the network, i.e., the observable facts.

In the work of theorization, the ideal would be that one single structure of the connections would predict the stable states observed in reality.

Unfortunately, it is easy to see that in most cases we are unable to reach that goal and even to come close to it, except under particularly favorable conditions. This is due to the fact that the number of possible connection structures is generally much larger than the number of possible states.

For example, for 5 interconnected units the number of states is of the order of $2^5 = 32$ if every unit can be in either one of 2 states, and it is $3^5 = 243$, or $4^5 = 512$, etc., if each unit can be in 3 states, 4 states, etc. However, the number of connections is 25 and the number of possible connection structures, i.e., of different networks with the same 5 units is of the order of 3^{25} if we assume that a connection can be activating, inhibiting or non-existent. 3^{25} is approximately 10^{12}, i.e. there are one thousand billion different ways to interconnect 5 units, in other words, one thousand billion different theories can be built with the same 5 elements.

Thus, we have, on the one hand, a few hundred states and, on the other, billions of different possible structures. As a result, a large number of network structures, different without being redundant, predicts the same stable states. In other words, a large number of different theories predicts the same observable facts. This is what is meant by underdetermination of the theories by the facts.

In order for a discrete automata model to have some chance of not being underdetermined by the observed facts, it is easy to show that if q is the number of different states of each unit, it must be much larger than the number p of different values that each connection can take. In the above mentioned example p was assumed to be 3, but it can be larger if different weights are given to the connections. More precisely, one must have $q>p^N$ where N is the number of elements in the network, since the number of states is q^N and the number of connection structures is p^{N^2}. Cases where this condition would be fulfilled would come close to situations where we can assume continuous variables, represent them by sets of continuous differential equations, and test them accurately because we would have access to large numbers of observable facts.

In principle one can think about several ways to reduce the underdetermination. One is looking for more empirical data; they would provide direct evidence on the network structure itself, making some of the previously accepted features unrealistic.

Another way would be to refine the theories by making additional, more specific predictions. This would allow one to define and look for new relevant facts, e.g., not only attractors of dynamics but transients, i.e., sequences of states the system goes through until it reaches one of its attractors. It may happen that a new experimental technique provides access to such sequences of states. Thus, the range of observable behaviors to be correctly predicted would increase, and this would allow one to demand more from the competing models.

However, it is not always the case that enough empirical data may become available to allow for the complete reduction of the underdetermination to one single theory. In view of the immense numbers mentioned above, the more heterogenous coupled phenomena must be taken into account by the model, so that the larger the minimum number of automata needed, the less likely we are to succeed. In fact, underdetermination of the theories is what complexity is about; and its interpretation in terms of automata networks shows that complexity is not an "all or nothing" property: natural phenomena are not either reducible to a complete analytical or computable unique description or mysteries where nothing can be understood rigorously. Most biological systems, when studied in physiological in vivo conditions, lie somewhere in between, and that is what makes them complex, i.e., neither fully and without a doubt understood, nor intractable.

The above example of a 5 automata network is taken from a model in immunology (Atlan & Snyder 1989; Cohen & Atlan 1989), and it shows that already small numbers of interacting units are enough to generate such underdetermination. It is neither necessary to reach the billions of interacting neurons of a mammalian brain, nor to resort to psychology and to the usual neuro-philosophical mind-body problem.

However, let us remember that it is in psychology and cognitive sciences in general that large automata network computation has become very popular in recent years, as a very crude simulation of the behaviour of neural networks. (As a matter of fact, it is an extension of the classical work of McCullough and Pitts on neural nets in the forties which has been made possibles today thanks to the availability of computing facilities.)

That is why most people view such networks as models of cognitive systems, i.e., not as theories which account for facts but as the cognitive systems which build the theories, like our brains. Thus, if we change our viewpoint and look at these networks as models of our cognitive systems whereby we perceive, understand and communicate, the perspective is completely changed. What appeared to us as a weakness in the theories, i.e., their underdetermination by the facts, now appears as a positive and advantageous feature of our cognitive neural networks. The same phenomenon, namely, the fact that the number of different network structures is much larger than the number of their states, now has a different meaning. It is transformed into a positive property of robustness and structural stability which allows for very different neural nets to reach identical stable states. To the extent that we are willing to recognize some validity to the neural net computation model of our brains, in spite of its oversimplification and underdetermination, this property of stability is a very important one for the functioning of our cognitive systems. Indeed, we have every reason to believe that both genetically and epigenetically determined structures in our brains lead to very different connection patterns for different individuals, at least as far as the details of the connections are concerned. And the neural net computation model shows us that different neural networks with different histories may nevertheless reach similar stable states. This may allow us to share some identical stable patterns in our different cognitive systems, and thereby to more or less understand one another. Thus, what appears as an irreducible complexity of our cognitive systems from the viewpoint of the underdetermination of our theories about their structure and function, may also be what leaves room for a possible intersubjectivity.

INTENTIONAL SELF-ORGANIZATION

In all these models of self-organization, we always take something for granted, namely, the meaning of the pattern which emerges and stabilizes, and the function it performs.

However, as we have begun to discuss above, this question of meaning is what makes all the difference between natural systems which we observe from the outside, and human systems or man – made systems which we can observe, at least partially, from the inside.

Thus, one cannot avoid taking into account the viewpoint of the observer (obviously not in a subjective sense but in the sense usual in physics of the objective conditions of observation and measurement). The notions of microscopic and macroscopic structures are relative to the level of observation. Consideration of a network as a device to recognize meaningful patterns on the basis of self-generated criteria is an interpretation. Similarly, saying that the "Soar" program "understands" or "knows" is an interpretation. More precisely, it is a projection of our own experiences of pattern recognition and understanding of the observed behavior of the network. Therefore, we must make another distinction based on the capability of not only creating what appears to be a non-programmed meaningful function but also of interpreting or understanding it as a function, as if it were purposely performing it. In other words, we must consider a third type of self-organizing systems, able to generate projects in the sense that they are capable of setting a goal for themselves and achieving that goal. Let us call this capacity " intentional self-organization".

I would therefore like to suggest how, in principle, a general model of self-organization, not only in the strong sense, but also exhibiting some degree of intentionality, could be constructed on the basis of the known properties of neural networks. This approach is coherent with the fact that we can also observe some kind of intentional behavior produced by neural networks different from our brains, i.e., brains of other animals not too different from ours, such as those of certain primates. In effect, it seems to be the case that some chimpanzees, who have often performed certain gestures having no apparent functional meaning, discover the effects of these gestures and reproduce them in an apparent purposeful manner in order to reproduce these effects. As in humans, the fabrication of primitive tools oriented towards their future utilization can be understood only as a certain capacity to implement projects and to determine behavior that we have every reason to qualify as intentional. A relatively simple machine, capable of creating a project, and thus reproducing simple intentional behavior, could function as follows.

Let us go back to our random boolean networks where we have seen that a purely causal succession of states produced by the dynamics of local mechanical interactions between the elements may acquire some meaning in the eyes of an external observer. This happens, for example, when such a succession of states is stabilized into a final state where the network does something which, to the observer, appears as a function. Then, this succession of states, a posteriori, appears to the eyes of the observer as a sequence of means used to produce the final state with its function as an end.

Let us assume now that this sequence of states is stored in memory as it is. Then everything happens as if what is stored were a procedure, since the final state is stored—together with what it is doing—as the end of this sequence. In addition, let us assume that a neural network can function as an observer of itself—something which we do not yet know how to do in artificial networks but which seems to be performed apparently by the reticular formation of our brains.

Now it may happen that the stored final state is retrieved, in an associative way, by a different sequence of states which is itself triggered by an eventual perception of the effects of that state—i.e., its function as interpreted by the network because of its observation of itself. Then what would be retrieved would be not only the stored final state, but the whole procedure which has been stored together with it.

As an example, let us consider the final state in the sequence of states of the brain of a primate associated with the modification of the shape of a dead bone which allows its accidental use as a tool by accidentally making a hole in the ground. As a model of that sequence of brain states and of the final one, let us consider the transient and the attractor, respectively, of the dynamics of an automata network. The digging of the hole is seen as a non-programmed effect, a consequence of the attractor, just as the pattern recognition was a non-programmed consequence of the attractor in our random boolean network. Then the whole sequence of states, including the image of a hole in the ground is being stored in memory. Eventually, another perception or representation of a hole in the ground will retrieve, by associative memory, the same final state "modification of a bone" as was produced previously. The "hole in the ground" will then be interpreted as a consequence of the "modification of the bone", although it is the cause of the retrieval of the network state corresponding to this modification. In addition, it is not only that state which is retrieved but the whole sequence of states which led to it for the first time. Thus it is the whole sequence of states which is now interpreted by the network as a procedure, the meaning of which is to fabricate a tool in order to dig a hole in the ground.

Eventually, each time this sequence is retrieved from memory by association with various stimuli, it will bear the same meaning of an intentional behavior appearing as the achievement of a project directed towards and by the future.

Whereas everything which happens in the network is always the result of a mechanical causal succession of states and its storage in memory, the inversion between past and future is possible because the whole sequence of causes and effects has been stored, including its last consequence as a representation of a sequence of symbols in memory.

This allows for the inversion between the effects and the cause in the representation. What was the end of a procedure and its consequence in the actual action before it was stored is now transformed into a cause in the representation. This inversion contains no mystery, since it is made possible by the elimination of real time which results from the mere fact of memorization. As soon as a temporal sequence is stored in memory, the temporal order from past to future is transformed into a symbolic order in which the actual direction of time has disappeared, since the whole sequence is there, at the same time, in memory. (Every event in the ordered sequence can be assigned an index in order to keep the initial order, but the index is symbolic and not really temporal.) Thus, it can be retrieved all at once and triggered by a stimulus associated with the action which was in fact the last consequence of the sequence of events when it was performed in real time. That is how what happens at the end in the temporal succession of causes and effects can be perceived as an initial stimulus, triggering the retrieval and implementation of the same sequence of causes and actions now interpreted as a procedure to reach that end, thus achieving a projected goal.

If we knew how to build a machine capable of observing itself, it seems to me that nothing could prevent us from building a purely causalist, mechanical model of our capacity to make simple projects, and of what seems to be a similar capacity already observable in great apes.

A further step could be reached if the process of memorization and interpretation of procedures were itself memorized. This would provide a model for our experience of our capability to make projects in general, independent of this or that specific project. In other words, this would provide a mechanical model for what we experience as our consciousness of intentionality.

As I suggested many years ago (Atlan 1979), conscious will and intentionality should be analyzed as secondary to the interaction between two different primary processes: consciousness as memory of the past on the one hand, and unconscious self-organizing processes on the other. Basically, consciousness would be mere recording, storage and retrieving of past events, whereas innovation and adaptation to newness with the creation of new struetures and functions would be produced by unconscious self-organizing processes. On top of these two basic primary processes, and secondary to their interactions, conscious will and creative intentionality would be interpretations of memorized products of previous self-organization, in such a way that previous unplanned consequences of actions are transformed into possible causes for the repetition of similar—though not necessarily identical—sequences of actions.

THE MEANING OF SIMULATION AND THE SIMULATION OF MEANING

Assuming that all the technical difficulties can be overcome and that such models of intentional self-organizing systems can be practically achieved, we are left with an ultimate question which is basically the classic question of the semantic capacities of Artificial Intelligence programs. If algorithms are assumed to be capable of simulating all our cognitive capacities, are they also able to understand the meaning of what they are doing? Searle's "Chinese room" thought experiment (Searle 1980) seems to give a negative answer to this question. Since it was published, it has triggered a whole pro and con literature. The discussion is still going on (see, for example, Churchland & Smith-Churchland 1990; Hofstadter & Dennett 1981; Searle 1990; Shannon 1990) and will apparently go on for a long time until an agreement is reached about what it is, for us, to understand the meaning of something.

So far, whether a computer program is a mere simulation of our understanding no more related to the real one than a computer simulation of digestion to actual digestion, or whether it actually understands as well as it actually computes, is decided on the basis of what is our opinion of human understanding. The situation concerning understanding as one of our cognitive capacities is different from that concerning our capacities to compute; and that is the source of confusion and the reason why the controversy will probably go on until we know better what it is, for us, to understand. In the case of computation, everybody agrees that the computer programs not only simulate some of our cognitive capacities to compute, but that they actually compute. The situation is not the same concerning our capacities to understand. If we tend to associate understanding with our capacities to compute and use logical symbols, then we would tend to look at the computer as capable of understanding. If we tend to associate understanding with a distinct function of ours, then we would tend to look at the computer program as a mere simulation of that function. In addition, everyone seems to agree that present-day computer programs cannot fulfill the minimum requirements (i.e., the Turing test) to be considered as capable of understanding. Therefore, the strong A.I. arguments against Searle (who assumed himself that a computer program had already passed the test) are based on speculations about what future computers would be capable of doing. The fact that parallel computing machines have entered into the picture seems to strenghten the strong A.I. argument. However, as far as the basic theory of programming is concerned, it does not matter whether a computer is working serially or in parallel, as a Turing machine working in series on one single tape is equivalent to a Turing machine working on several tapes in parallel. Genuinely new achievements could be reached only if the notion of programming itself were changed, for example, by making systematic use of randomness and of different possible kinds of interactions between different levels of organization. It is true that automata network computation may help to produce such changes. Anyway, the discussion suffers from a lack of clarity about what a computer program may be in the future as well as about our understanding—similarly our feeling, our consciousness, etc.—is "in reality".

Our work on the emergence of classification procedures in boolean networks may add something to the discussion in that it is based precisely on a switch from the notion of programming to that of self-organization in the strong sense including the emergence of non-programmed properties. In the case of the limited models of creation of meaning in a primitive sense, Searle's negative answer is strengthened, since it is clear that the network does not understand what it is doing and that one can talk about meaning because of our capacities of interpretation as human observers only.

The main difficulty with this negative answer being extrapolated to all other kinds of self organization, including intentional ones, is that it seems to put our internal experience of semantics and intentionality outside of the physical world into a kind of absolute dualist philosophy (although Searie explicitly rejects such a stand). On the other hand, partisans of strong A.I. often use the functionalist argument which is not devoid of dualism in that it considers the hardware (in our case the biochemical nature of the brain) to be irrelevant to its logical and psychological performances, so that the latter could be reproduced by software running on different machines. From this point of view, neural network computation may be in a better shape than classical programming. It seems to allow more intricate forms of interaction between levels of organization, including between software and hardware, and its physical organization makes it more similar than classical computers to networks of real neurons.

SOPHISTICATION AS MEANINGFUL COMPLEXITY AND HERMENEUTIC CAPACITIES

In the task of unifying neural network computation and our experience of meaning and intentionality, a last step would be to find a new organizational principle which would be unique to our semantics and intentionality in that it would explain our unique capacity of interpretation, i.e., of projecting and giving meaning to everything—either an intentional or a non-intentional, purely causal meaning, depending on the rules of the games that we set for ourselves.

Such a principle could be based on the idea that a model of intentional self-organization, such as the one we discussed, could be generalized even more. We started from the simple capacity of creating a projection by converting the consequence of a sequence of events into the trigger of a procedure; then we suggested a capacity to memorize the memorizing of procedures as a possible substrate for the consciousness of intentionality.

What is missing now is a capacity not only to memorize procedures with their meaning, and then the process of memorization itself but also to modify the meanings of the procedures under almost any circumstances which modify the procedures themselves. The basis of this idea is that actual meaning is not going to be found as an intrinsic property of a sentence or an obiect. As some psychologists have already noted (Shanon 1990b) it seems to be generated by an act of interpretation which takes place at the interface between the observer and the observed. However, observation and even self-observation are not sufficient. That is why the mere existence of the reticular formation in the brain of many vertebrate species other than humans—in the human brain it seems to be responsible for self-awareness—may lead us to conclude that those animals have a capacity for self-observation without necessarily having semantic and intentional capacities as we feel we have. In order to work like an apparently infinite source of interpretations, self-observation or awareness must not be a mere passive memory device. It must be connected to one or several self-organizing devices able to indefinitely produce novelty which it will indefinitely interpret with a new meaning.

From the point of view of the theory of algorithm complexity, such a capacity of interpretation could be assigned to a special type of algorithms, defined as capable of generating infinite objects with a seemingly infinite "sophistication". Contrary to the classical measures of complexity, sophistication is a measure of meaningful complexity. Whereas the classical theory of algorithmic complexity assumes that computer programs are meaningful, it does not account for what constitutes their meaning. The usual measure of their complexity is based only on the possibility to reduce their length without taking into account the more or less intricate or intelligent aspect of their task. Similarly, the estimates of natural complexity based on probabilistic information theory are measures of statistical unexpectedness and not of any kind of meaningful complexity (Atlan 1988). Sophistication was defined as the meaningful part of the classical complexity of an algorithm (Atlan & Koppel 1990; Koppel 1988) by drawing a formal distinction between the program (which defines a structure with a given meaning) and the data (which merely specify a particular instance out of many members in the class of obiects sharing the same structure). Among other properties, this definition allows one to discriminate between the maximum classical complexity of random strings and their zero sophistication. Although objects of infinite sophistication can be formally defined in precise terms, no real algorithmic procedure could, by definition, generate one (Koppel 1988; Koppel & Atlan 1990). As such, they have the peculiar property of being neither recursive nor random, which is what we expect from meaningful things to which new meanings are constantly assigned upon constant, unexpected reorganizations. If the sophistication of the sequences of states in our brains would be large enough to approximate infinity—especially in view of the communications between different brains in space and time with the help of accessory memories and computing devices—this would explain their apparent ability to create indefinitely new meanings by interpretations. Then, if such machines, producing behaviors of apparently infinite sophistication, could ever be made by A.I., especially in the form of multilevel, partially random architectures of automata networks, perhaps the question could be raised as to whether they would experience the same feeling of meaning and intentionality as we do.

Prof. Henri Atlan
Hadassah University Hospital
P.O.B. 12.000
il 91120 Jerusalem, Israel

REFERENCES

Atlan H. (1979), Entre le cristal et la fumée, Paris, Seuil.
Atlan H. (1987), "Self-Creation of Meaning", "Physica Scripta", 36: 563-576.
Atlan H. (1988), "Measures of Biologically Meaningful Complexity", in: Measures of Complexity (L. Peliti and A. Vulpiani, eds.), Lecture Notes in Physics, 314, Berlin, Springer-Verlag, pp. 112-127.
Atlan H. (1989), "Automata Network Theories in Immunology: Their utility and their underdetermination", "Bulletin of Mathematical Biology", 51 (2): 247-253.
Atlan H., Ben Ezra E., Fogelman-Soulie F., Pellegrin D. & Weisbuch G. (1986), "Emergence of Classification Procedures in Automata Networks as a Modelfor Funchonal Self-Organization", "Journal of Theoretical Biology", 120: 371-380.
Atlan H. & Koppel M. (1990), "The Cellular Computer DNA: Program or Data", "Bulletin of Mathematical. Biology", 52 (3): 335-348.
Atlan H. & Snyder S.H., "Simulation of the Immune Cellular Response by Small Neural Networks", in: Theories of Immune Networks (H. Atlan & I. R. Cohen, eds.), Berlin, Springer-Verlag", pp. 85-98.
Churchland P.M. & Smith-Churchland P. (1990), "Could a Machine Think?", "Scientific American", January, pp. 26-31.
Cohen I. R. & Atlan H. (1989), "Network Regulation of Autoimmunity: An Automaton Model", "Journal of Autoimmunity", 2: 613-625.
Grossberg S. (1990), "Self-organizing Cognitive and Neural Systess", in: Twelfth Annual Conference of the Cognitive Science Society (eds.), Hillsdale N.J., L. Erlbaum, pp. 1027-32.
Grossberg S. & Rudd M. (1990), "Self-organizing Neural Networks for Visual Form and Monon Perception", Twelfth Annual Conference of the Cognitive Science Society (eds.), Hillsdale, N.J., L. Erlbaum, pp. 1032-1034.
Hofstadter D.R. & Dennett D.C. (1981), The Mind's I, New York, Basic Books, pp. 373-382.
Hofstadter D.R. (1985a), "Analogies and Roles in Human and Machine Thinking", in: "Metamagical Themes" (ed.) New York, Basic Books, pp. 547-602.
Hofstadter D.R. (1985b), "Waking up from the Boolean Dream, or, Subcognition as Computation", in: "Metamagical Themes" (ed.) New York, Basic Books, pp. 631-665.
Hopfield J.J. (1982), "Neural Networks and Physical Systems with Emergent Collective Computational Abilities", Proceedings of the National Academy of Science (USA), 79, pp. 2554-2558.
Kauffman S.A. (1984), "Emergent Properties in Random Complex Automata", "Physics", 10D: 145-156.
Kohonen R. (1984), Self-Organization and Associative Memories, Berlin, Springer.
Koppel M. (1988), "Structure", in: The Universal Turing Machine: A Half Century Survey (R. Herken, ed.), Oxford, Oxford Univ. Press, pp. 435-452.
Koppel. M. & Atlan H. (1991), "An Almost Machine-Independent Theory of Program-Length Complexity, Sophistication and Induction", "Information Sciences", 56, 23-33.
Mitchell M. & Hofstadter D. R. (1990), "The Emergence of Understanding in a Computer Model of Concepts and Analogy-making", "Physica" D., vol. 42, pp. 322-334.
Newell A. (1990), Unified Theories of Cognition, Cambridge Mass., Harvard Univ. Press.
Searle J.R. (1980), "Minds, Brains and Programs". "The Behavioral and Brain Sciences", 3: 417-457.
Searle J.R. (1990), "Is the Brain's Mind a Computer Program?" "Scientific American", January, pp. 20-25.
Shanon B. (1990), "Consciousness", "Journal of Mind and Behavior", 11 pp. 137-152.
Shanon B., "Are Connectionist Models Cognitive?" "Philosophical Psychology", in press

Waldrof M. (1988), "Soar: A Unified Theory of Cognition?", "Science", 241: 296-298.
Weisbuch G. (1986), "Mini review: Network of Automata and Biological Organization", "Journal of Theoretical Biology", 121: 255-267.
Weisbuch G. & Atlan H. (1988), "Control of the Immune Response", "Journal of Physics". A, Math. Gen., 21: L189-192.

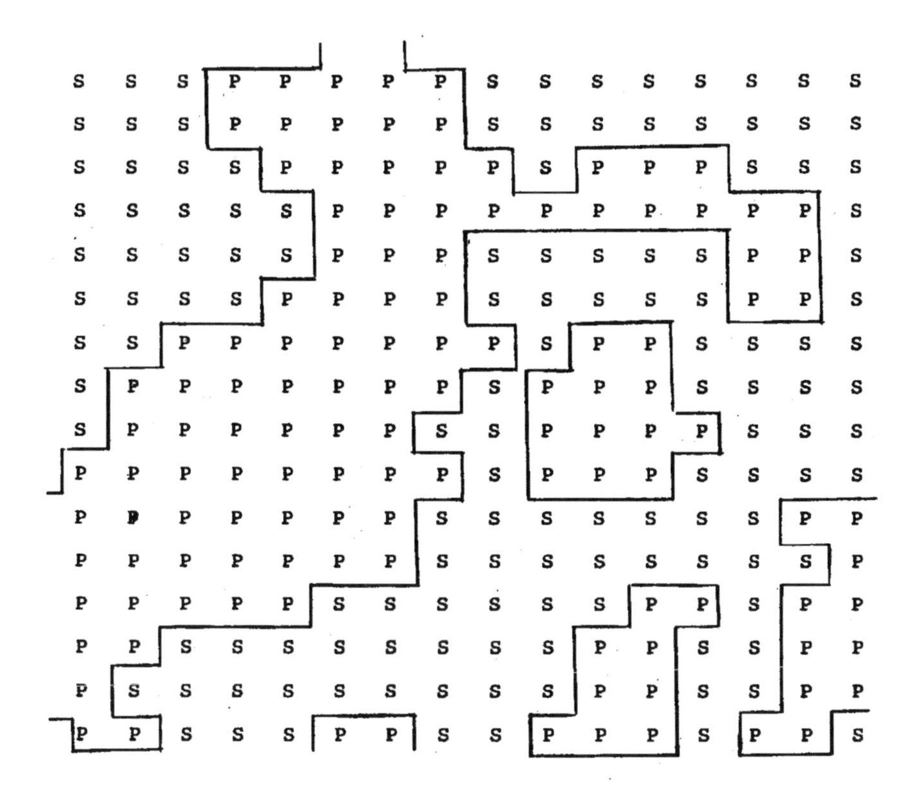

Fig. la - Emergence of macroscopic structures.
The elements are connected in such a way that each of them receives two binary inputs from two of its nearest neighbors and sends the same output to the two others after computation. Each row and column is closed on itself. At every step of the computation every element computes its output from its two inputs by means of one of the 14 possible 2-variable binary functions (leaving out the two constant functions). The functions are randomly distributed on the elements of the network and do not change during the computation.
The figure represents the states of the elements when the network has reached one of its attractors and do not change during the computation. S stands for stable, P for Periodically oscillating with a relatively short cycle length.

2 2 2 0 0 0 0 0 0 0 0 0 0 2 2 2

2 2 2 0 0 0 0 0 0 0 0 0 0 2 2 2

2 2 2 0 0 0 0 0 0 0 0 0 0 2 2 2

2 2 2 2 2 0 0 0 0 0 0 0 0 0 0 0

2 2 2 2 2 0 0 0 0 0 0 0 0 0 0 0

2 2 2 2 0 0 0 0 0 0 0 0 0 0 0 2

2 2 0 0 0 0 0 0 0 0 0 0 0 0 2 2

0 0 0 0 0 0 0 0 0 0 0 0 0 0 0 0

0 0 0 0 0 0 0 0 0 0 0 0 0 0 0 0

0 0 0 0 0 0 0 0 0 0 0 0 0 0 0 0

0 0 0 0 0 0 0 0 0 0 0 0 0 0 0 1

0 0 1 0 0 0 0 0 0 0 0 0 0 0 0 1

0 0 1 0 0 0 0 0 0 0 0 0 0 0 1 1

0 0 0 0 0 0 0 0 0 0 0 0 0 0 1 1

1 0 0 0 0 0 0 0 0 0 0 0 0 0 1 1

0 0 0 0 0 0 0 0 0 0 0 0 0 0 0 2

Fig. lb - Emergence of function.
The arrow indicates an element which serves as an input device for a recognition channel. After the network has reached its attractor (Fig. la), binary sequences are imposed on the input element and perturb the state of the network. The state of the elements indicated as 0 is the same (either S or P) as in the unperturbed attractor. The elements indicated as 2 were stable in the unperturbed network and are destabilized by the sequences imposed on the input element. The elements indicated as 1 were oscillating in the unperturbed network and are stabilized by a Cb55 of sequences imposed on the input element; they serve as output for the recognition channel. The sequences which stabilize them are said to be "recognized".

* * 0 * * * 0 * * * 0 * * * 0 * * * 0 * * * 0 * * * 0 *

Fig. 2 (see text).

JOHANN GÖTSCHL

SELF-ORGANIZATION: EPISTEMOLOGICAL AND METHODOLOGICAL ASPECT OF THE UNITY OF REALITY

(I) THE PARADIGM OF SELF-ORGANIZATION: TOWARDS A NEW MODELLING OF REALITY

The attempt to take the concept of reality as it has been developed by modern science - in particular by physics - as a means of orientation shows that the physical concept of reality is, compared to non-physical concepts, highly specific and differentiated. If one searches for a structure in the edifice composed of the sequences of physical theories, one finds three dimensions of a theoretical construct "physical reality".

These three dimensions, which will be analyzed in the following and examined as to whether they hold any implications for the subject-object relations, mark the change in the structure of subject-object relations. Taking natural sciences as they are understood today as a basis the attempt will be made to show that the following epistemological insight is valid and/or that there are well-founded arguments in its favour.

General epistemological insight: The perception of physical reality, as it is represented in the theories of self-organization, implies a categorial change in subject-object relations.

It is taken for granted today that the characteristic feature of the physical form of the perception of reality is invariance. Currently, this invariance is more distinctly developed in physics than in any other empirical science. Basically, the significance of this invariance lies in the fact that the invariant principles of construction are the highest form of objectivity. These invariants lead to a higher degree of differentiation of subject-object relations. The highest degree of differentiation is reached by the theoretical elimination of the perceiving subject or consciousness from the physical representation of reality. The objectivation accomplished by the formation of invariants separates subject from object. It is only now that the important concepts of objectivity and truth assume a deeper semantic content. The formation of invariants transforms subjectivity, which is a source of physical theories, into a cognitive level of scientific-physical objectivity. Against this background the structure of the subject-object relations does not seem to pose any problems, as the perceiving subject - consciousness - does not occur in the theoretical physical representation of reality. On the other hand, it is evident that with the advancement of physical knowledge it becomes indispensible to understand that the perceiving subject as an integral part of reality must be understood without any reduction. This, however, is only rendered possible by developing the science of the self-organization of reality (matter, nature). This new science constitutes a concept of physical reality which reveals the categorial shift of subject-object relations.

A. Carsetti (ed.), Functional Models of Cognition, 143-166.

If one searches for a classification of those representations of reality which exist in the form of physical theories and serve to reconstruct the new determinability of the subject-object relations, one finds three levels. One has to proceed from a general construct "physical reality" which mirrors both the prephysical and the physical aspects of physical reality. In order to trace the connection between "hypercritical physical reality" and the categorial change in the "subject-object relations" one may proceed from the following postulate:

Postulate 1: All our knowledge of principles of reality (matter, nature) stems from physical theories. Ultimately, everything that is singular needs to be deductively ordered by a principle in order to assume the character of reality.

Postulate (1) essentially says that there was and still is a prephysical understanding of reality which may currently be expressed quite well by means of a representation in terms of the theory of evolution. After all, evolutionary epistemology contributed substantially to the realization that the prephysical perception of nature is not accidental but basically constituted by the determinants "mutation, selection", "combination" and "recombination".

Thus, postulate (1) maintains that there are good reasons for introducing a general construct for physical reality. Against this background and taking into consideration the most important theories the various forms of representation of physical reality can be identified. This identification is also the key for identifying the renewal of the subject-object relations which appear due to the increase in scientific knowledge. The philosophical reconstruction of the physical theories brings to light at least two aspects: (1) the emergence of a categorial renewal of the subject-object relations by means of the theory of self-organization; and (2) an expanded differentiation of the plural structure of reality, also by means of the theory of self-organization. This fact can be expressed in postulate (2):

Postulate 2: The sequence of the central physical theories (NT), (RT), (QT), (TSO) (Newtonian theory, theory of relativity, quantum theory, theory of self-organization) leads to a representation of reality which results in a pluralization of the conceptions of reality.

The above mentioned physical theories are of decisive significance for achieving the aim pursued here, i.e., to find arguments in favour of the central hypothesis (H), as they permit the identification of different ontologies and their implications. A philosophical reconstruction of the theories leads to the following result: A substance ontology corresponds to (NT), a functional ontology to (RT) and (QT), and - which is of utmost importance here - a process ontology which points to the revolutionary categorial change in subject-object relations.

These relations can be summarized in the following diagram:

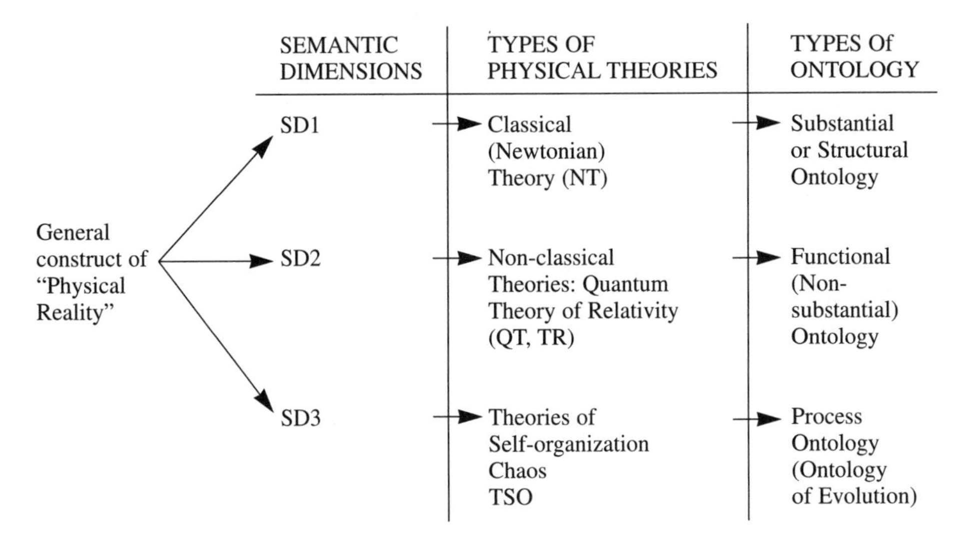

Diagram - 1 -

Two brief interpretations of this diagram suffice to demonstrate the significance of (SD3) for the identification of the connection between hypercritical physical realism and subject-object relations.

INTERPRETATION 1

(1) (SD2) constituted a *heuristic* of physical thought, which essentially maintains: Only the further development of quantum theory and the theory of relativity can lead to a better understanding of both the extra-human and the human (relations between genotype and phenotype) types of phenomena. According to this heuristics it is for example possible to develop a quantum theory of the human brain which is understood to be a network unit of neuronal and cognitive processes.

(2) Inspite of the great difference between (SD1) and (SD3) their common feature is that they both represent macro theories. This common feature, however, is of no major significance for the identification of the subject-object relations. All the more important are the differences. (SD3) differs - this is decisive for the argumentation in favour of the initial hypothesis (H) - in some aspects of fundamental importance both from (SD1) and from (SD2). After all it was the findings of Prigogine[1], Mandelbrot[2] and others and the phenomena they discovered which led to a tentative but categorial renewal of the subject-object relations. The key terms of the theories of self-organization state the following:

(i) The complexity of a system is not primarily defined by a large number of components, but the dominant factor of determination is the recursive networking of the components. The order of a system in terms of the theory of self-organization and/or chaos is determined by the mechanism of recursion and the principle of self-similarity. Recursivity and self-similarity means that, although a system is determined by what has happened before, it is not predeterminable.

(ii) There is a sensitive dependence of system formation on the initial conditions. The decisive fact is that each state of the system is the initial condition for its next state.

What implications do (i) and (ii) hold in terms of our search for arguments in favour of the central hypothesis (H)? Only in a first step do the theorems of recursive networking and historicity (not predeterminability) permit what was not possible for (SD1) and (SD2), i.e., the constitution of such a concept of physical reality from which a structural transfer to extra-physical zones seems to be possible. Now, step by step, a concept of "hypercritical physical realism" may be developed, on the basis of which (TSO) may be transferred to processes not exclusively physical.

INTERPRETATION 2

(1) The semantic dimension (SD1) is completely indifferent or neutral as far as the subject-object relations we are looking for are concerned. Physical realism corresponds to the substance ontology where an "ultimate" object is postulated and modelled as invariant.

(2) The semantic dimension (SD2) is no longer completely indifferent or neutral as regards the subject-object relations we are looking for. This is explained by the fact that quantum theory and the theory of relativity cause the dissolution of the classical concept of substance or object and thus the constitution of a physical realism corresponding to a functional ontology. Here we notice for the first time that (SD2) enables us to see the outline of self-reference in terms of the theory of nature.

(3) Only the semantic dimension (SD3) permits the identification of self-reference in terms of the theory of nature, thus leading to a categorial change in subject-object relations compared to (SD2) and - to a greater extent - in comparison with (SD1). This is due to the fact that (SD3) represents a hypercritical physical realism which basically says: The understanding of nature and self-understanding are related terms.

By means of three postulates of realism[3] we can now achieve:

(i) An explication of the type of physical realism that is structured by physical theories;

(ii) An explication of the connection between a type of realism on the one hand and the subject-object relations on the other hand.

We can now add one step to diagram 1, as is shown in the following:

Types of Theories	Types of Ontologies	Types of Realism (R1, R2, R3)
Classical Physics: Newtonion Theory	Substance Ontology	R1: The object (nature) exists independently of the subject (man), and though it is categorially different from the subject it may be perceived by the subject.
Non-classical Physics: Quantum Theory, Theory of Relativity	Functional Ontology (Non-substantial Ontology)	R2: The object (nature) may be perceived by the subject (man) only insofar as the subject can in all its manifestations be shown to be an integral part of the object.
New Physics: Theory of Self-organization	Process Ontology	R3: Subject and object may be perceived as coherent constructs. They are neither categorially different, nor do they form one complete whole. Only in the light of a reconstruction of history do subject and object present themselves as a whole composed of the history of nature and civilization.

Diagram - 2 -

Each of these three postulates of realism will be briefly commented on to provide the foundation for investigating more thoroughly the structure and dynamics of the "man-nature-frame".

Comment on (R1): Though all three postulates of realism are still inherent in 20th century thought, we notice that (R1) is increasingly pushed into the background and might be wholly replaced in the near future. The main reason why (R1), rather implicitly, has to give way lies in its contents expressing the central element of Newton's conception of nature.

The subject (man) does not exist within Newton's conception of nature - there is no room for the subject. (R1) precisely states that, although man can perceive nature (the world), the perceiving subject cannot occur in this world in a consistent form. To put it differently: within the Newtonian world (nature) the interpretation of the perceiving subject is instrumentalistic.

Comment on (R2): (R2) represents what is part of the mainstream of thought in the second half of the 20th century: the philosophical and scientific conceptions of nature gradually seem to realize the traditional concept of harmony between man and nature, i.e., the harmony between subject and object. Reflecting upon the current situation of man, it becomes evident that man aims at achieving harmony with nature. It is, however, equally obvious that currently there is neither a philosophical nor a scientific conception of nature which can stringently prove that the perceiving subject (man) can, in all ist manifestations, be understood as an integral part of the object (nature).

Comment on (R3): This postulate represents in the most striking manner both the philosophical and the scientific knowledge of the present. (R1) has not only been generally criticized in the 20th century, but concrete alternatives for (R1) have been discovered which are above all expressed in the quantum theory and currently in the theory of self-organization. Particularly in the second half of the 20th century it has become evident that a shift in categories has occurred: the "problem of nature" has turned into a problem of "subject-object relations".

On the basis of diagram 2 the subject-object relations we are searching for can now be defined more explicitly. The process ontology yielded by the theory of self-organization has the following characteristics suggesting an important conclusion:

(i) The development potential of matter is not limited. The emergence of consciousness (subject) as a new and specific form of the process of evolution is no longer as mysterious as it used to be.

(ii) The structure of matter is not exclusively symmetrical. The mechanism of recursion and the principle of self-similarity make it superfluous to look for extensions of quantum theory and the theory of relativity in order to scientifically deal with self-organizing or chaotic systems.

(iii) The new heuristics:

Very slowly a process of inversion becomes apparent. The complexity of organic matter and, beyond that, the complexitiy of the unity of neuronal and cognitive processes is x times greater than the complexity of inorganic matter. The inorganic world is pictured by models of the organic world and not vice versa, as it used to be the case.

Which conclusion do (i), (ii) and (iii) suggest? Without going into further detail it seems to be highly probable that the following holds true:

Postulate 3: Subject and object are not necessarily categorially different or separate, they are - seen from the point of view of the new hypercritical physical conception of reality - coherent constructs.

(TSO) established the preconditions for the development of a process ontology that is adequate for our time. This process ontology, which should result in a more uniform and consistent representation of subject-object relations, has the following philosophical implications:

(i) (TSO) opens up a way to considerably improve man's self-interpretation in terms of the theory of nature: (TSO) facilitates self-reference.

(ii) With (TSO) a level of convergence between the "sciences of matter (nature)" and the "sciences of man (mind)" has been reached which permits a first insight into the degree of coherence between object and subject.

The philosophical and ontological investigation of (TSO) clearly shows that the self-perpetuating development potential of material systems is - under specific conditions - greater than the behaviour caused by the respective set of forces outside the system. To put it differently: The empirical facts "cause" that the description or reconstruction of the material world does not exclusively point to an ontic world structure, but that the diversity of material events experienced by the perceiving subject corresponds to a section of the world or reality within the general process dynamics. The ontology of the TSO gives a precise insight into the process dynamics which necessarily led to the definite and complete withdrawal of the substance ontology and to a modal expansion of the functional ontology. This withdrawal reduces familiar problems insofar as the differences and connections between the prebiotic and biotic, the biological and cultural categories are no longer at issue. Due to the fact that (TSO) leads to an empirical, scientific in-depth understanding of specific evolutionary processes in general, any reasons for trying to explain organic manifestations by inorganic causes, social effects by psychical causes etc. cease to exist. In the language of the theory of science we might say that (TSO) does not solve those problems which have defied any such effort up to now, but that the issues vanish. Although (TSO) has emerged as a "science of matter (nature)" and is currently still classified as such, due to its message this conception holds important implications for the formation of a new basic understanding of the "science of man (self-consciousness)" as well as for the relationship between these two fundamental types of science. One of the major aspects of (TSO) is the fact that the description and explanation of pattern or structure formation reaches even into the human genetic code and thus, progressively, into the higher configurations of the formation and development of life. This means that a new and existentially significant level of cognition is reached which shall be expressed in the form of a postulate:

Postulate 4: The potential for description of (TSO) implies that reference of (TSO) is not only represented by the object (nature) opposite to the perceiving subject but to a certain extent by subject-object relations[5].

Postulate (4), which expresses the connection between reference and self-reference, leaves the classical version of the subject-object relations far behind. The statements pertaining to (TSO) represent the first empirical scientific piece of theory which says that the relationship between man and nature or the subject-object relations may neither be thought to be comprehended by means of a structural nor by a functional ontology but by a process ontology only. Proceeding from an epistemological and methodological reconstruction of (TSO) one can at least identify the following characteristics, presented in this paper.

(II) CONVERGENCE BETWEEN "NATURAL SCIENCE" AND "SOCIAL SCIENCE"

New knowledge derived from the theory of self-organization, (TSO), of specific real systems constitutes a new understanding of nature and culture, or society and implies a new self-understanding. Containing in its core the principles of emergence, self-production and creativity of new forms or structures, (TSO) leads to the possibility of a categorially new, highly differentiated process-evolutive concept of reality and thus breaks up the traditional scientific classifications and different kinds of scientific ontology. (TSO) has imbued the notional categories of configuration and transformation of material systems with new significance. For, according to (TSO), those principles that lead to the emergence of dissipative structures are principles which, under specific conditions, lead to the occurrence of a separate self-organizing dynamic in material systems. This separate dynamic is described by the principles of energetic and informational openness and the operational closeness as well as conditions for the instability and stability of systems. These principles of separate dynamics concerning material systems serve as the foundations for a more unified and process-evolutive conceptualization of reality which no longer solely refers or is restricted to material systems.

This process-evolutive concept of reality no longer holds to the configurations of the inorganic, organic or social which require categorial differentiation. Contrary to what has been hitherto understood by this, it is here intended to mean the interwoven functions of 1) understanding, 2) explanation, 3) predictability and 4) realization. (TSO) leads us to a model of reality in which the framework for the compatibility of determinism and specific unpredictable developments in material systems are shown in the concept of bifurcation, i.e., in the concept of the unity of linearity and non-linearity in the evolution of systems. A compatibility of determinism and uncertainty, the linking of linearity and non-linearity have thus become the new prerequisites for the construction of a universal, process-evolutive understanding of reality.

However, a generalization or synthesis of the classical theories of the natural and social sciences which could bridge the apparent categorial gap between material systems on the one hand and the social systems on the other hand is still not possible. This is so because the subject matter which is fundamentally posited as ontological correlate in the "natural sciences" and as unalterable prerequisite in the "human sciences" is not represented through self-organizing characteristics. New breakthroughs, however, have been made with (TSO). New prerequisites for a more unified description of the multiplicity of reality are coming into being: (TSO) represents a new unity in reality's multiplicity.

Up to now - and in a manifest way since Newton - scientific development has changed. New theories have occurred, but methodology and ontology have remained more or less the same. (TSO) has drastically changed the situation; theories, methodology and ontology are appearing in a new interconnected way. With the help of the philosophical reconstruction of (TSO) the ontological correlate of classical natural science is gradually being recognized more accurately. This correlate of the classical physical theory (CPT) is a substance-based ontology which determines the subject matter. The four familiar principles of the classical physical theory (CPT), summed up here in philosophical reconstruction, are as follows: (1) the principle of the invariance of mass, (2) the principle of linear causality, (3) the principle of limited potentiality of matter, (4) the principle of reversibility (principle of symmetry), and (5) the principle of the categorial separateness of matter and mind, object and subject (principle of non-coherence of object and subject).

These five well-known principles mirror what can be named substantial ontology. On the physical level of terms this paradigm corresponds to Newton's concept of reality, the two central characteristics being those of (i) mass, which is understood to be constant (invariant), and of (ii) movement which describes dynamic segments of reality but cannot explain the origination, stability or instability of new forms and structures. It falls short of supplying exactly those scientific characteristics for a process-evolutive concept which are in fact provided by (TSO).

Although the Theory of Relativity and Quantum Theory differ in many respects, as far as the conceptualization of reality is concerned they both contribute something new: a high degree of de-substantialization with the result that all five characteristics of classic substantial ontology are specifically suspended to a degree. This de-substantialization takes place essentially within the conceptual framework of the Theory of Relativity by means of the matter-energy equivalence relations, together with the field theory representation of reality, and in the conceptual framework of Quantum Theory by means of transforming the notional object together with the interdependence of subject-object relations. For this reason, it is helpful here to speak of functional ontology. This functional ontology is, in fact, a far-reaching process concept of reality. But here, too, the relevant characteristics of process-evolutive processes of self-organization are missing.

Against this background the central hypothesis in this piece of work can be formulated using (TSO) in three versions which become increasingly more succinct. The second version is intensified to the effect that a tendency towards suspension of an apparent categorial difference between material systems and social systems is allowed for. The third version of the central hypothesis refers to a tendency towards transcending the mind-matter duality hypothesis.

Central hypothesis (1st version):

(TSO) results in a radical improvement of a **more unified conceptual representation** of reality. Material systems and social systems can increasingly be represented in a more unified way. This **more unified conceptual representation** of two apparently categorially different realities investigates analogies more deeply to the effect that these analogies are eliminated through a process-evolutive, universalizing modeling of reality.

(TSO) allows for the identification and an increasingly differentiated explanation of presuppositions necessary for a new classification of science with radical consequences for the construction of a "General theory of reality". What are these prerequisites? The central point is that it is not claimed that (TSO) has found new characteristics of material systems which can simply be added to the characteristics which material systems have so far been stated to have. The generating principles state that reality must be modeled along a dominant configuration and transformation of elements and structures with its own dynamics. There are characteristics that lead to the generation of coherent structures or shapes, respectively, in material systems, and which are, within certain limits, an immanent part of a system. The notion of material systems (or subject type) constituted by the principles of (TSO) represents material systems in a way where areas of reality which have up to now appeared to differ categorially, e.g., material processes and social processes, can be increasingly bridged theoretically in a process-evolutive way without reductionism.

The development of (TSO) gives us the opportunity of more strongly differentiating scientifically constituted conceptions of reality or their corresponding ontology. The different kinds of ontology constituted by science can be divided into the following minimal classification:

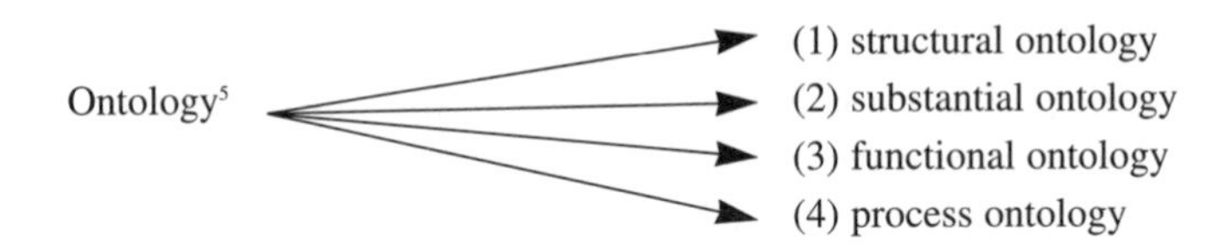

For a brief interpretation of these four predominant scientifically constituted types of ontology we can say that:

1) is essentially constituted by mathematics, logic and pure system theory
2) is essentially constituted by the Classical Physical Theory (CPT)
3) is essentially constituted by the Theory of Relativity (RT) and Quantum Theory (QT)
4) is essentially constituted by self-organization (TSO)

What results from the above in terms of an identification of relations between "natural science" and "social science" in general, and in terms of a proof of a convergence between the two in particular? From an epistemological standpoint, three insights are of importance. These are: (i) (TSO) functions as a bridging science, (ii) (TSO) functions as a science of integration, and (iii) (TSO) functions as a universalizing science. What are the most important characteristics?

The various efforts undertaken in the course of the 20th century have shown that all reductionist attempts to create a more unified model of reality, of material and social processes fail. This failure is mainly owing to the fact that insufficient differentiation has been made between reduction in the sciences and philosophical reductionism. Reduction is indeed indispensable; but reductionism is not. The attempt to trace all phenomena back to logical/mathematical and purely system-theoretical structures failed. Even the less ambitious attempt to trace everything back to physicalist structures has failed. Essentially, there are three reasons for this unsuccessful outcome. (TSO), however, opens new opportunities which do not altogether solve the problem of reduction but bypass it, so-to-speak. The failure of this program of reduction can be traced back to three reasons:

1) It is not possible to provide a satisfactory empirical interpretation of logic, mathematics and pure system theory.

2) There is too high a complexity in attempting a unified notional representation or even a homogenization of material and social structures.

3) The dynamics of material and social structures under certain conditions are too high.

What is new is that (TSO) has brought about fundamental changes for all three structural features by supplying the basis for this change by bypassing 1) using the method of mathematical computer simulation (taking into account fractals). 2) and 3) can be bypassed by the central part of (TSO), namely, by the concept of dissipative structures. (TSO), which is increasingly becoming an immanent component of mathematical computer simulation, enables more unified terms to be apprehended from the complexity and dynamics of material and social structures. The linking of mathematics and computers into a new type of processing in the form of computer simulation represents a new methodological element. In computer simulation an interwovenness of material, notional and visual dimensions of thinking about reality in reality occurs. (TSO) is, at the same time, the theory of complex and dynamic systems and gradually identifies those characteristics for process-evolutive material systems which show a strong affinity with the characteristics displayed by social systems. This has produced more justified reasons, for the first time in the history of science, for examining the social structures by means of conceptual frameworks of material structures and, vice versa, examining the material structures by means of social structures. This possibility - although it still needs working out in detail - is leading to a new understanding of the contexts between "material structures" on the one hand and "social structures" on the other hand. Put more strongly, the following supposition can be better justified:

Central hypothesis (2nd version):

(TSO) supplies two things, particularly in the context of computer simulation support: (i) a tendency towards ontological neutralization and simultaneously (ii) an empirical adequacy which enables a more unified representation of material and social structures.

These two characteristics - far-reaching ontological neutrality and, at the same time, empirical adequacy - make it possible to arrive, from mere analogies between "material structures" on the one hand and "social structures" on the other hand, at evolutionary process models which enable, in turn, a more unified notional representation of material and social phenomena. This scientific advance to evolutionary process models is supported by the following familiar characteristics of systems which have been identified by (TSO). It is mainly these characteristics which qualify (TSO) to manage without recourse to physicalism and biologism.

(TSO) demonstrates that:

1) All real systems are open systems. Far from equilibrium, dissipative, chaotic and self-organizing processes can occur. As sequences of dynamic patterns they are of a process-evolutive character.

2) Systems can, under certain circumstances, create self-determinations where they influence themselves. This can lead to levels of configurations which cannot be reduced to other levels. There are limitations for system reductions.

3) In dissipative, self-organizing and irreducible processes there is no necessity to postulate the general existence of hierarchic principles of formation and development.

4) Systems can have the structure of self-referentiality and recursivity, i.e., the dominance of the internal coherence of a system reduces and relativizes the meaning of factors external to the system for the process-evolutive unfolding of a system.

5) System-immanent characteristics of co-operation and co-evolution lead to a further differentiation of identity (self-identity) or autonomy of the system.

On the basis of 1) to 5) a decisive characteristic can be formulated in order to establish a unified process-evolutive modeling of material and social systems in a better way:

6) (TSO) provides a new stipulation of the relationship between macro and micro levels by using circular determinations. This means that the elements of a system cause the regularities (pattern, orders), and the regularities cause the order of the elements. This makes it rational to assume that system elements have a certain stability.

Point 6) shows on the basis of 1) to 5) that (TSO) is gradually enabling a notional representation of physical and social systems in a more unified way. This tends to lead to the elimination of the classic problem of reduction especially in view of the fact that there is no reduction of classical physics, (RT) or (QT) - or the Theory of Evolution - (TSO). Neither is there a reduction in the other direction, but rather other characteristics of systems are identified by (TSO), which gives the terms "object" and "reality" a new meaning. That is to say that the ontological correlate of (TSO) is a process-evolutive concept of reality or of objects. Two things are becoming possible with (TSO): (i) a more unified foundation of the natural sciences (physics, biology), something which is already well developed and, going beyond that, a more unified foundation of our understanding of material and social systems.

Put in a simplified diagram:

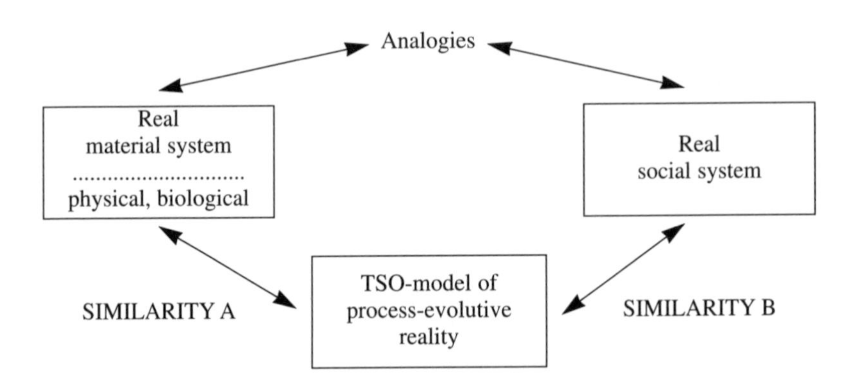

The philosophical reconstruction of (TSO) validates:

$$\text{Similarity A} \approx \text{Similarity B}$$

- which means a reduction is based on mere analogies or, put differently: the construction of "process-evolutive models of reality" tends to substitute for the use of analogies. Here, the outline of a change in paradigm takes place.

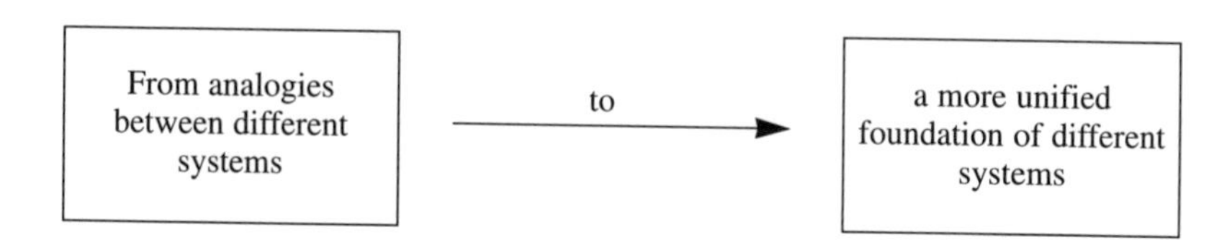

The approach which establishes: similarity A ≈ similarity B, has been well documented over the last few years. (TSO) has shown the general context of cosmic, biotic-ecological and cultural (social, economic) self-organization. The central point here is, as is well-known, the inner context between irreversibility (arrow of_time) and self-organization (negative entropy as creator of order).

The philosophical reconstruction of (TSO)´s area of validity shows, however, that one important limitation caused by an as yet unsolved problem needs mentioning. This problem results from the fact that, according to (QT), a physical system must be understood as the totality of material processes with specific interconnections of micro and macro levels. Through (QT), (TSO) appears as a specific macro theory. The validity of (TSO) for objects and processes belonging to the quantum world has not yet been proved, although currently there is no good reason that raises doubts about (TSO)'s representation of the quantum world. Working through this aspect reveals that the similarities between physical and social systems and the thus implied reduction - and possibly even elimination - of mere analogies of quantum-theoretical representations of physical systems and the representations of social systems by (TSO) might contain a limit. As far as the (QT) representation of material systems is concerned, the thesis of a tendency towards a unified representation of physical and social reality can currently not be claimed. This means that similarity A is, for the time being, limited to the molecular dimension. Since being both a macro and a micro theory, (TSO) can not currently represent the (QT) dimension; it follows that certain implied assumptions as to a unified representation of nature, man and society must be made for heuristic reasons. The most important implicit assumption, supported by a high plausibility, is as follows: the theoretically appreciable chances of developing (QT) model-theoretically in such a manner that social processes could thus be represented by it, appear to be low at the moment. Much higher are the chances of developing (TSO) model-theoretically so as to represent the (QT) dimensions of material reality.

(III) COGNITION, COMPLEXITY AND SELF-ORGANIZATION

The simultaneous increase in ontological neutrality and empirical adequacy lead to a more unified conceptual basis for the construction of a "General theory of reality". This reveals that (TSO) has brought with it a new structuring for the existing canon of science. This new order consists, in its main features, in accounting for the above-mentioned three-fold functionality:

1) (TSO) as a bridging science.
2) (TSO) as an integrating science.
3) (TSO) as a universalizing science through forming complexity, forming convergence between the individual sciences, new stipulation of the relations between inner and outer world by means of establishing links between the individual sciences.

The outlines, especially of the third function for the scientific canon, can be seen from the fact that the transition supplied by the complexity of (TSO) from substantial and functional ontology to a process-evolutive ontology is leading to a new definition of the relationship between evolution and cognition. This new definition admits of a sharpening of the central hypothesis in the form of a third version. In the following, it will be shown how this sharpened hypothesis is a viable alternative to the dualistic conception of reality.

Central hypothesis (3rd version):
(TSO) as a theory of the complexity and dynamics of reality identifies and represents the existence of new affinities between evolution, self-organization and cognition. These affinities lead to progress in overcoming the dualistic matter-mind hypothesis.

Epistemologically speaking, this hypothesis can be expanded by the fact that it becomes recognized with the help of (TSO), that, from a certain notional level of development, the scientific knowledge of the outer world (knowledge of objects) represents the potential knowledge of the inner world (self-knowledge). A process-ontological understanding of the world, of nature and matter mediates in a non-reductionist manner between the various levels of structure (material and social ones) by means of a continuously improving knowledge of the conditions for the occurrence of various levels of structure. Now it is becoming distinguishable that the classical canon of scientific classifications according to which, for ontological reasons, a differentiation had to be made between natural and social sciences can no longer be maintained. Although the ground for knowledge of the unity of natural and cultural history had been prepared even before the time of (TSO), it has only been through (TSO) that considerable progress in this direction has been made. It is owing to (TSO) that the coupling processes between material dynamics and social dynamics can be identified and explained. The classical scientific differentiation according to "given" objects - according to a given ontology - has apparently been left behind. The "sciences of nature" and "sciences of man" constitute a synthesis, even though there is not, at this point, a name for it. Let me call it here the "General theory of reality". This developmental tendency mainly leads to the fact that a recourse to the philosophically reconstructed (TSO) reasons that the classic differentiation between the natural sciences and social sciences is tending to expire.

The central hypothesis in the third version, in particular, is supported by three characteristics of (TSO) which indicate the way to overcome the dualistic model of reality.

Characteristic A:
(TSO) functions as a theory of complexity.
Characteristic B:
(TSO) functions as a theory of convergence forming between the sciences.
Characteristic C:
(TSO) functions as a theory of interweaving of external reference (natural sciences) and self-reference (psycho-social sciences).

These three functions - complexity, convergence-forming and interweaving – have a tendency to replace the dualistic concept of reality with a process-evolutive monistic model of reality. It is becoming evident that the ontology of substance is completely failing, and that the ontology of function is partly failing when it comes to the central category. What they both lack are modal categories for a theoretical modeling of realities in terms of a process-evolutive system on various levels of structure. This deficit will only be corrected by (TSO) because the latter shows the possibilities and limits within which systems organize themselves and emerge. With this approach, (TSO) leads, for the first time, to a substantial scientific overcoming of the matter-mind concept of duality. A brief analysis of the three characteristics should help to explore the limits and possibilities of overcoming the matter-mind concept of duality.

Re A) (TSO) as a theory of complexity:

Scientific developments of a more recent date have identified one tendency: increasingly more scientific results function as bridging sciences leading to a weakening of the matter-mind dualism. There are mainly two tendencies which were first developed independently of (TSO) but which have been expanded theoretically through (TSO). Let me make two brief preliminary remarks about this here.

Preliminary remark (1): The fast elaboration of the contexts between genotype and phenotype led to the recognition that human consciousness can only be modeled, in a more empirically adequate manner, as a specific emergent configuration or structure of neural processes. The whole spectrum of biological sciences of the apprehending subject-from macro-biology to the most highly differentiated molecular biology and neuro-biology, from neuro-physics, neuro-chemistry and neuro-linguistics to neuro-psychology – reveals itself as an integrative, universal integration of the subject into the context of nature. This integration transfers both the natural term and the term of the social or cognitive system to the categorial representation, interpretation and modeling frameworks of a naturalistic, complex system term.

Preliminary remark (2): It is also from a modeling of nature in the categorial framework of terms of complex naturalistic systems that new possibilities have arisen which have now become well-known. This is, generally speaking, computer simulation on the basis of material and social (symbolic) systems, namely, mathematics, informatics and technology. These have made it evident that a technology (computer) constructed on the basis of natural laws can represent system processes which grant insight into its universalizable degrees of freedom. This, too, is a new form of ordering of the apprehending subject (the "self") into a complex naturalistic system term.

These two preliminary remarks explain the assumption that it is the complexity of (TSO) which leads to progress in the attempt to eliminate the dualistic hypothesis of reality. In a summarized way, the level of complexity of (TSO) can be seen from the following characteristics which are of central importance for a philosophical analysis in order to overcome the hypothesis of duality.

A philosophical reconstruction and philosophical expansion of (TSO) has led to a new, more general term of object or reality because the ontological correlate of (TSO) consists in a process-evolutive conception of object and reality. In comparing the term "reality" with that used in classical physics (CPT), Quantum Theory (QT) and the Theory of Relativity (RT), the following result has been arrived at:

Reality concept (CPT) ≠ reality concept (QT, RT) ≠ reality concept (TSO)[6].

(TSO) constitutes a reality term which leads to a series of epistemological or ontological implications. Tied to_this first is that (TSO) means a unity "behind" the multiplicity of material and social systems. This unity in multiplicity increases the representational and explanatory values of concepts according to which the differentiated and interwoven structural levels of biological and social complexities have evolved from material structural levels. This essentially means that (TSO) categorially expands the cognition achieved up to now without, however, being incompatible with (RT) and (QT). The essence of (TSO) is that complex systems can increase in degrees of freedom by themselves. This cognition is a categorial new step in scientific development since, as has been shown earlier, (TSO) - in contrast to, and expansion of, (RT) and (QT) - allows for a decisive twofold reference owing to its ontological neutrality and empirical adequacy:

(i) a reference to material structural levels
(ii) a reference to cognitive structural levels

Expressed diagrammatically:

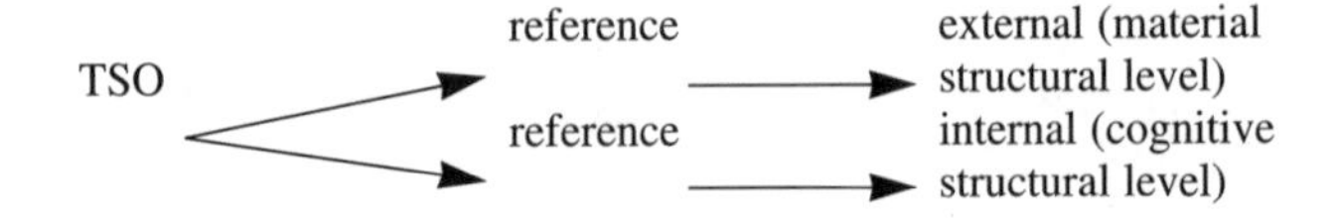

This twofold reference structure of (TSO) can be given the following philosophical interpretation which shows several further aspects. With the help of the complex potential of description and explanation of (TSO), we can show that the reference of (TSO) is represented not only by the object which is opposite the apprehending subject but, to a certain degree, by the relations between subject and object. This indicates that, with the rise of (TSO), a structural interwovenness of object, knowledge of object and self-knowledge can be identified. Against the background of a twofold reference potential of (TSO), the following more complex approach can be taken as a starting point for a conceptual unfolding of the central hypothesis (third version!):

Complex approach: a scientific theory (Ti) as conceptual framework shall be categorized as more complex than a scientific theory (Tj) provided (Ti) can represent more differentiated elements as regards co-operative and evolutive structures as ordered totality than (Tj).

This complex approach as a necessary but not sufficient criterion can, by means of a diagram, point out the new principles for the construction of a "General theory of reality".

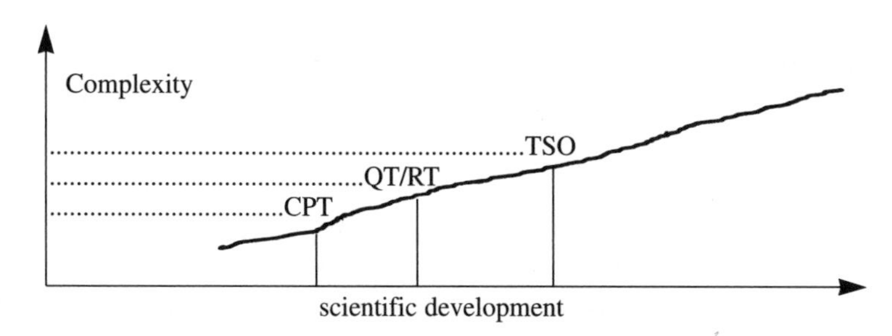

This complex[7] approach reveals that, in scientific development, it took (TSO) to advance to a more unified theoretical-notional representation of material and social (cognitive) structural levels. Thus, complexity (comp) in the sense of the above complex[8] approach increases as follows:

$$\text{Comp (CPT)} < \text{Comp (QT, RT)} < \text{Comp (TSO)}.$$

This result, which is founded mainly on the principles of a very strict ontological neutrality and empirical adequacy on the one hand, and - closely connected - of the twofold reference potential on the other hand, allows for implications which signal the following familiar transitions. These transitions form the categories for promoting a construction of a "General theory of reality" as a symbiosis of material and socio-cultural structural levels. These transitions are mainly:

- Naturalistic understanding of systems instead of idealism
- Monism instead of idealism
- Holism instead of particularism
- Emergentism instead of (classical) determinism
- Self-organization instead of organization from outside

All this reveals a better and better justification of the central hypothesis. (TSO) as a naturalistic theory of complexity offers the possibility of an augmenting naturalistic-systematic self-interpretation of man. A unification of the dimensions of nature, man and society is developing into a scientific option.

Re B) (TSO) as a naturalistic system-science:

The twofold reference structure of (TSO) identified by means of this philosophical reconstruction can be made more explicit by two further characteristics. So far, analyses have come up with possibilities which establish the following: the reference potential of (TSO) is such that there is the possibility of a more unified notional representation of material and social (cognitive) structural levels.

Expressed in a small diagram, we can say:

In order to make this aspect clear, two levels of analysis can be used. One level concerns the specific contexts between socio-cognitive and physico-material structural levels. The other level of analysis refers to a brief comparative treatment of the dominant indicators.

Re Analysis level 1:

In order to be able to understand the external (material) world a division must be made, according to one classical methodological and ontological model of cognition, of subject and object. To understand the external (material) world means to produce symbolic and experimental interactions and aggregations with the material world. This leads to the occurrence of specific social and cognitive phenomena, such as identities, forms and totalities. An ontological assumption which receives its more precise meaning from (TSO) is at hand. It is that there is a definitive correlation between the state of the socio-cognitive world on the one hand, and the physico-material world on the other hand. "Understanding", for instance, represents the state of correspondence between the socio-cognitive and the physico-material system. Thus, understanding, on the basis of (TSO), apparently represents the existence of a dynamic equilibrium of the cognitive system and, simultaneously, the existence of a dynamic dis-equilibrium of the neural material system. Even though these contexts will require further theoretical exploration, they already vaguely suggest that and how socio-cognitive structural levels are linked with material (neural) structural levels and through which parameters the socio-cognitive and material structural levels can be identified as interdependent totalities.

The twofold reference potential of (TSO), namely the reference to material and social (socio-cognitive) structural levels, changes the meaning of self-understanding, of "self". For "self" becomes evident as a specific totality of material and socio-cognitive structural levels by categorization through (TSO).[9]

Put into a small diagram:

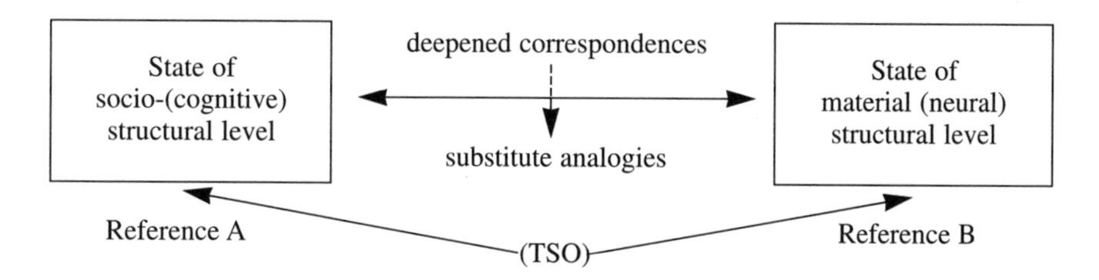

Reference type A (internal reference of (TSO)) and reference type B (external reference of (TSO)) make it possible to say that the fundamentals are about to get one step closer towards a "General theory of reality". (TSO) as naturalistic system-science is tending towards making possible something that, up to now, has been unthinkable: a sociological or human-theoretical access to material structural levels, to the physical world and to nature.

Re Analysis level 2:

A deepening of the compatibility with tendencies towards ontological neutrality and empirical adequacy, such as the twofold reference structure of (TSO) which is strongly connected with it, can be achieved with the help of a comparative analysis of scientific criteria.

The analysis so far has revealed that (TSO) has supplied a new theoretical model of reality. What is central to this model is the process-evolutive modeling type through which a more unified foundation of material and socio-cognitive structural levels is possible. With the philosophically reconstructed (TSO) the preconditions for the construction of a "General theory of reality" are becoming increasingly identifiable. This means that the theorems of philosophically reconstructed (TSO) which allow for a system of cognition to be founded,

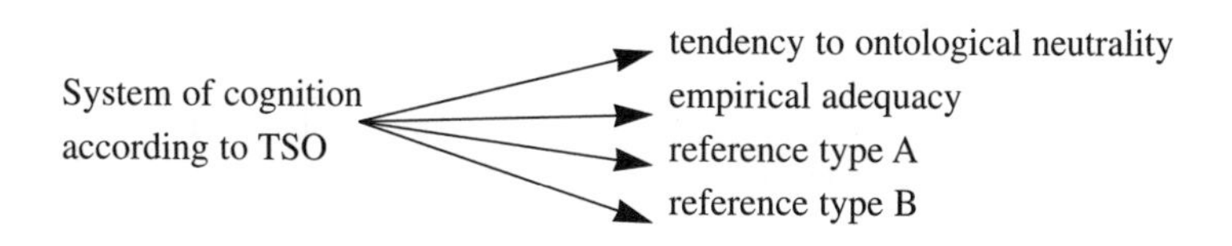

mature to be a heuristic, fruitful core of scientific development for both the "natural" and the "social" sciences.

As a diagram:

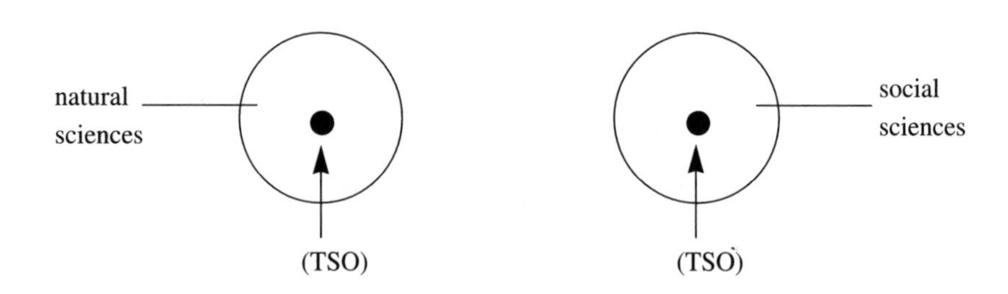

The possibility, appearing in outline, that the theories of philosophically reconstructed and philosophically expanded concepts of (TSO) are developing into an immanent component of the natural scientific approach in the dimension of a socio-cognitive access to natural phenomena and, vice versa, a socio-scientific access to natural phenomena, reduces the familiar problems of physicalism and biologism, analogism and projectism.

Let us make a choice of central characteristics (criteria) which distinguish the types of science of the 20th century, i.e.

Type A: (NSC) ❍ Natural sciences
Type B: (SOSC) ❍ Social sciences
Type C: (TSO) ❍ Sciences of self-organization.

Scientific criteria	NSC	SOCS	TSO
1) Clarity	yes	yes	yes
2) Consistency	yes	yes	yes
3) Reproducibility	yes	no	no
4) Predictability	yes	no	no
5) Objectivity	yes	no	yes
6) Reducibility	yes	no	no
7) Truth	yes	no	yes
8) Historicity	yes	yes	yes
9) Quantifiability	yes	no	yes
10) Sensitive dependence on initial conditions	no	yes	yes
11) Recursivity	no	yes	yes
12) Self-referentiality	no	yes	yes
13) Creativity	no	yes	yes
14) Knowledge of not-knowing	no	yes	yes
etc.			

On comparing (NSC) and (SOSC) we find at least 11 incompatibilities among 14 comparative indicators - which underlines the principal direction of development. On comparing (SOSC) and (TSO), only three incompatibilities are detected.

What results from this for the complex questions according to the new presuppositions relevant to the construction of a "General theory of reality"? The answer to this question by means of a 2nd level analysis has been enriched by one aspect, at least: the philosophical reconstruction of (TSO) shows that (TSO) as a naturalistic and systematic science on the one hand, and as a social science on the other hand, reveals interesting affinities (convergences).

Re C) (TSO) as a basis for identifying a linked rule:

The deeper exploration into the structural formation and structural dynamics of material and socio-cognitive structural levels leads to the insight that the level of scientific development which is represented by the categories of (TSO) culminates in one central item of knowledge: knowledge of the external, material world has a specific meaning as knowledge of the internal world. In other words: on the level of categories of the philosophically reconstructed (TSO) it becomes apparent that each perception of the external world represents a potential perception of the internal world, i.e., (TSO) as the nucleus of our perceptual production of the external and internal world facilitates the philosophically reconstructed (TSO) and scientific self-reference. Thus, it also becomes clearer that material and social structural levels, "matter" and "mind", are coherent constructs, i.e., foundations for the construction of a "General theory of reality" are becoming evident.

It can be seen from two characteristics that this is a further decisive aspect:

(i) More and more "new" sciences allocate themselves in between the classic types of "natural sciences" and "social sciences". The range between these two scientific types is becoming more and more dense.

(ii) More and more sciences occur in linkage.

Aspects (i) and (ii) are becoming more easily explicable by means of the system of cognition of (TSO). This increasing density and linkage between material and socio-cogntive structural levels can be seen, among others, from the rapid developments in the following "more recent" sciences. No claim is made as to the completeness of the list. Even a selected extract shows that the two extremes "matter" and "mind" which, according to the hypothesis of duality, make necessary a fundamental discontinuity are subject to paradigmatic change. Risking an overstatement, the assumption could be made that the scientific development of the classic discontinuity of "matter" and "mind" will transform itself into a continuity.

In a diagram it looks like this:

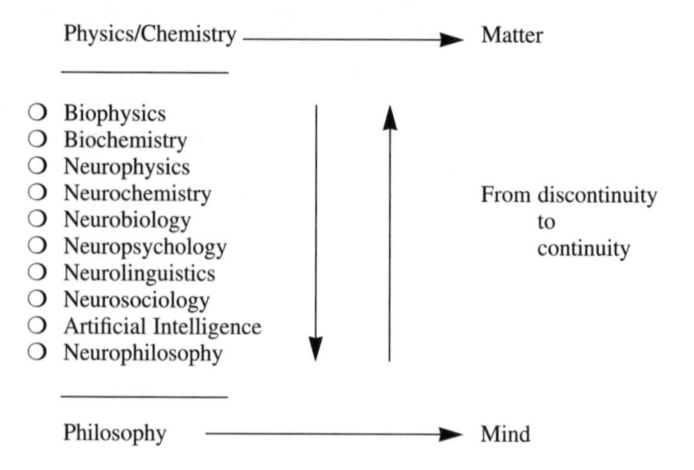

When the decisive aspect of development is focussed on, it can be seen that this state of a tendency to a continuum between mind and matter is about the linking of sciences. In a diagram, it looks like this:

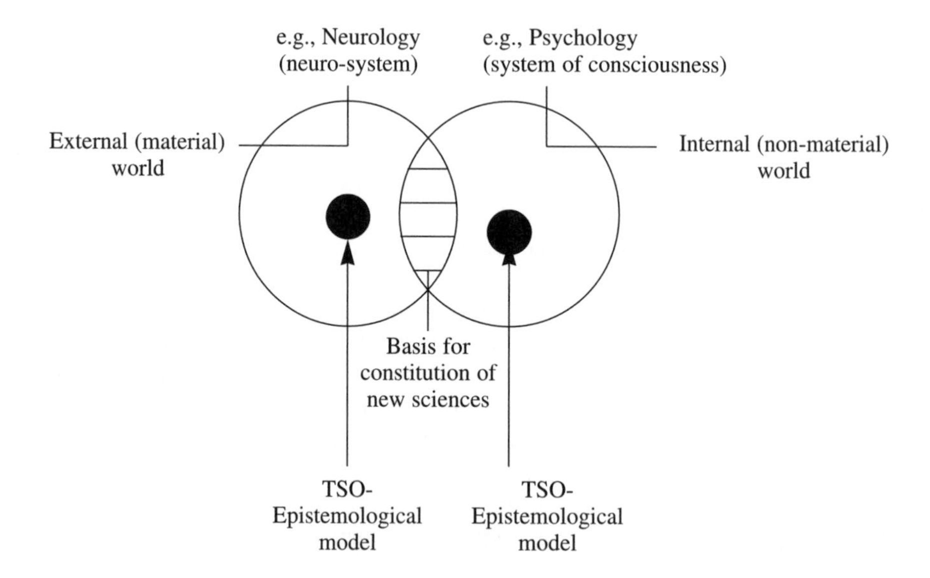

The diagram illustrates one important aspect: the process-evolutive modeling of reality by means of the (TSO) system of cognition facilitates the constitution and identification of links between individual sciences. This type of linking rule shows once more that the scientific conditions improve for establishing a "General theory of reality" in which the apparently categorial differentiation between matter and mind will, in any case, not be a scientific necessity any more.[10]

Conclusion 1:
This philosophical exploration of (TSO) has outlined the possibility of a "General theory of reality"[11]. The ultimate aim of such a theory will, however, not be achieved for some time, since it needs to be shown that the human thinking process is consistent as a real process and can theoretically be imbedded in reality.[12]
Conclusion 2:
This philosophical exploration of (TSO) - from which the (TSO) system of cognition can be developed - reasons that nature and society, mind and matter are coherent constructs.
Conclusion 3:
(TSO) argues to a considerable extent that a higher ontological affinity, and probably even the same ontological status, will have to be allocated to nature and man.

Prof. Johann Götschl
L. Boltzmann Institut
für Wissenschaftsforschung
Mozartgasse 14 - A- 8010 Graz, Austria

NOTES

[1] Prigogine, I.: Order out of Chaos. London 1984.
[2] Mandelbrot, B.: The Fractal Geometry of Nature. New York 1977.
[3] Götschl, J.: "Philosophical and Scientific Conceptions of Nature and the Place of Responsibility". In: "La Nuova Critica", Nuova Serie 7-8. Roma 1987, pp. 5-22.
[4] Götschl, J.: "Zur philosophischen Bedeutung des Paradigmas der Selbstorganisation für den Zusammenhang von Naturverständnis und Selbstverständnis". In: Krohn, W., Küppers, G. (Eds.): Selbstorganisation - Aspekte einer wissenschaftlichen Revolution. Vieweg Braunschweig 1990, pp. 181-199.
[5] To be understood as scientifically constituted models of reality. These models as different kinds of ontology are understood here to mean the result of philosophical reconstruction of scientific theories. For more details compare Götschl, J., "Philosophical and scientific conceptions of nature and the place of responsibility", in: "International Journal of Science Education", 1990, pp. 288-296 und Götschl J., "Hypercritical physical realism and the categorial changes in the subject-object-relations", in: "La Nuova Critica", Roma 1991, pp. 5-19.
[6] For details compare 1. Götschl, J., "Philosophical and scientific conceptions of nature and the place of responsibility", in: "La Nuova Critica", 1988. 2. Götschl, J. , "Hypercritical physical realism and the categorial changes in the subject-object relations", in: "La Nuova Critica", 1991. 3. Götschl, J. , "The role of physics in Erwin Schrödinger's philosophical system", in: Erwin Schrödinger. Philosophy and the Birth of Quantum Mechanics, M. Bitbol and O. Darrigol (eds.), Editions Frontières 1992. 4. Götschl, J. and Leinfellner, W., "Erwin Schrödinger's world view. The role of physics and biology in his philosophical system", in: What is Controlling Life? 50 years after Erwin Schrödinger's What is Life?, E. Gnaiger, F.N. Gellerich and M. Wyss (eds.), Modern Trends in BioThermoKinetics, Vol. 3, Innsbruck: University Press, 1994.
[7] Götschl J., "Hypercritical physical realism and the categorial changes in the subject-object-relations", in: "La Nuova Critica", 1991, pp. 5-19.
[8] Simon, H.A., "The architecture of complexity", in: "Proceedings of the American Philosophical Society" 106, 1962, pp. 467-482.

[9] Götschl, J., "Physics in Erwin Schrödinger's Philosophical System", in: Philosophy and the Birth of Quantum Mechanics, M. Bitbol/O. Darrigol (eds.), Editions Frontières, Paris, 1993, pp. 121-142.
[10] Leinfellner W., "The change of the concept of reduction in biology and in the social sciences", in: Centripetal Forces in the Sciences, Radnitzky, G. (ed.), New York, 1989, pp. 55-77.
[11] Leinfellner, W., "The new theory of evolution - a theory of democratic societies", in: Götschl, J. (Ed.): Revolutionary Changes in Understanding Man and Society. Scopes and Limits, Kluwer Academic Publishers, Dordrecht, 1995, p. 149-189.
[12] Leinfellner, W., "The new theory of evolution - a theory of democratic societies", in: Götschl, J. (Ed.): Revolutionary Changes in Understanding Man and Society. Scopes and Limits, Kluwer Academic Publishers, Dordrecht, 1995, p. 149-189

REFERENCES:

Bohm, D. [1984], Wholeness and the Implicate Order, London: Ark Paperbacks.
Carrier, M., Mittelstraß, J. [1991], Das Leib-Seele-Problem und die Philosophie der Psychologie X, Berlin: de Gruyter.
Erpenbeck, J. [1993], Wollen und Werden, Universitätsverlag Konstanz.
Geyer, F. [1991], The Cybernetics of Complex Systems, Salinas, CA: Intersystems Publications.
Götschl, J. [1991], "Hypercritical physical realism and the categorial changes in the subject-object-relations", in: "La Nuova Critica", pp. 5-19.
Götschl, J., Leinfellner, W. [1994], "Erwin Schrödinger's world view. the role of physics and biology in his philosophical system", in: What is Controlling Life? 50 years after Erwin Schrödinger's What is Life? E. Gnaiger, F.N. Gellerich, M. Wyss (eds.), Modern Trends in BioThermoKinetics, Volume 3, Innsbruck University Press, pp. 23-31.
Götschl, J. [1989], "Philosophical and scientific conceptions of nature and the place of responsibility", in: "La Nuova Critica", pp. 5-22. For an abridged version of this paper with stronger emphasis on the coupling of material and non-material structural levels, see: Götschl, J. [1990], "Philosophical and scientific conceptions of nature and the place of responsibility", in: "International Journal of Science Education", pp. 288-296.
Götschl, J. [1992], Erwin Schrödinger's world view. The dynamics of knowledge and reality, Dordrecht: Kluwer Academic Publishers.
Götschl, J. [1990], "Zur philosophischen Bedeutung des Paradigmas der Selbstorganisation für den Zusammenhang von Naturverständnis und Selbstverständnis", in: Selbstorganisation - Aspekte einer wissenschaftlichen Revolution, Krohn, W., Küppers, G. (eds.), Braunschweig: Vieweg, pp. 181-199.
Götschl, J. [1986], "Zum Subjekt-Objekt-Problem von transzendentaler und evolutionärer Erkenntnistheorie", in: Transzendentale oder evolutionäre Erkenntnistheorie, Lütterfelds (ed.), Darmstadt: Wissenschaftliche Buchgesellschaft, pp. 285-306.
Hörz, H. [1994], Selbstorganisation sozialer Systeme, Lit Verlag, Münster/Hamburg.
Jantsch, E. [1979], Die Selbstorganisation des Universums, München: Deutscher Taschenbuch Verlag.
Luhmann, N. [1985], Soziale Systeme. Grundriß einer allgemeinen Theorie, Frankfurt: Suhrkamp.
Nicolis, G., Prigogine, I. [1977], Self-Organization in nonequilibrium systems: from dissipative structures to order through fluctuations, New York: Wiley.
Nicolis, G., Prigogine, I. [1989], Exploring complexity systems, New York: Freeman.
Penrose, R. [1989], The emperor's new mind, Oxford: University Press.
Prigogine, I., Stengers, I. [1984], Order out of chaos, New York: Bantam.
Riedl, R., Huber, L., Ackermann [1991], "Rational versus ratiomorphic strategies in human cognition", in: "Evolution and cognition" 1, pp. 71-88.
Schrödinger, E. [1958], Mind and matter, Cambridge: University Press.
Searle, J.R. [1992], The rediscovery of mind, Massachusetts Institute of Technology, Massachusetts.
Whitehead, A.N. [1967], Science and the modern world, New York: The Free Press.
Whitehead, A.N. [1978], Process and reality: an essay in cosmology, New York: The Free Press.
Wuketits, M.F. [1984], Concepts and approaches in evolutionary epistemology, Theory and Decision Library, Boston: Kluwer Academic Publishers.
Zeleny, M. [1985], "Spontaneous social orders", in: The science and praxis of complexity, The United Nations University, pp. 312-327.

FORTUNATO T. ARECCHI

HOW SCIENCE APPROACHES THE WORLD: RISKY TRUTHS VERSUS MISLEADING CERTITUDES

SUMMARY

The scientific revolution can be seen as an innovation within the linguistic system currently adopted in ordinary languages. Extraction of quantitative features from observations via suitable measuring devices means that the words of science are numbers, and the connecting syntax is a set of mathematical rules. Once a general law is available (as, e.g., Newton gravitation) all consequences can be worked out in a purely deductive way. This characteristics of modern physics displays two orders of drawbacks, namely, Gödel undecidability of deductive procedures, and intractability of computer modelings of complex situations.

The way out of such a crisis consists in an adaptive strategy, that is, in a frequent readjustment of the rules suggested by the observed events. This way, the language has no longer fixed rules, hence it appears as semantically open. Semantic openness implies a re-evaluation of the notion of "truth" as recognition of the essential role of some external features in orienting the cognitive procedures, against "certitude", i.e., self consistency within a deductive procedure.

OUTLINE

1. INTRODUCTION

In a previous essay (Arecchi 1992) I have compared the two main lines of human investigation, the deductive one ("top-down") characteristic of the scientific procedure (Carnap) and the inductive one ("bottom-up") characteristic of the argumentation procedure (Perelman). It was shown that in fact the two procedures are not mutually exclusive. Indeed the attempt to outline a purely deductive science, consisting in the logical construction of a set of theorems yielding a "complete" description of experimental obser-

A. Carsetti (ed.), Functional Models of Cognition, 167-187.

vations, fails in face of two main drawbacks, namely, *undecidability* and *intractability*. Thus it is necessary along the deductive chain to "open" the theory to "extrinsic" elements, i.e., elements not included in the set of axioms.

In this essay I explore how scientific language emerges from ordinary language. I shall show that, in both cases, a foundation can not be found "intrinsically", that is, embedded within the same dynamical flow which characterizes the language. Such an intrinsic foundation would amount to a certitude criterion implying merely the self consistency of the syntactical chain. On the contrary, we shall see that elements extrinsic to the flow play a crucial role, providing a semantic aperture which represents the truth value of the linguistic formulation.

Whence the contraposition of "truths" versus "certitudes" in the title.

In accord with the old definition of "veritas = adaequatio intellectus et rei", truth denotes the adjustment of the flow characterizing a sequence of linguistic utterances following the suggestions from the context.

In contrast, "certitude" ("clear and distinct ideas" in Descartes' formulation) denotes the correctness of the syntactic procedure. We shall show that a correct syntax does not provide in general a unique result, but rather leads to a large number of distinct consequences, most of which are never verified.

We shall further demonstrate a way to recover a unique result. If the syntactic flow, intended as the chain of deductions generated by a set $\{\alpha\}$ of axioms, is supplemented by a set of external influences that we call $\{A\}$, then the increasing multiplicity is reduced and eventually a unique outcome is reached. Let us call "truth" the collection $\{A\}$ of external events which genetically shapes the representation of an observed reality by biasing the syntactic flow $\{\alpha\}$ towards the description of that same reality. Then the truth value $\{A\}$ seems tailored upon the specific language $\{\alpha\}$, and it is subjected to change into $\{B\}$ as we adopt a different description ruled by another set $\{\beta\}$ of axioms. At a metalinguistic level, we can draw two parallel chains $\{\alpha\}$, $\{\beta\}$, $\{\gamma\}$ ..... of different axioms (different scientific descriptions of the same phenomenon) and discover that the corresponding syntactical chains must be supplemented respectively by the sets $\{A\}$, $\{B\}$, $\{C\}$...... of external agents breaking the inner symmetries of the deductive processes.

Should we say that the truth ($\{A\}$, $\{B\}$, or $\{C\}$) depends upon the selected language? In a set-theoretical sense, this would be the right answer and thus truth, as self consistency, would be a matter of the chosen language.

However we shall recur to an adaptive strategy, whereby the rules are adjusted in the course of the cognitive process rather than pre-assigned (Arecchi et al. 1994). An adaptive strategy means that the data collected in a sequence of observations are not part of the same set described by a single formal language (e.g. $\{\alpha\}$), but they should be matched by different languages in so far as we have readjusted the rules (i.e., the syntax) in the course of the observation, in order to optimize the recognition. If we still want to make use of the standard formulation, we should say that after having examined at a metalevel all possible theories $\{\alpha\}$, $\{\beta\}$... with all possible sets of constraints $\{A\}$, $\{B\}$...., we succeed in building an optimal path among all of them, which is no longer language-specific. This procedure is exempt from the two drawbacks of *undecidability* and *intractabil-*

intractability. However, even though the general conditions of success of an adaptive strategy have been stated, and many examples worked out, the formulation is still in terms of ordinary language. This is after all the status of any new scientific line before it is formalized by a metatheoretical procedure.

2. THE LANGUAGE OF SCIENCE

The word of the ordinary language is polysemic. In fact it does not represent univocally an object per se, but rather the object embedded in different contexts, which means an infinitely large variety of different situations that we call "events". Thus the same bare object can be part of different events as it appears in different contexts (Fig. 1a).

The same noun "event" is ambiguous, in so far as it implies a clear separation between observed thing and observer. It was a tenet of naive realism that the observer, as a self consistent entity, could classify the "events" without perturbing them. This requires already a given degree of abstraction, since at first sight we barely perceive such a separation. Furthermore, after the critique by Ernest Mach and the discussions put forth by the development of quantum mechanics, there is no clear cut separation between observer and event. On the other hand, the two separations "observer-event" and within the event "object-context" do not appear as arbitrary linguistic features. Already a cat playing with a ball has separated it from the legs of the table or from the carpet, and if in presence of a dog, the cat puts itself at a clear distance from the dog.

Let us recall the birth of Galileo's scientific revolution. He explicitely limited the scientific endeavour to the "quantitative affections" rather than trying to "grasp the essences". To catch the importance of such a limitation, let us recall that in "logocentric" cultures such as the Mediterranean ones, language is based upon polysemic words which do not have an univocal meaning but are adaptable to different situations. In Fig. 1a the horizontal axis is a semantic space where each point represents an object embedded in a specific context. Usually the same word can be attributed with different degrees of appropriateness to different events (object plus context). The different attributions have a different probability, as we can see from a perusal of the historical dictionary of a given language where the probabilities are given as a histogram of frequencies of occurrence in the literary texts of that language. In fact, the histogram is a finitistic approximation, a kind of coarse-grained structure due to the limited number of available texts. If, however, we consider the everyday use, the continuous probability curve is more appropriate since the context includes the observer with his (her) own mood of the moment which nobody would dare to reduce to a countable number of states.

This is a rather crucial feature, at variance with that of an artificial cognitive agent (a collection of detectors feeding the input of a universal computer). Such an artificial observer would extract a histogram, thus justifying a finitistic approach, whereas finitism seems to be excluded from the human everyday experience, as we reflect on the variety of nuances which qualifies a poem, or even a private conversation.

Whence the problem of interpretation, that is, of what is the right meaning to be attributed to a word; within the wide support subtended by its probability distribution. In Indo-European languages, a quasi-univocal, or narrow range space of meaning is obtained by a "filtering" operation, applying to the word a sufficient number of attributions or specifications, as sketched in the figure.

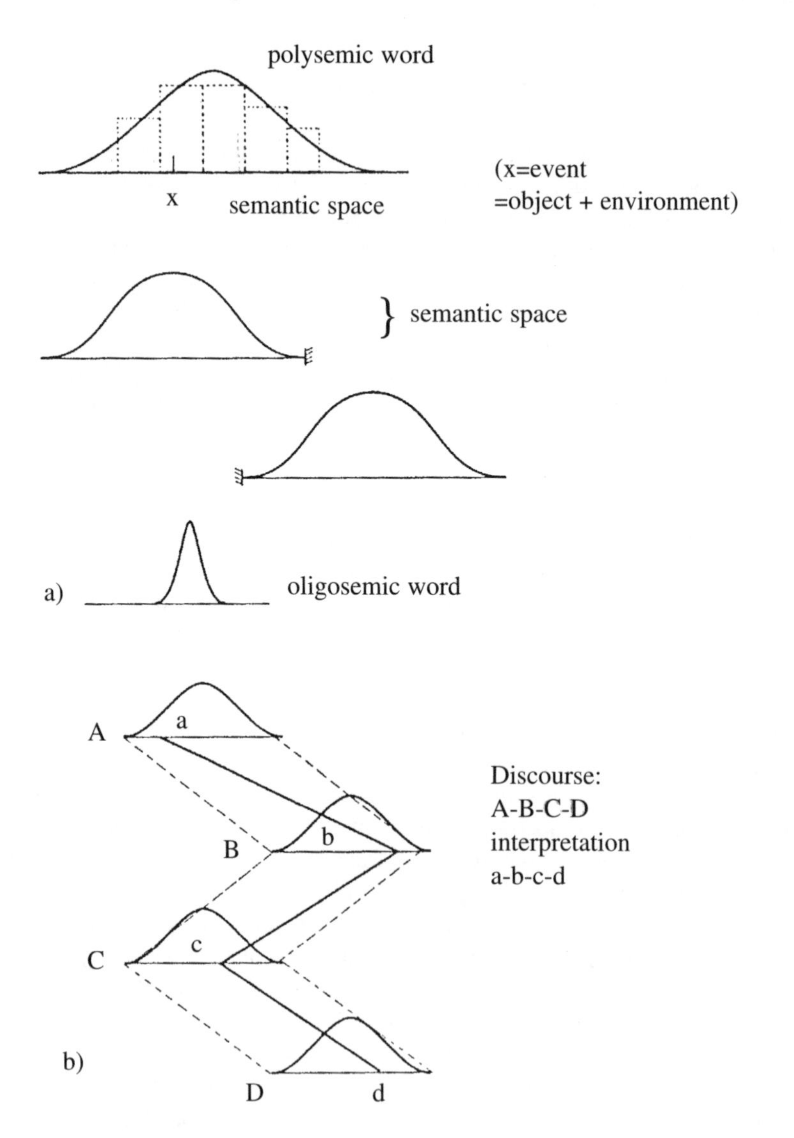

Figure 1. a) Polysemic word represented as a probability curve over a semantic space. Constraining the word by further attributions narrows its semantic range, that is, reduces the probability spread on the semantic space ('oligo-semic' word). A discourse, as the connection of different semantic spaces, yields a wider river bed allowing for different interpretations. (From the point of view of a linguist this sketch is a rough caricature, since it reduces the semantic narrowing to a pre-contextual session).

For instance: the English word "snow" has fourteen equivalents in the Eskimo language. Each of these terms corresponds to "snow" plus a suitable addition of further specifications in order to obtain fourteen different meanings corresponding to a more finely grained perception of such a phenomenon. In this way, a definition by "genus and specific difference" implies an elaborate system of stipulations where highly polysemic terms are filtered down to narrow range connotations on which most people can agree. A discourse, seen as a flow of different words connected by grammatical rules, appears as

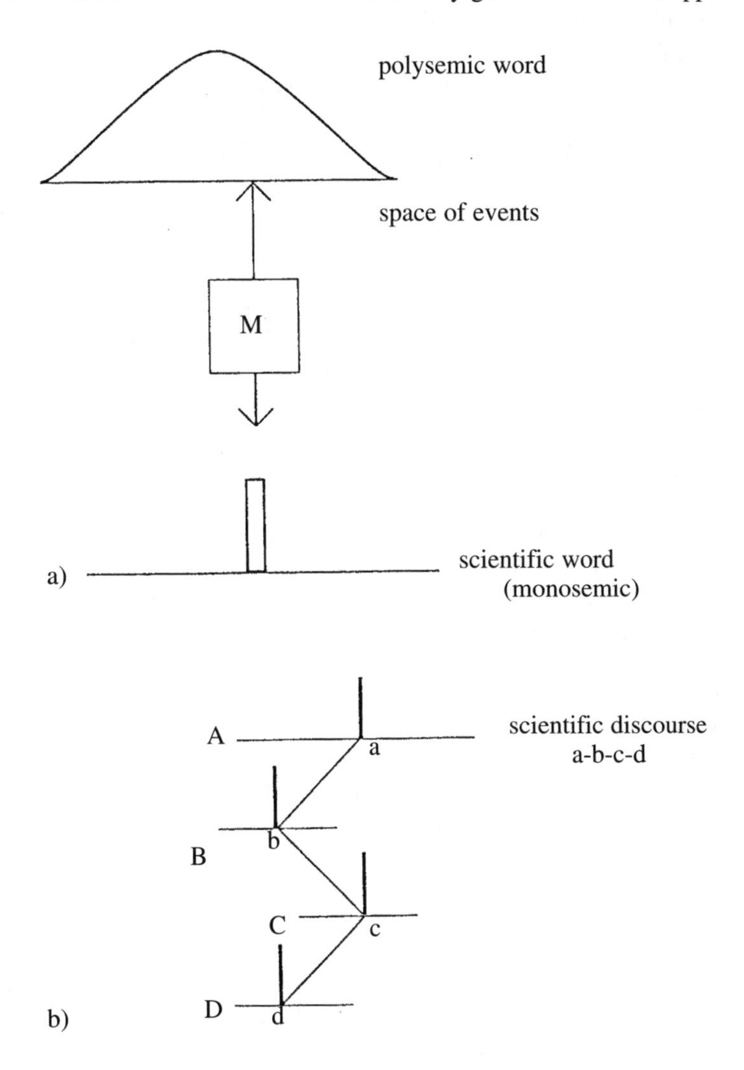

Figure 2. (a) A fixed measuring apparatus selects a univocal denotation, attributing to the event a membership property to a suitable set. (b) A scientific discourse is a chain of necessary connections between univocal terms, thus leading to a unique pattern.

a wide river-bed within which everybody can cut out a different interpretation (Fig. 1b). As is well known, there is no unique sense of a given discourse, but one must recur to other sources of information besides the text itself, in order to narrow the semantic range of each term. A self-consistent reading of a text, whereby each word is specified by all others, is somewhat illusory.

The problem of interpretation is common to all languages (painting, music, etc.). The same problem arises in the translation from one language to another.

The philologist tries to extract the most "faithful" meaning of a text by reconstructing the cultural and psychologic history of the author and by looking for other elements besides the flow of words. Even so, this is a relatively poor help for the personal interpretation by a new reader who strongly affects the context by his (her) own mood and cultural bias, thus shifting the position of each word in its semantic space with respect to the "most probable" one found by the philologist.

It is a common experience that rereading a poem or listening again to a certain piece of music may lead to a completely different attribution of sense (i.e. interpretation) depending on the different psycological situation. The observer has changed the context, and hence shifted the position of each term within its individual probability distribution, with respect to a previous approach to the same text.

Such ambiguities of the ordinary language were discussed at length by many Renaissance philologists and were well known to Galileo. He provided a way out of the ambiguities through his suggestion of "naming" an event via the number extracted by a suitable measuring apparatus M applied to the event itself (Fig. 2a). This procedure apparently provided univocal meanings since it filtered out a single denotation, clearing away all the context. As an example, a physicist does not speak of a "table" but of the "weight" or the "length" of the table. In this new language, the syntactical rule connecting two words becomes a mathematical relation connecting the output numbers from two measuring apparatuses related to two "objects".

The words of the new language have sharp probability distributions, and the experimental recipe (set of measurements) which generates a description of an object makes it possible to classify the object as the member of a given set in Cantor's sense.

In the early ages of modern science, two different attitudes have grown regarding the nature of the relations between physical events. On one side, the relations were regarded as abstracted from series of measurements, that is, "read" in the book of Nature. On the other side, they were put forward as theoretical inventions (think of Newton's gravitation law) to be confirmed by their experimental results.

Whatever the value attributed to the laws, they are mathematical relations between the numbers resulting from measurements which provide a solid framework for any scientific description in terms of well established existence and uniqueness theorems. Thus, the flow of a scientific discourse consists of sharp, necessary connections among point-like objects of different semantic spaces, corresponding to different measurements, as shown by the solid line in Fig. 2b. That solid line seems to represent a great progress as compared to the wide river-bed of Fig. 1b. It means that the scientific language is free from interpretational ambiguities. Ambiguities may arise, though, when a chain of scientific statements is translated into ordinary language. In such a case, any precise scientific concept has to be expressed by everydays words which reintroduce a certain amount of metaphorical content, excluded by the Galilean procedure from the scientific terms. Giving up with conversational terms, one should get around a scientific discourse in an univocal, non ambiguous, way.

However, a reverse use of Church's thesis introduces a new type of ambiguity considered recenty in nonlinear dynamics.

Church's thesis (1936) says that any function which can be calculated, that is, any logic predicate which is decidable, is a general recursive function and vice versa. In other words, two universes whose objects are sharply defined as members of Cantor sets can be mapped one into the other. It is reasonable to draw the following correspondences of Church's type, independentely of the different semantic values:

I) general recursive functions;

II) any formal theory based on a finite set of concepts and axioms;

III) sequence of operations on a universal computer (Turing machine);

IV) dynamical systems (discrete objects evolving by dynamical laws).

Church's mapping among the four universes (I-IV) holds in so far as in all cases individuals are defined in a set-theoretical sense.

In view of this correspondence, results derived in one of the above areas are applicable to the other ones.

Church's thesis has been used to constrain universal computers (III) by the Gödel theorem (II) (Webb), and dynamical systems (IV) by algorithmic complexity (III) (Wolfram). Vice versa, I exploit well-known results in nonlinear dynamical systems (IV) to introduce relevant limitations of the non-ambiguity of a scientific discourse (II).

If Fig. 1b is meant as a dynamical flow in the course of time, we know from the theory of deterministic chaos (Arecchi 1985) that a minimal uncertainty in the location of the initial condition results in large deviations, increasing exponentially with time. By Church's thesis, a similar instability can occur within a scientific discourse. One might object that in the dynamical case the initial coordinate is in general a real number and the uncertainty arises from its truncation to a finite number of digits. The point is that even Galileo's procedure of Fig. 2a does not define sharply one point of a semantic space, since the apparatus M has a given amount of context dependence, either because of environmental effects or, in the quantum case, because of the non-commutativity of position and momentum of the pointer. This last consideration was recently developed by Aharonov, leading to the concept of "weak measurement" (Aharonov and Vaidman, 1990).

Thus, as asymptotic instabilities affect a dynamic flow, similarly they affect the development of a formal deduction. One might wonder if the formal deduction can lead to an even larger spread in the semantic space than that allowed by ordinary language. In the next section, making further use of Church's thesis, we will find even stronger limitations to the pretension of a formal language to provide complete and non ambiguous descriptions of our scientific experience.

3. AN INNER CRITIQUE OF SCIENCE

Modern science has eventually reached a deductive structure (Carnap), in the sense that a theory - in order to be predictive - must anticipate the effective behaviors of the world throught a purely conceptual development. Indeed, a theory is considered successful if it is a "compressed" description of the world, that is, if the length in bits of the words stating the initial assumption is much shorter than a detailed description of the events themselves. From this point of view, a physical theory is not different from an abstract mathematical theory. It becomes a model, that is, it acquires semantic values whenever we interpret the objects of the theory as elements of reality (Tarski).

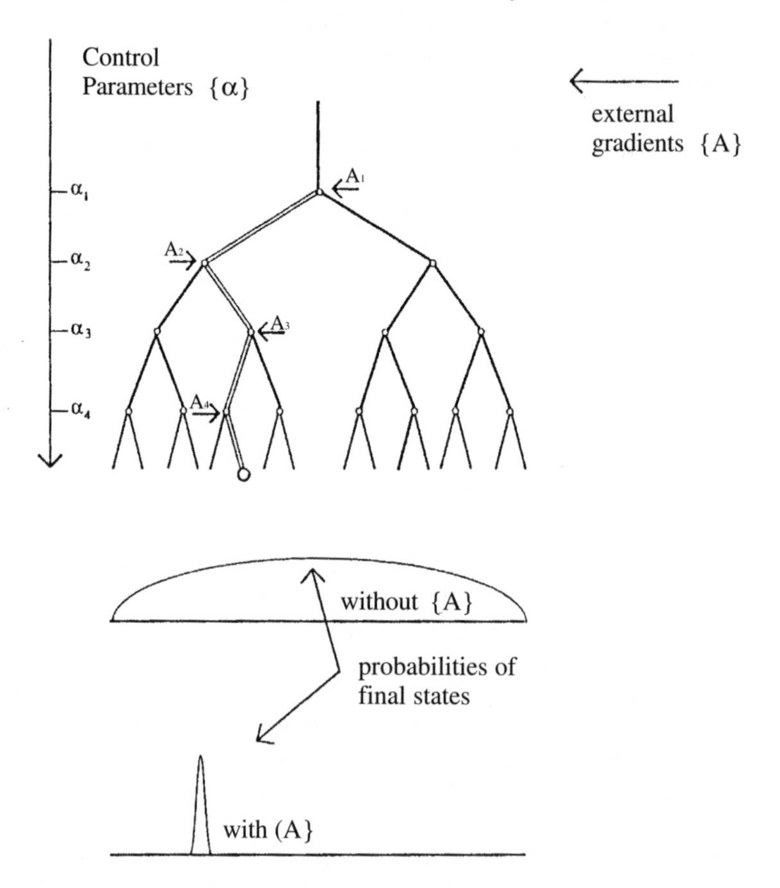

Figure 3. Bifurcation tree in onlinear dynamics. As a control parameter α is tuned through different values, novel steady states appear. By tuning a from α_0 to α_N, the system goes from 1 to 2^N different states. We call dynamical complexity such an ambiguity. It is responsible for the failure of reductionism. In the absence of external gradients, all final outcomes are equally probable. We call "organization" the occurrence of just one event out of 2^N. This implies that at each α_i an external agent A*i* has broken the bifurcation symmetry.

Therefore a scientific theory must be considered as a set of primitive concepts (defined by suitable measuring apparatuses as M of Fig. 2a) related by axioms. The deductions of all possible consequences (theorems) provide predictions which have to be compared to the observations. If the observations falsify the expectations, then one tries different axioms. Axioms look like mental inventions, justified only a posteriori through their consequences. This motivates the incommensurability of theories (Kuhn) or methodological anarchism (Feyerabend).

Furthermore like any formal theory, the deductive process should be affected by Gödel's undecidability in the sense that it should be possible to build a well formed statement where the rules of deduction are unable to decide whether that statement is true or false.

A finitistic approach consists in limiting any sensible theory to a finite set of theorems for which a sufficient number of axioms excludes any undecidability (Webb). In Sec. 2 we have already criticized the finitistic pretension of the scientific language, after analyzing the spread in semantic space due to the quantum resolution of the pointer of the measuring apparatus. We will provide later an even stronger criticisms of finitism.

Thus, *undecidability* does emerge as a ghost within any scientific theory.

Besides that, a second drawback is represented by *intractability*, that is, by the exponential increase of possible outcomes among which we have to select the final state of a dynamical evolution. Fig. 3 sketches a bifurcation tree familiar to computer scientists when they have to perform a complex calculation with branch points implying multiple choices of the type "if-then". Due to my professional bias, I rather prefer to consider the tree as the bifurcation tree of a complex non-linear dynamics.

As said in (Arecchi, 1992) entangled decision trees are the results of successive bifurcations which characterize the evolution of a complex system consisting of many interacting elementary constituents.

The problem of *complexity* emerges from the attempt to reduce the description of a piece of world to the mutual interactions of some atomic components. As an alternative we may use different scientific descriptions for different aspects of reality as shown in Fig. 4.

A particular set M of measuring apparatuses selects a given level of reality and defines a formal language for it (i.e., a language connecting the outputs of M). If we now choose a different set M_j, then we have a different description. The two descriptions are related by metaphor, that is, by elements of pre-formalized knowledge which carries information about the context surrounding the object, or, to express it more clearly, about the specific quantitative aspects of the object which is picked up by the measurement.

Going back to the reductionistic construction of a reality out of its constituents, we find an exponentially high number of possible outcomes, when only one is being observed. This means that, while the theory, that is the syntax, would give equal probability to all branches of the tree, in reality we observe an *organization process* whereby only one final state has a high probability of occurrence.

It is here necessary to recall some descriptive elements of the bifurcation of the stable branches of a dynamics for different settings of a control parameter. These are the necessary ingredients of any complex dynamics (Nicolis and Prigogine 1989, Arecchi 1992).

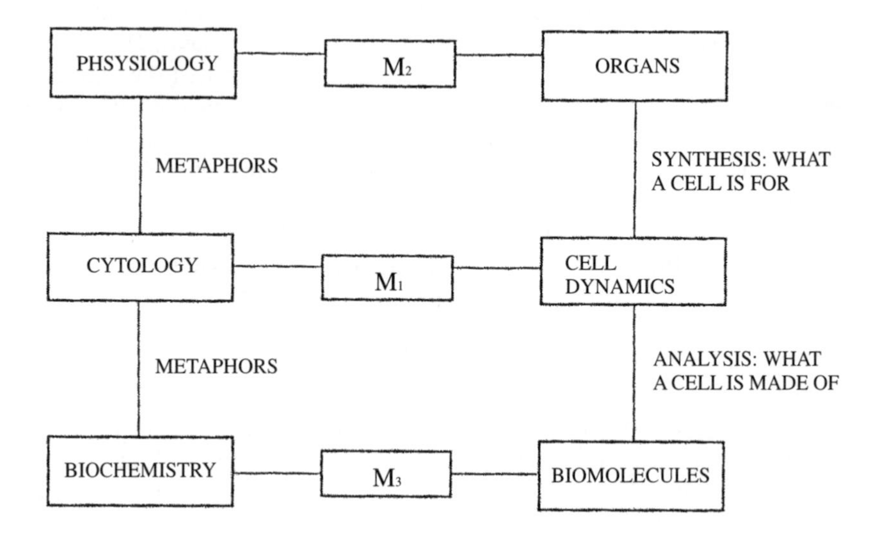

Mi : MEASURING APPARATUS WHICH CHARACTERIZES A SPECIFIC SCIENCE (i=1,2,3)

Figure 4. A living organ (e.g., a heart) is scientifically approached on different levels, depending on different apparatuses *Mi* giving rise to different sciences. Within each science, the univocal words are defined by the corresponding *Mi*. To communicate among different sciences, one must rely on metaphors.

Notice that intrinsic bifurcations of a dynamical theory display specific symmetries (Fig. 5a). Only external gradients break this symmetry (Fig. 5b). However a model ad hoc can incorporate broken symmetries by postulating that already atoms are equipped with suitable selection mechanisms. This, of course, sounds rather artificial. It is the way to build specific $\{\varepsilon\}$ theories which include from the outset superselections leading to a unique outcome without the introduction of external gradients.

Thus, during the course of a dynamical evolution, either because some control parameters $\{\alpha\}$ are tuned from the outside to assume different values, or because internal feedbacks change the $\{\alpha\}$ set in the course of time, starting from some initial conditions we expect an exponential increase of final states. Whence the dynamical interpretation of the decision tree of Fig. 3 originally introduced in computer science.

Whenever there has been *organization*, this means that at each bifurcation vertex of Fig. 3 the symmetry was broken by an agent external to the system under investigation.

We can thus stipulate the following:

i) A set of control parameters

$$\alpha_1, \alpha_2, \ldots. \alpha_N = \{\alpha\}$$

is responsible for successive bifurcations leading to an exponentially high number (of the order of 2^N) of final outcomes. If the system has no boundary effects (considered to be of infinite size), then all outcomes have comparable probabilities (the probability of their occurrence is a smooth function) and we call complexity the impossibility of predicting which one is the state we will observe at the end of the chain of bifurcations.

ii) A set of external forces

$$A_1, A_2, \dots A_N = \{A\}$$

applied at each bifurcation point break the symmetries, biasing toward a specific choice and eventually leading to a unique final state.

We are in the presence of a conflict between (i) syntax represented by the set of rules (axioms) $\{\alpha\}$ and (ii) semantics represented by the set of external agents $\{A\}$. The syntax provides 2^N legal outcomes. But if the system is open to the external world, the presence of which is expressed by $\{A\}$, then it leads to a unique final outcome. Once the syntax $\{\alpha\}$ is known, the final result will be an assertion that the set of external events $\{A\}$

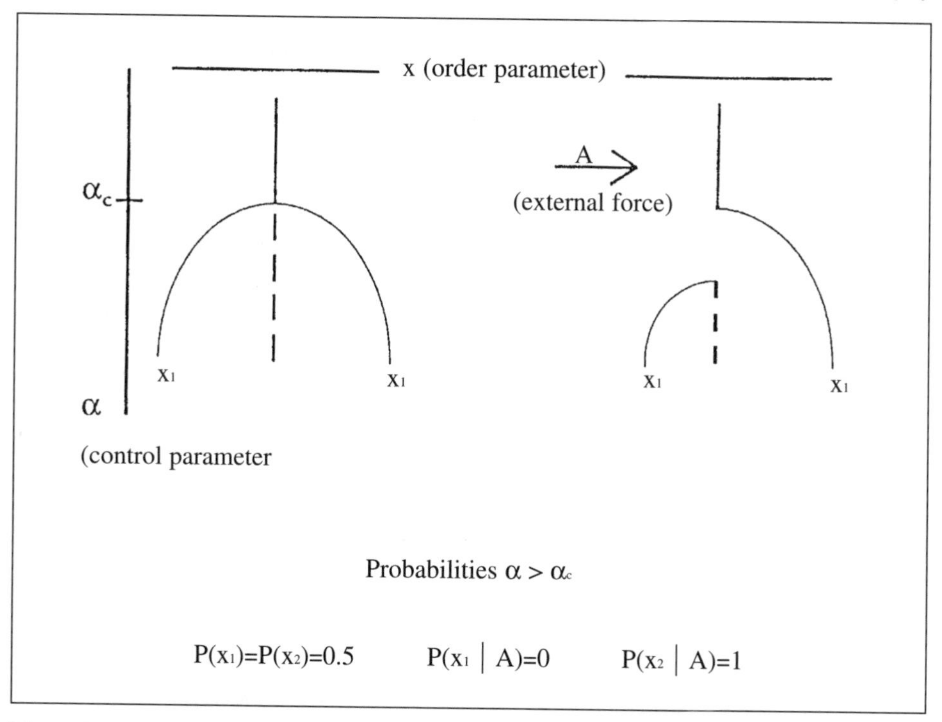

Figure 5. Examples of bifurcation diagrams. The dynamical variable x (order parameter) varies horizontally, the control parameter α varies vertically. Solid (dashed) lines represent stable (unstable) steady states as the control parameter is changed. Left: symmetric bifurcation with equal probabilities for the stable branches. Right: asymmetric bifurcation in the presence of an external field A. If the gap introduced by A between right and left branch is wider than the range of thermal fluctuations at the transition point α_c, then the right (left) branch has probability 1 (0).

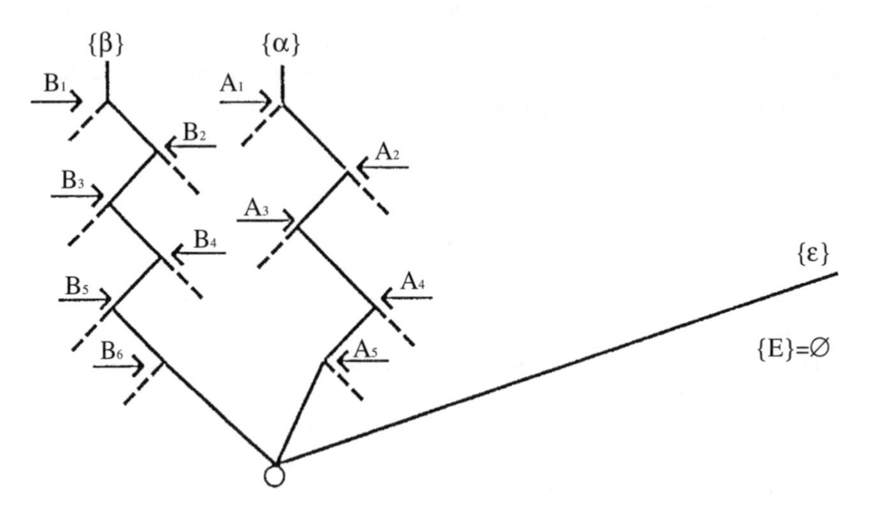

Figure 6. Different theoretical models may explain the same final state. The backward reconstruction of the dynamical path will then retrieve different classes of external agents.

must have occurred. Therefore, we can take {A} as the element of reality in which our system is embedded.

We define "certitude" as the correct application of the rules {α}, and "truth" as the adaptation to the reality which is expressed by {A}.

However, the same final outcome could be reached by a different set of rules {β}. In such a case, retracing the new tree of bifurcations, we would reconstruct a set {B} of external agents. Thus it seems that truth, {A} or {B}, is language dependent!

Furthermore, the "emergence" of organization means that we can even build a set of axioms {ε} which succeeds in predicting the correct final state without external perturbations, that is, {E} = ∅ (Fig. 6). This is indeed the pretension of the so-called "autopoiesis", or "self-organization" (Krohn et al. 1990), to which I have opposed the term "hetero-organization" (Arecchi 1992).

In fact, as stressed in the Introduction, an adaptive procedure discussed in the next Section will disclose an optimal trajectory built into the language space by a change of rules as a compromise among all formal theories {α}, {β} .... {ε}, with a corresponding optimized adaptation to reality made of pieces of {A}, {B}....

4. LIMITATION OF THE {E} THEORY (SELF-ORGANIZATION)

We have concluded the preceding Section stating that in general, among all possible theoretical models {α}, {β} ..., there may be a privileged one, {ε}, within which the organization "emerges" in a natural way, without having to recur to a non empty set {E} of external perturbations (in other words, organization is "self-organization").

From a cognitive point of view, the theory {ε} can be regarded as a "petitio principi", a tricky formulation tailored for a specific purpose and not applicable to slightly different situations, more or less like Ptolemaic epicycles which provide very accurate predictions for the future positions of the bodies of the solar system, but only from a particular point of view. In fact, a space traveler away from Earth must build a new "ad hoc" theory to describe planetary motions as seen in any new observation frame. Anyway, {ε} theories are highly respectable. Rather than explicitely listing the elements of reality, as, e.g., {A} for {α}, the user of language {ε} has already exploited the elements of reality at a pre-formalized level, and has made good use of them in planning {ε}.

These pieces of knowledge which precede axiomatization have received different names like "abduction" (Peirce), "tacit dimension" (Polanyi) or "common sense" (Livi). Some of them have been embedded as universal tools either in our genetic heritage, or during infancy in our learning age. This seems to be the cognitive valence of Jung's archetypes" (cfr. von Franz). A critical survey of these useful "preliminaries" is contained in the work by Bateson (Bateson 1972).

The choice of {ε} axioms is a clever achievement acceptable only if it duly acknowledges the elements of the underlying reality upon which it has been built. However, these pre-scientific pieces of information are no longer explicitely mentioned in {ε} theories, thus fostering the misleading belief that a self-organized approach is always workable ("autopoiesis", cfr. Krohn et al. 1990).

We might accept {ε} theories if there were a single one capable of explaining wide classes of different phenomena, but so far they have been purposefully planned to explain a very small class of observations.

Due to the close correspondence between a cognitive interpretation (IV) and a dynamical interpretation (II) of the bifurcation tree of Fig. 3, the cognitive strategy leading to an {ε} theory has a dynamical counterpart. This corresponds to some interacting species (e.g. chemicals) which under suitable boundary conditions evolve without need for secondary bifurcations, thus reaching a well-defined organized state. Situations of this kind are fortunate occurrences which in general must be actively searched, since they are rather uncommon.

Whether complex phenomena as "life" or "intelligence" are the outcomes of a single {ε} dynamics or rather the result of competition which has involved many crucial external elements as {A, B...} is a matter of historical reconstruction, a kind of paleontological investigation which may imply a lot of bifurcations toward the past, and thus may not provide a unique answer. Complexity works also backwards in time!

5. AN ADAPTIVE STRATEGY IMPLYING SEMANTIC OPENNESS

Inferring models from given data is affected by many different changes in representation. Any such change affects the semantic content of the resulting model, as we have seen in the previous section by comparing models {α}, {β}... and the corresponding semantics {A}, {B}... .

Let us see why we should change models while doing an investigation. When we start with a given partition of our space of events, any trajectory is coarse-grained as a sequence of symbols representing the cells encountered (see Fig. 7).

After the partition the trajectory is coded into a discrete sequence, as the example of Fig. 7 shows. The symbol sequence depends on the chosen alphabet; therefore, the descriptive language, and hence the model, changes as one changes the partition.

For a given partition, we are unable to distinguish points within the same box. However we know from deterministic chaos that close-by points within the same box can generate exponentially diverging trajectories. Since the aim of our model is to make predictions of the future once we have observed the system for a while, then chaos is reducing the predictive power to a small time interval beyond the last observation.

Thus we decide to perform a finer partition and have to replace the previous code with a different one. We have changed the "words" and hence the language. This change of model towards increasing resolution requires however exponentially increasing execu-

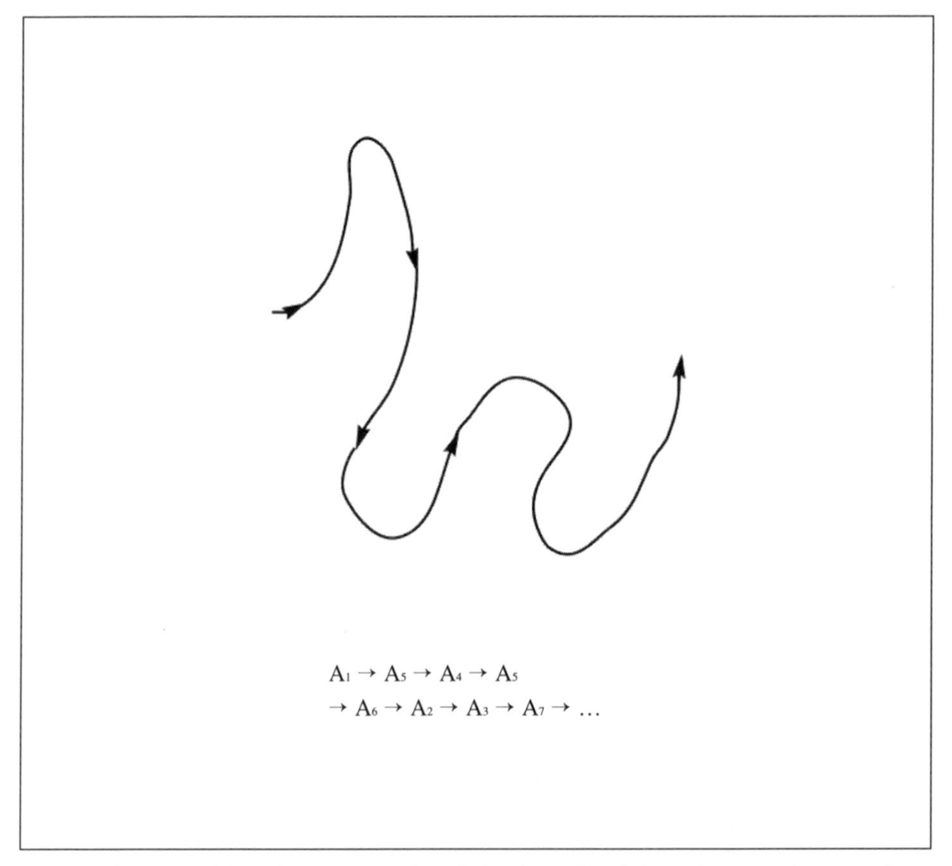

$A_1 \rightarrow A_5 \rightarrow A_4 \rightarrow A_5$
$\rightarrow A_6 \rightarrow A_2 \rightarrow A_3 \rightarrow A_7 \rightarrow \ldots$

Figure 7. Coarse graining of a state space into finite size cells Ai. A continuous trajectory reduces to a sequence of discrete symbols A_1 A_5 A_4 A_5 A_6 A_2 A_3....

tion times. As soon as the computation time is longer than the effective time of evolution, we are no longer able to develop a model since a straightforward simulation is the best we can do (alas without any predictive power).

An adaptive strategy (Arecchi et al. 1994) consists in a frequent series of observations and in a comparison of the spread away from the average trajectory.

We readjust the resolution depending on how fast the irregularities grow. In this way, we are coding with a variable grain (coarse in regions where chaos is low, very fine in regions where there is a fast increase of irregularities). In general terms, we are building sequences of words with different alphabets, and therefore with different rules.

The adaptation demonstrated in (Arecchi et al. 1994) refers to a coarsening of the time axis, introducing a stroboscopic observation time adjustable along the indication coming from the same data sequence. However any representative space can be parsed at variable resolution, even the discrete space of possible dimensions, as done by the embedding technique (Arecchi 1985).

The example of (Fig. 8a) refers to a data set embedded in a three-dimensional space, and to individuals which have to be discriminated because they may correspond to classes of useful or harmful features, If we limit space to a two-dimensional view, the two classes of data are intertwined in such a way (I) that no useful discrimination criterion emerges on that projection x-y. Without resorting to the full three-dimensional representation, by an adjustment of the point of view, a new projection x-y (II) is found which allows the complete separation of the two classes, providing a truth which was lost in (I).

Going back to the probability distribution in a semantic space, as sketched in Fig. 1, we can represent a collection of different scientific approaches to a given reality as a set of sharply defined characteristic functions which are zero everywhere, except in the small region selected by the measuring apparatus.

The group physiology-cytology-biochemistry illustrated in Fig. 4 will appear as three rectangular windows located in different regions of semantic space (Fig. 9). The sharp boundaries correspond to the precise definition characterizing the property of a given set. The exchange among the windows can be expressed in the terms of each theory; but they are used beyond their range of formal validity. In this sense, they are used as metaphors.

A fuzzy logic introduced in recent years (Zadeh 1987) allows for smooth distributions in semantic space, thus providing overlaps among terms centered at different positions (Fig. 9b).

At variance with the above "fixed logic" approaches, the adaptive strategy appears as a "moving logic" approach. Indeed, during the cognitive process, as one adjusts to the increasing series of data with better approximations, the scientific description shifts its center and becomes narrower as in the sequence 1 - 3 of Fig. 9c. Of course, the final position depends upon the motivation of the observer in selecting a specific point of view. Going back to the example of Fig. 4 if the emphasis is on biomolecules rather than on organs, then the adaptive process will be 1 → 2' → 3' (dashed lines) rather than 1 → 2 → 3 (solid lines).

It is a matter of reflection to realize that similar strategies are common to any language,

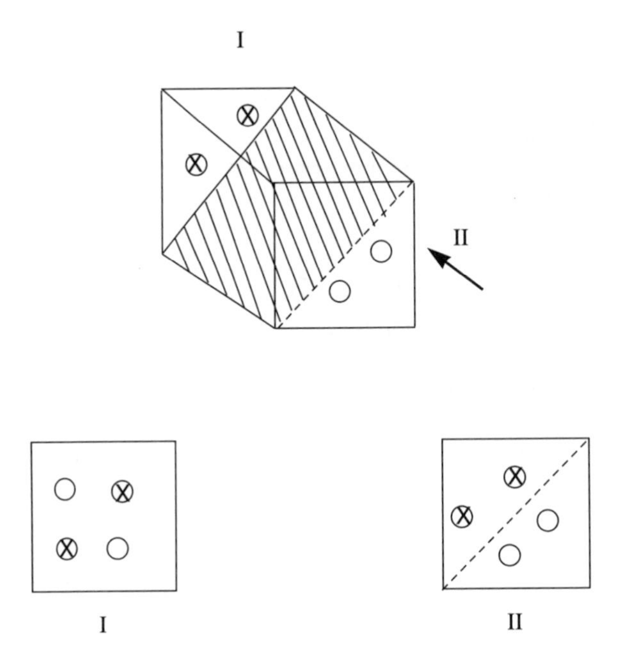

Figure 8. Choice of the most convenient point of view. Two types objects (empty and crossed circles, respectively) ar confined to disjoint regions, as shown in the three dimensional picture. A two-dimensional projection may give a wrong (I) or right (II) answer.

from common speaking to music and painting. Interpretation is never a neutral affair, but it is always biased by a personal selection mechanism, as is well known by everybody who has been exposed to an aesthetic experience. That is why different performances by the same orchestra may be so different!

In conclusion, applying adaptation to human cognition, we can consider any well established concept as a compromise between globality (genus) and detail (species). This is represented in Fig. 10 as the crossing point of two conflicting requirements in a diagram of latitude versus resolution.

The exact position of the crossing point C depends on the slopes of the two lines which are influenced by the past experience and the cultural background of each private subject. In fact the exact location of C is irrelevant and it represents a private affair. The way C is reached and hence named is however expressible in terms of objective operations, so that agreement among different observers is achievable. Any discussion whether my "red" is equal to your "red" is nonsensical. The important point is that we agree in com-

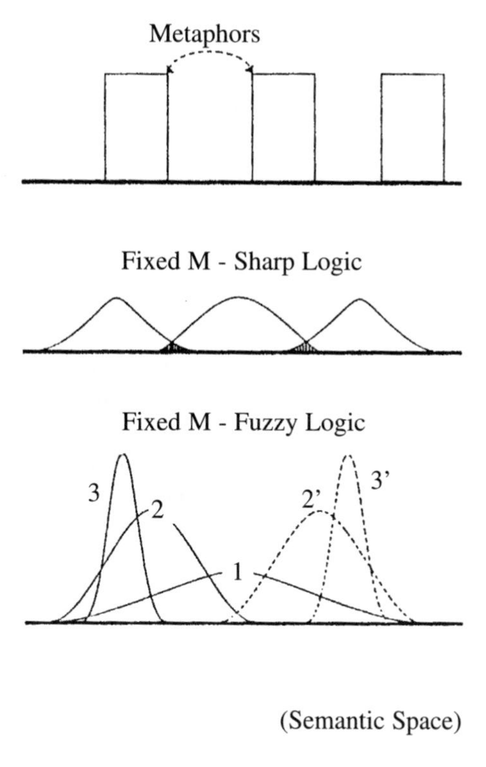

Figure 9. Fixed measuring apparatuses giving rise to different sciences. Intercommunication is based on metaphors (upmost figure) unless, as in fuzzy logic, one accepts ill defined terms with overlaps (center figure).
When the measurement is no longer locked to a fixed position of semantic space, nor to a fixed resolution, then the scientific knowledge can cover, with different degrees of detail, different areas of the semantic space, allowing information exchange within a unique language.

paring e.g. Titian's or Veronese's red with the help of some scale of appreciation based on which we can build and transfer opinions. I do not consider this as a subjective limitation, but as the real core of truth as "adaequatio intellectus et rei". Reading the deep discussion by St. Thomas Aquinas in "Quaestiones de Veritate" (Thomas Aquinas 1992) I do not find any element of disagreement with what I have presented above.

By the way, this adaptationist description of cognition is also the reason why we observe stable features around us. In fact, once deterministic chaos was explored (Arecchi 1985) it appeared so ubiquitous, that people started wondering how we could talk of anything. If we assume an active role and interact with a chaotic system by frequent observations we can intervene at the onset of any new bifurcation and stabilize a wanted state by tiny perturbations (Ott et al. 1990). In fact, a dynamical aspect of the bifurcation tree of Fig. 3 is that a chaotic dynamics can be seen as a multiple bifurcation cascade folded on itself and allowing for the onset of many regular orbits which, however, survive for a short time only (i.e. which are unstable) and are then replaced by another one. We can avoid de-stabilization and fix any one of the orbits if we start with small corrections once we have reached the desired position in bifurcation space.

Nowadays we think that such a control strategy is at the root of any stable feature. A many-body system can be conceptually splitted into subparts, each one described by rather simple nonlinear dynamics giving rise to chaos. Chaos is stabilized by the interaction with the rest of the system.

Then, why cannot we outline a global picture which includes the chaotic subsystem plus its surrounding as a single non-chaotic entity? Yes, indeed we can, but at a different level of description (see the ladder of Fig. 4) and in so doing we loose the advantage of the reductionist approach. Reductionism, on the other hand, is "economic" because it relies on a smaller number of fundamental ingredients, in accordance with Occam's precept: "Entia non sunt multiplicanda praeter necessitatem".

Once again, globality and detail are in conflict, as seen in Fig. 10.

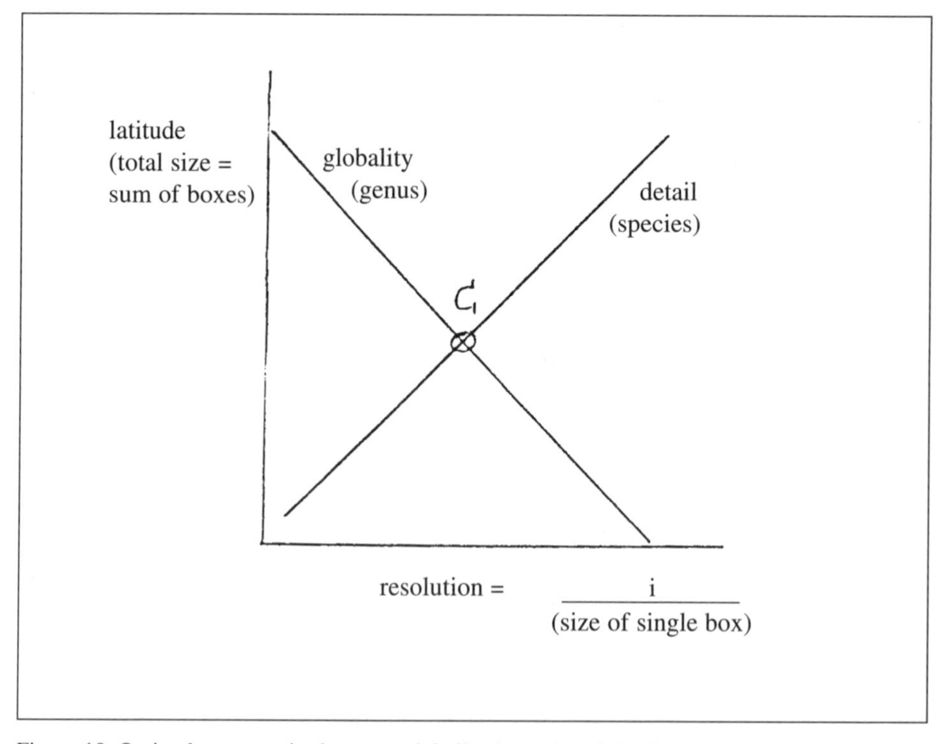

Figure 10. Optimal compromise between globality (genus) and detail (species).

6. CONCLUSION: RISKY TRUTHS VS. MISLEADING CERTITUDES

In this conclusion we discuss the truth value of scientific statements on the basis of the considerations of the previous Sections.

In the scientific endeavour, we select a quantitative feature by application of an apparatus M at a particular point of the semantic space (Fig. 2a). This means choosing not only what to measure, but also the measurement protocol (that is the rules of use of M).

The emerging description of reality represents an observation "from one point of view". Such is the sense of scientific truth "from a point of view" as discussed by Agazzi (Agazzi 1974).

It is important to notice that, as expressed by Bridgman (Bridgman 1966) the definition of the scientific concepts is *operational*, that is, determined by M and by the corresponding instructions for its use.

Due to the variety of possible M's, we must justify at a metascientific level *why* we have selected that M rather than another one.

This is a general semantic question dealing with the role of those elements of reality which are preliminary to one particular program (Pierce's abduction, Polanyi's tacit dimension, Jung's archetypes, etc.).

A purely pragmatic attitude consists in stressing the power of a technology as its actual truth. In fact, any technology is the mature fruit of a rather complex bifurcation tree. The identification of truth with a technological skill is equivalent to isolating the final portion of the bifurcation tree, without considering its early branches. But this amounts to overlooking not only the semantic values (truth) but also the preliminary syntax, thus risking to violate some rules which are a condition of the effective success of the technological procedure.

Many advanced technologies (bio-manipulations, chemical or weapons experiments), may be criticized on a purely procedural basis because they are defective in their obedience to the canons of the scientific endeavours as outlined above, not to mention ethical considerations which are beyond the scope of this essay.

Once we have criticized the equation "truth=technological power", we must reconsider whether truth has any sense within the scientific program.

In section 5 we have seen that adaptation is a slalom among different sets of rules $\{\alpha\}$, $\{\beta\}$,... under the guidance of a preferential set of external elements $\{T\}$ which has non zero intersections with $\{A\}$, $\{B\}$... but does not coincide with either one of them. By $\{T\}$ I denote an open set which may be re-adjusted; see the notion of collection below.

In front of the truth problem we can assume three attitudes, namely,

i) assume the adaptive strategy and its associated reality set $\{T\}$ as a kind of privileged reference frame. Indeed, being the result of an optimization process, it appears more appropriate than any particular theory.

ii) Consider the truth problem as a metatheoretical problem that belongs to a metalevel. At this metalevel, the set of all sets of truth values $\{A\}$, $\{B\}$,... has to be considered as the truth, but with the stipulation that any individual set makes sense only if associated with the corresponding theory.

iii) A more fundamental approach is being developed by E. De Giorgi and his collaborators at the Mathematics Department of the Scuola Normale Superiore, Pisa. De Giorgi does not confine himself to questioning the power intrinsic to any specific theory $\{\alpha\}$ or $\{\beta\}$ etc., he even questions the set-theoretical approach to the fundamental concepts of physical description. In Fig. 2 we have seen that the sharp definition provided by the measuring apparatus M wipes out the smooth probability measure in semantic space, providing a sharp connotation which allows to classify any observed entity as an individual of an appropriate set. Whence the set-theoretical character of all modern sciences, with the consequent antinomies of modern logic after Cantor, Frege, Russell, Gödel etc., transferred into the heart of the scientific language.

By a suitable re-introduction of *qualities*, besides the *quantities*, De Giorgi escapes from the limitation of sets. He calls the new groups of objects *collections* where membership in these collection is no longer as strictly determined as whenever it is defined by a finite number of precise stipulations as for *sets*. This degree of smoothness seems to me like going back to the polysemicity of ordinary language. Hence "Epimenides, the Cretan, says: all Cretans are liers" is no longer an antinomy, since Epimenides is not bound to be always a liar, but a liar in general, even though sometimes he can even tell the truth!

In conclusion, going back to the title of this paper, the fall of the myth of *certitude* (Barone) may lead to a devaluation of the power of science in reading the world. *Truth* appears instead as providing a grasp of a reality of which we are a part and that we do not invent, but by which we are conditioned. However, *truth* appears as a work in progress, a kind of program without end, which develops in the course of time, and which does not permit fancy predictions (these are just outcomes of a powerful syntax). In fact, we must recognize that a new event has modified our expectations only when it has occurred.

Such a continuous compromise between the power of theory and its limitation by the irruption of reality seems to be the essence of our human condition. It was central to classical philosophy and it was lost after Descartes, in a delirium of mental power which has led to the final triumph of nihilism in recent philosophy.

I hope to have demonstrated that the present scientific speculations strongly lean towards an ontological foundation.

Prof. Fortunato Tito Arecchi
Istituto Nazionale di Ottica
L. Enrico Fermi, 6
50125 Firenze, Italy

REFERENCES

E. Agazzi, Temi e problemi di filosofia della fisica, Ed. Abete, Roma, 1974 (paragrafo 50).

Y. Aharonov and L. Vaidman, "Properties of a quantum system during the time interval between two measurements", "Phys. Rev." A41, 11 (1990).

F.T. Arecchi, "Caos e ordine nella fisica", "Il Nuovo Saggiatore" (S.I.F.), Vol. 1, n. 3, pp. 35-51 (1981).

F.T. Arecchi, "A critical approach to complexity and self-organization", "La Nuova Critica" I-II, Quaderno 19-20, pp. 7-39, (1992).

F.T. Arecchi, G.F. Basti S. Boccaletti and A. Perrone, "Adaptive recognition of a chaotic dynamics", "Europhys. Lett." *27*, 327 (1994).

F. T. Arecchi, "Complexity, complex systems deal adaptation" F. T. Arecchi, A. Farini (eds) Lexicon of complexity, Firenze, 1997, pp. 7-41.

F. Barone, "Il mito della certezza", "Civiltà delle Macchine" XVIII, n. 3, pp. 55-59 (1970).

G. Bateson, Steps to an ecology of mind, Ballantine Books, New York, 1972.

P.W. Bridgman, The logic of modern physics: an introduction to the philosophy of science, Basic Books, New York, 1966.

R. Carnap, Philosophical foundations of physics: an introduction to the philosophy of science, Basic Books, New York, 1966.

E. De Giorgi, "I fondamenti della matematica" (unpublished lecture notes), Scuola Normale Superiore, Pisa 1995.

P.K. Feyerabend, Against method: outlines of an anarchist theory of knowledge, New Left Books, London 1975.

W. Krohn, G. Küppers and H. Nowotny, Selforganization, portrait of a scientific revolution, Kluwer Academic Publishers, (1990).

T.S. Kuhn, The structure of scientific revolutions, Univ. of Chicago Press, Chicago, 1962 (tr. it. Einaudi, Torino 1976).

A Livi, Filosofia del senso comune, ed. Ares, Milano 1990.

G. Nicolis and I. Prigogine, Exploring complexity, W.H. Freeman, New York, 1989.

E. Ott., C. Grebogi and J. Yorke, "Controlling chaos", "Phys. Rev. Lett.", *64*, 1196 (1990).

C.S. Peirce, Collected papers, Vol.s I-VI, ed. by C. Harshorne and P. Weiss, Harvard University Press, Cambridge, Mass. 1931-35, Vol.s VII and VIII, Ed. by A. W. Burks, same publisher, 1958; see also: W. R. Hanson, Patterns of discovery: an inquiry into the conceptual foundations of science, Cambridge University Press, Cambridge, 1958.

C. Perelman and L. Olbrechts-Tyteca, Traité de l'argumentation. La nouvelle rhétorique P.U.F. Paris (1958).

M. Polanyii, Personal knowledge: towards a post-critical philosophy, Routledge & Kegan Paul, London 1958.

A Tarski, "The concept of truth in formalized languages" in: Logic, semantic, mathematics: papers 1923-1938 by A Tarski, Clarendon Press, Owford, 1956.

Thomas Aquinas, Quaestiones disputatae - De Veritate (q. 1-20) 2 volumes (Latin text and Italian translation), E.S.D., Bologna 1992 (see in particular: q. 1: The truth, q. 10: The mind, q. 15: The reason).

M.L. von Franz, Psyche und Materie, Einsiedeln (C.H.) 1988.

J.C. Webb, Mechanism, mentalism and meta-mathematics (an essay on finitism) D. Reidel, Dordrecht (1980).

S. Wolfram, "Undecidability and intractability in theoretical physics", "Phys. Rev. Lett"., 54, 735 (1985).

L.A. Zadeh, Fuzzy sets and applications (selected papers), J. Wiley, New York, 1987.

KARL SVOZIL

ON SELF-REFERENCE AND SELF-DESCRIPTION

CONTENTS

1. THE CARTESIAN PRISON

Let us assume that our world is discretely organized, and that it is governed by constructive, i.e., effectively computable, laws [1]. By that assumption, there exists a "blueprint", a complete set of ruls or laws governing the universe. This seems unlike mathematics for which Gödel, Tarski, Turing and others proved that no reasonable (i.e., strong enough and consistent) formal system will ever be able to prove all true well-formed statements. Indeed, Chaitin showed that certain mathematical entities are as random as a sequence produced by the tossing of a fair coin [2, 3]. Hence, let us contemplate the assumption that, when it comes to an enumeration of laws and initial values, nature is finitely "shallow" while mathematics is infinitely "deep" [4].

One might think of such a world as an ongoing computation based on a finite-size algorithm; or, equivalently, as the derivation of successive theorems of a formal system. It is very much like a virtual reality, a computer-generated universe.

Any observer who is embedded in such a system [5, 6, 7] is naturally limited by the methods, devices and procedures which are operational therein. Such an observer cannot "step outside" [8] of this "Cartesian prison" and is therefore bounded to self-referential perception. Can one give concrete meaning to this "boundedness by self-reference?" Indeed, a research program is proposed here which is capable of the formal evaluation of bounds to self-reference. This program is based on a recursion theoretic re-formulation of physics. It may result in a paradigm change concerning the perception of indeterminism in physics.

2. SELF-DESCRIPTION BY SELF-EXAMINATION

Is it possible for a system to contain a "blueprint", a complete representation, of itself? This question has been raised by von Neumann in his investigation of self-reproducing

A. Carsetti (ed.), Functional Models of Cognition, 189-197.

automata. With such a "blueprint" it should be possible for the automaton to construct replicas of itself [9, 10, 11].

To avoid confusion, it should be noted that it is never possible to have a finite description with itself as proper part. The trick is to employ *representations* or *names* of objects whose code can be smaller than the objects themselves and can indeed be contained in that object (cf. [11], p. 165).

Gödel's first incompleteness theorem [12] stating its own unprovability is such an example [9].

Another example is the existence of a description p of length $|p|$ whose algorithmic information content $H(p) = |p| + H(|p|) + O(1) = |p| + ||p|| + |||p||| + ... + O(1)$ exceeds the length of its code. Intuitively, this can be interpreted as representing algorithmically useful information (e.g., coded in the program length, in the length of the program length, in the length of the length of the program length, ...) which is not contained by an immediate interpretation of the symbols of the string alone [15].

2.1 FIXED-POINT THEOREM OF RECURSION THEORY

Kleene's fixed-point theorem of recursive function theory states that, given any total function f, then there exists an index i such that i and $f(i)$ compute the same function; i.e., $\varphi_i = \varphi_{f(i)}$ [10, 11]. One application of the fixed-point theorem is the existence of self-reproducing machines and, therefore, the existence of intrinsically representable system "blueprints" [7].

This is an indication that it is at least possible to represent all the "finite-size" laws governing the system within such a system. A second aspect, which was the motivation of von Neumann to study self-reproduction, is the possibility for living systems to reproduce.

2.2 RECURSIVE UNSOLVABILITY OF THE RULE-INFERENCE PROBLEM

A totally different problem is the question how, if ever, a system can obtain such a blueprint by mere self-inspection. Two considerations yield the impossibility of such an attempt for the general case. The first one is connected to the recursive unsolvability of the rule-inference problem [17, 18, 19, 20]. The second one, which will be discussed below, is connected to the disruptive character of self-measurement [7].

Even without self-reference it is impossible to guess the law governing an effectively computable system. Assume some particular (universal) machine U which is used as a "guessing device". Then there exist total functions which cannot be "guessed" or inferred by U. One can also interpret this result in terms of the recursive unsolvability of the halting problem: there is no recursive bound on the time the guesser U has to wait in order to make sure that the guess is correct.

2.3 COMPUTATIONAL COMPLEMENTARITY

Self-reproduction by self-inspection usually presupposes an unchanging original. In the general case, this is again impossible because of disruptive effects. To put it pointedly: self-measurement exhibits (paradoxical) features strongly resembling complementarity. An idealized self-referential measurement attempts the impossible: on the one hand it pretends to grasp the "true" value of an observable, while on the other hand it has to interact with the object to be measured and thereby inevitably changes its state. Integration of the measurement apparatus does not help because then the observables inseparably refer to the state of the object and the measurement apparatus *combined*, thereby surrendering the original goal of measurement (i.e., the measurement of the object). These considerations apply to quantum as well as to classical physics with the difference that, disregarding "interaction-free" measurements, quantum theory postulates a lower bound on the transfer of action by Planck's constant $\hbar$.

Computational complementarity is not directly related to diagonalization and is a second, independent source of indeterminism. It is already realizable at an elementary pre-diagonalization level, i.e., without the requirement of computational universality or its arithmetic equivalent. The corresponding machine model is the class of finite automata, characterized by a finite number of states and a finite number of input and output symbols [21, 22].

Computational complementarity is based on the following observation [23]. Whenever the experimenter is part of the system, any measurement of one particular system feature causes an interaction with the observed object. This interaction may induce a transition of the observed object which results in the impossibility to measure another, complementary, observable and *vice versa.*

An automaton (Mealy or Moore machine) is a finite deterministic system with input and output capabilities. At any time the automaton is in a state q of a finite set of states Q. The state determines the future input-output behavior of the automaton. If an input is applied, the machine assumes a new state, depending both on the old state and on the input, emitting thereby an output, depending also on the old state and the input (Mealy machine) or depending only on the new state (Moore machine). Automata experiments are conducted by applying an input sequence and observing the output sequence. The automaton is thereby treated as a black box with known description but unknown initial state. As has already been observed by Moore [23], it may occur that the automaton undergoes an irreversible state change, i.e., information about the automaton's initial state is lost. A second, later experiment may therefore be affected by the first experiment, and vice versa. Hence, both experiments are incompatible. In this setup, the observer has a qualifying influence on the measurement result insofar as a particular observable has to be chosen among a class of non-co-measurable observables. But the observer has no quantifying influence on the measurement result insofar as the outcome of a particular measurement is concerned. This gives rise to the Copenhagen interpretation of automaton logic.

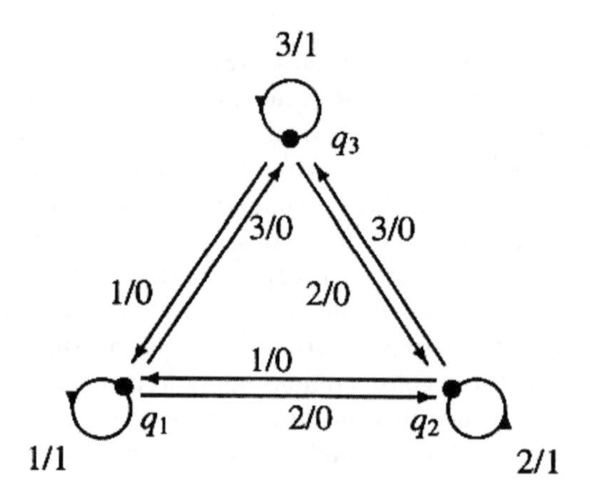

Figure 1: Quantumlike Mealy automaton featuring computational complementarity. Arrows indicate transitions. A/B should be interpreted as "input of symbol A, output of symbol B".

I shall illustrate this fact by an example. Consider the automaton whose transition diagram is drawn in Fig. 1. Assume that the initial value of this automaton is unknown. The task is to get information about the initial value by input-output experiments. The automaton is constructed to undergo a transition into the state which corresponds to the input; e.g., input of symbol "3" will steer the automaton into state three, no matter in which state it was before. In case of a "hit", it responds with output symbol "1", any failure to guess the state results in the output of symbol "0". Complementarity manifests itself in the following way: one has to make choices between the input of symbols "1", "2", "3"; corresponding to definite answers whether the automaton was in state "1", "2" or "3" (output "1") or nd; i.e., "2 or 3", "1 or 3" or "1 or 2" (output "0"). In the latter case, i.e., whenever the automaton responds with a "0" (for failure), one has lost information about the automaton's initial state, since it will have made a transition into the state "1", "2" or "3" for sure.

Since any finite state automaton can be simulated by a universal computer, complementarity is a feature of sufficiently complex deterministic universes as well. To put it pointedly: if the physical universe is conceived as the product of a universal computation, then complementarity is an inevitable and necessary feature of its operational perception. It cannot be avoided. It may serve not only as a constructive re-interpretation of quantum complementarity, but as an indication of the principal impossibility of self-reproduction by self-inspection.

3. FORECAST

Let us, for the moment, disregard the impossibility to find all laws of nature and assume that such a complete enumeration is presented to us by some oracle. What can we make out from that?

Imagine statements of the form, *"fed with program x and input y my computer will output z"*, or *"at time t the system will be in state xyz"* or, *"on May 2nd of next year there will be sunshine in Vienna; a wind will blow from northwest at 5 km/hour"*. As a consequence of the *recursive unsolvability of the halting problem* [24] and related speed-up theorems, such statements are undecidable. Indeed, there exist uncomputable observables even for computable systems whose "laws" and "input parameters" are completely determined. In particular, no effective computation can predict the behaviour of an arbitrary computable system in any "reasonable" (i.e., computable) time. Stated pointedly, in general there does not exist any "computational shortcut", no optimization with respect to time which would allow a forecast of the "distant future". – A "speedup" of a calculation is generally impossible.

This blocking of speedup theorems interpretable as forecasts applies even to observers which are outside of the system. It becomes even more dramatic when rephrased in terms of self-referential prediction. The following argument resembles Zeno's paradox of "Achilles and the Tortoise" [25]. K. Popper has given a similar account [26], based on what he calls *"paradox of Tristram Shandy"*. Think of the attempt of a finitely describable "intelligence" or computing agent to understand itself completely. It might first try to describe itself by printing its initial description. (It has been argued above that there is nothing wrong with this attempt *per se*, and that there indeed exist automata which contain the "blueprint" of themselves). But then it has to describe itself printing its initial description. Then it has to describe itself printing its printing its initial description. then it has to describe itself printing its printing its printing its initial description ... *ad infinitum*. Any reflection about itself "steers" the computing agent into a never-ending vicious circle. In a sense, "in the limit of an infinity of such circles", the agent has completed the task of complete self-comprehension. Yet for any finite time, this cannot be achieved.

4. FOR CONSISTENCY, INFORMATION IS HIDDEN FROM INTRINSIC OBSERVERS

I cannot go too deeply here into the "mind-boggling" features of quantum mechanics. But there is a peculiarity which seems rather interesting: quantum mechanics allows for nonlocal correlations without violating causality. That is, two spatially separated events which correspond to an entangled [27] state can show the tendency of coordination and adjustment which cannot be explained (locally) classically [28, 29, 30]. On a purely phenomenological level, this means that if one event happens, then also the other event (e.g., a click in the particle counter) happens, but there is hardly any classical way of explaining these coincidences (despite nonlocal hidden parameter models) [31, 32, 33].

Another related, truly mind-boggling feature of quantum mechanics was pointedly stated by Wheeler and is well known as the *delayed choice experiment* [34]. There the future has, expressed in classical terms, a qualifying influence over the past. All this can be well phrased in terms of the standard Copenhagen interpretation of quantum mechanics. And yet it seems that indeed it is not unreasonable to assume that, in view of quantum mechanics, the structural organization of space-time events has to be reorganized; in addition to the revisions made necessary by relativity theory.

Can we utilize these nonlocal coincidences and perform faster-than-light signalling? The answer to this question is unknown. It is true that, since the single events which coincide occur unpredictably and cannot be systematically controlled by any action at a distance, no EPR-type faster-than-light telegraph can be built. Shimony's thesis of *peaceful coexistence* [35] between quantum mechanics and relativity theory relates to that quantum feature and states that quantum mechanics violates *outcome independence* - by coincidences of arbitrary spatially separated events - but it conforms to *parameter independence* - by the impossibility to control spatially separated events through some parameter of a measurement apparatus.

Interpreted rigidly, Shimony's principle of peaceful coexistence, however, is not the weakest statement which would catch the overall consistency of phenomena. We may require that even *faster-than-light signalling via nonlocal quantum correlations is allowed, provided it is impossible to construct any time paradoxes or other inconsistencies.*

I shall give as an example a "high-tech" version of Nick Herbert's FLASH-device (cf. [37] and subsequent refutations [38]). The device would be capable of circumventing the Heisenberg uncertainty principle, i.e., complementarity, by the introduction of a "super-observer" (equivalent to a dualistic model). Let us assume that we want to measure the spin of a single electron in directions x and $y \neq x$ *at the same time.* (This is impossible due to complementarity). In purporting to do so, we proceed by (i) measuring the electron spin in x-direction, then (ii) copy and (iii) keep a record of this measurement result, then (iv) reconstruct the wave function, then (v) measure the spin in y-direction. From all the steps taken, only step (iii) - keeping a record of the "first measurement" of the electron spin in the x-direction - is forbidden. Because in order to be able to reconstruct the wave function for the "second" measurement in the y-direction, quantum mechanics requires that we have to eliminate *all* physical information about the measurement result in the x-direction. Thereby, reversibility and one-to-one computation is enforced [39, 40]. Keeping a record of the "first measurement" of the electron spin in the x-direction would clearly contradict this requirement. One possible way of circumventing complementarity and thereby of escaping the "Cartesian prison" of the physical world would be a dualistic, more generally, hierarchical, model, in which physical information is transferred into a sphere - spiritual or otherwise accessible to a "super-observer" - in which the copying and recording of this information has no physical correspondent [41]. In this model, the state (wavefunction) of the electron is reconstructed in purely physical terms, but the "super-observer" retains a "secret, obscure" knowledge of the "first measurement". After the "second" measurement of the electron spin in the y-direction, the "super-observer" can claim knowledge of the electron spin in the x-direction; but no physical measurement exists to test this claim.

Despite the merely speculative character of the above scenario, it should not be possible to produce any paradoxical situation by, for instance, faster-than-light back-to-the-future signalling (one may still allow non-paradoxical tasks). Indeed, in order to preserve consistency, information may be hidden from intrinsic observers.

5. NEW PHENOMENA AND OPEN WORLDVIEW

Is there a moral to be learned from the various forms of uncertainties to which an intrisic observer is exposed even in a system which is effectively computable on a step-by-step basis?

I believe, yes. The moral is to maintain a humble open-mindedness with regarde to new ideas, very much in the sense of Lakatos [42]. What will turn out as progressive or degenerative research program, no contemporary peer knows.

We do not know how "deep" The Great Beyond is. Historic examples like electricity show that the Beyond of the past is today's technology. And it is very likely that there is plenty "undiscover'd country" waiting for us to be discovered, from whose bourn the traveller will hopefully return wiser and will not lose the name of action.

Prof. Karl Svozil
Institut für Theoretische Physik
University of Vienna
Wiedner Hauptstrasse 8-10
A-1040 Wien, Austria

REFERENCES

[1] In contradistinction, contemporary physical theories are expressed in terms of continua: time, position, momentum, wave amplitudes, ... The very notion of continuum embodies indeterminism insofar as "almost all" (i.e., with probability 1), elements of continua are (Martin-Löf/Solovay/Chaitin) random. Physical chaos, if it exists, is the necessary consequence of this fact. In these models, indeterminism is "put-in" from the very beginning. There is no reasonable machine representation and no conceivable "explanation" corresponding to such models. They (together with classical, non-constructive, mathematics) are irrational at heart, and, while interesting in many respects, we shall not dwell deeper on this subject.

[2]C. Calude, Information and Randomness - An Algorithmic Perspective (Springer, Berlin, 1994).

[4] In this context, the terms "shallow" and "deep" refer to algorithmic information [2] rather than to Bennett's notion of "logical depth", cf. Ch. H. Bennett, "Logical Depth and Physical Complexity", in The Universal Turing Machine. A Half-Century Survey, ed. by R. Herken (Kammerer & Unverzagt, Hamburg, 1988). The apparent "paradox" that a complex phenotype originates from low-complex initial values and evolution is not paradoxical at all. Indeed, that the world appears complex by all means does not necessarily mean that its laws have a high algorithmic information content.

[5] T. Toffoli, "The role of the observer in uniform systems", in Applied General Systems Research, ed. by G. Klir (Plenum Press, New York, London, 1978).

[6] O.E. Rössler, "Endophysics", in Real Brains, Artificial Minds, ed. by J. L. Casti and A. Karlquist (North-Holland, New York, 1987), p. 25; Endophysics, Die Welt des inneren Beobachters, ed. by P. Weibel (Merwe Verlag, Berlin, 1992).

[7] K. Svozil, "Il Nuovo Cimento" 96B, 127 (1986); Randomness and Undecidability in Physics (World Scientific, Singapore, 1993); Quantum Logic, (Springer, Singapore, 1998).

[8] Archimedes (≈ 287-212 b.c.) encountered the mechanical problem "to move a given weight by a given force". According to Plutarch "Marcellus, 'that he declared ... that any given weight could be moved by any given force (however small)' and boasted that, 'if he were given a place to stand on, he could move the earth' [cited from T. Heath, A History of Greek Mathematics, Volume II' (Clarendon Press, Oxford, 1921), p. 18].

[9] J. von Neumann, Theory of Self-Reproducing Automata, ed. by A. W. Burks (University of Illinois Press, Urbana, 1966).

[10] H. Rogers, Theory of Recursive Functions and Effective Computability (MacGraw-Hill, New York 1967).

[11] P. Odifreddi, Classical Recursion Theory (North-Holland, Amsterdam, 1989).

[12] K. Gödel, 'Monatshefte für Mathematik und Physik 38', 173 (1931); English translation in [13] and in Davis, ref. [14].

[13] K. Gödel, Collected Works, Volume I, Publications 1929-1936, ed. by S. Feferman, J. W. Dawson, Jr., St. C. Kleene, G.H. Moore, R.M. Solovay, J. van Heijenoort (Oxord University Press, Oxford, 1986).

[14] M. Davis, The Undecidable (Raven Press, New York, 1965).

[15] Here "H(p)" is defined [2, 3, 16] as the length of the smallest program p* (in prefix code) which runs on a universal (Chaitin) computer and outputs p.

[16] M. Li and P.M.B. Vitányi, "Kolmogorov Complexity and its Applicatinos', in Handbook of Theoretical Computer Sciences, Algorithms and Complexity, Volume A (Elsevier, Amsterdam and MIT Press, Cambridge, MA., 1990).

[17] E.M. Gold, "Information and Control" 10, 447 (1967).

[18] D. Angluin and C.H. Smith, "Computing Surveys' 15, 237 (1983).

[19] M. Li and P.M.B. Vitányi, "Journal of Computer and System Science" 44, 343 (1992).

[20] L.M. Adleman and M. Blum, "Journal of Symbolic Logic" 56, 891 (1991).

[21] J.E. Hopcroft and J.D. Ullman, Introduction to Automata Theory, Languages, and Computation (Addison-Wesley, Reading, MA, 1979).

[22] W. Brauer, Automatentheorie (Teubner, Stuttgart, 1984).

[23] E.F. Moore, "Gedanken-Experiments on Sequential Machines', in "Automata Studies", ed. by C.E. Shannon & J. McCarthy (Princeton University Press, Princeton, 1956).

[24] A.M. Turing, "Proc. London Math. Soc." (2), 42, 230 (1936-7), reprinted in [14].

[25] H.D.P. Lee, Zeno of Elea (Cambridge University Press, Cambridge, 1936; reprinted by Adolf M. Hakkert, Amsterdam, 1967).

[26] K.R. Popper, "The British Journal for the Philosophy of Science" 1, 117, 173 (1950).

[27] E. Schrödinger, "Naturwissenschaften" 23, 807; 823; 844 (1935) [English translation in J.A. Wheeler and W.H. Zurek, eds., Quantum Theory and Measurement (Princeton University Press, Princeton, 1983), p. 152-167].

[28] A. Einstein, B. Podolsky and N. Rosen, "Phys. Rev." 47, 777 (1935).

[29] J.S. Bell, Speakable and Unspeakable in Quantum Mechanics (Cambridge University Press, Cambridge, 1987).

[30] D.M. Greenberger, M. Horne and A. Zeilinger, in Bell's Theorem, Quantum Theory, and Conceptions of the Universe, ed. by M. Kafatos (Kluwer, Dordrecht, 1989); D.M. Greenberger, M.A. Horne, A. Shimony and A. Zeilinger, "Am J. Phys." 58, 1131 (1990).

[31] A. Peres, "Am. J. Phys". 46, 745 (1978).

[32] A. Peres, Quantum Theory: Concepts & Methods (Kluwer Academic Publishers, Dordrecht, 1993).

[33] G. Krenn and K. Svozil, "Stronger-than-quantum correlations" (TUVienna preprint, 1995).

[34] J.A. Wheeler, "Law Without law", in J.A. Wheeler and W.H. Zurek, eds., Quantum Theory and Measurement (Princeton University Press, Princeton, 1983), p. 182-213.

[35] A. Shimony, "Controllable and uncontrollable non-locality", in "Proc. Int. Symp. Foundations of Quantum Mechanics", ed. by S. Kamefuchi et al. (Physical Society of Japan, Tokyo, 1984), reprinted in [36], p. 130; A. Shimony, "Events and Processes in the Quantum World", in Quantum Concepts in Space and Time, ed. by R. Penrose and C.I. Isham (Clarendon Press, Oxford, 1986), reprinted in [36], p. 140.

[36] A. Shimony, Search for a Naturalistic World View, Volume II (Cambridge University Press, Cambridge, 1993).

[37] N. Herbert, "Foundation of Physics" 12, 1171 (1982).

[38] W.K. Wooters and W.H. Zurek, "Nature" 299, 802 (1982); L. Mandel, "Nature" 304 188 (1983); P.W. Milonni and M.L. Hardies, "Phys. Lett." 92A, 321 (1982); R.J. Glauber, "Amplifiers, Attenuators and the

Quantum Theory of Measurement", in Frontiers in Quantum Optics', ed. by E.R. Pikes and S. Sarkar (Adam Hilger, Bristol 1986); C.M. Caves, Phys. Rev. D 26, 1817 (1982).

[39] R. Landauer, "Irreversibility and Heat Generation in the Computing Process", "IBM J. Res. Dev." 5, 183-191 (1961); reprinted in: Maxwell's Demon, ed. by H.S. Leff and A.F. Rex (Princeton University Press, 1990), pp. 188-196; R. Landauer, "Wanted: a physically possible theory of physics, in IEEE Spectrum" 4, 105-109 (1967); R. Landauer, "Fundamental Physical Limitations of the Computational Process; an Informal Commentary", in "Cybernetics Machine Group Newsheet" 1/1/87; R. Landauer, 'Computation, Measurement, Communication and Energy Dissipation", in Selected Topics in Signal Processing, ed. by S. Haykin (prentice Hall, Englewood Cliffs, NJ, 1989), p. 18.

[40] C.H. Bennett, "Logical Reversibility of Computation", "IBM J. Res. Dev." 17, 525-532 (1973); reprinted in: Maxwell's Demon, ed. by H.S. Leff and A.F. Rex (Princeton University Press, 1990), pp. 197-204.

[41] One may think of a virtual reality installation corresponding to the physical universe, and the player who is connected to it via a passive interface corresponding to the "super-observer".

[42] I. Lakatos, The Methodology of Scientific Research Programs (Cambridge University Press, Cambridge, 1978).

PART III

METHODS OF FORMAL ANALYSIS IN THE COGNITIVE SCIENCES

MOSHE KOPPEL AND HENRI ATLAN

SELF-ORGANIZATION AND COMPUTABILITY

In this paper we will use some ideas from the theory of computable functions in order to clarify certain imperfectly understood aspects of the theory of self-organizing systems [von Foerster 1960, Nicolis and Prigogine 1977, Eigen and Schuster 1979, Kauffman 1984].

Roughly speaking, a system is said to be "self-organizing" if:

(a) beginning from some random initial state it evolves towards some "organized" state, and

(b) the organization which ultimately evolves is not programmed but rather "emerges spontaneously".

Two examples of systems which are often cited as typical self-organizing systems are:

i. a cellular automaton in which a random array of 0's and 1's evolves into an organized array on the basis of simple deterministc local-interaction rules.

ii. a neural net with randomly-weighted edges on which backpropagation is used in order to evolve a weighted graph which can be used to distinguish elements of a group from non-elements.

Neural nets differ from the simple case of cellular automata, in that the dynamics are driven by a small set of randomly-selected elements of the set and another small set of randomly-selected non-elements of the set. Nevertheless, such nets are considered self-organizing, albeit in a weak sense [Atlan, 1992], since neither the actual ultimate edge weights nor the definition of the set to be identified are used to define the dynamics.

In attempting to define self-organization more precisely we must first define precisely what we mean by the term "organization". Moreover, we need to restate condition (b) above so that it does not depend on the contingent issue of whether the system was in fact programmed, but rather on the principled issue of whether the system could in fact be programmed. We will use computability theory to overcome these difficulties and thus to define self-organization within the context of that theory in as precise a fashion as possible.

In the next section we will use ideas from the theory of computation in order to give a formal definition of that quality which has been called "organization". We will then define a "true" self-organizing system as one which evolves more and more organization *ad infinitum.* We will show that any such system automatically satisfies the condition that the organization which evolves was not programmed because such a system is by its very nature not programmable.

While this result is satisfying in the sense that it captures our intuition for self-organization, it is also limiting in the sense that it means that true self-organization cannot be simulated using computational methods. In fact, according to the "physical Church-Turing thesis" that all natural phenomena are computable, true self-organizing systems could not exist in nature. Nevertheless, we will give a general method for simulating "bounded" self-orga-

A. Carsetti (ed.), Functional Models of Cognition, 201-209.

nizing systems. Such systems are based on a sub-routine which recognizes organization and thus are capable of evolving increasing organization by continually choosing the most organized array from among a randomly-generated selection.

PROGRAM-LENGTH COMPLEXITY AND ORGANIZATION

We wish to formally define the somewhat intuitive notion of "organization". Roughly speaking, we want to quantify the amount of planning which must have gone into the generation of a given string. For example, our measure should be greater for a string which is 1 in the ith spot precisely if i is prime, than a string which is 1 in the ith spot precisely if i is even; it should be greater, in turn for this string than for a string consisting only of 1's. Moreover, it should be greater for all of these than for a completely patternless string.

Let's first consider a measure which is not quite right but which will be useful for finding the right measure. The *program-length complexity* of an objects is the length of the shortest description of that object (in some appropriate language) [Chaitin 1975]. Thus, objects which are completely characterized by some simple property have low program-length complexity. Program-length complexity correctly orders the first three examples above: "all ones" is simpler than "evens" which is simpler than "primes". Where program-length complexity fails to capture the intuitive idea of organization desribed above is the case of patternless strings: objects which have no characterizing properties, and can therefore be described only by enumeration, are maximally complex, although they are minimally organized.

We will use the definition of program-length complexity as a basis for defining a quality called "sophistication" which we believe is the correct formal analogue of organization [Koppel and Atlan 1991]. The definition of sophistication uses the distinction between two different parts of a description of an object. The first part of the description of an object consists of the set of that object's properties. This set of properties constitutes the object's structure. This structure might in fact be common to a whole class of string; thus we regard a given structure as defining a class of strings. The second part of an object's description consists of the specification of the object from among the class of objects defined by its structure.

For example, the string S_0 = 00001100111100... might be described by first noting that for all odd-even pairs, the 2nd bit is the same as the 1st bit (that is, the structure of the string consists of the doubling feature) and then specifying the sequence of pairs 0010110... . Thus, if we were to use this description to predict the continuation of the string we would say that the continuation of the string will consist of pairs but we would not be able to specify which pairs.

A random string has no structure and therefore its description consists only of the second part - its specification from among the class of all strings. Thus, given a long random string we would not be able to make any reasonable predictions about its continuation.

The sophistication of an object can then be defined as the size of that part of the object's minimal description which describes its structure. By this definition, both simple objects and random objects have low sophistication.

We can formalize the notions of complexity and sophistication by showing that the two parts of a description correspond to the familiar notions, program and data.

Consider the set of all programs which take binary strings as input and produce binary strings as output. If running the program P on input D produces output S then we write P(D) = S. Many programs enter infinite loops for particular inputs. P is *total* if P(D) is defined for all D. If P is a total program and P(D) = S, then we say that the pair (P,D) is a *description* of S.

For a program P, let | P | be the length of P. In general, for a binary string x, let | x | be the length of x.

We define the *complexity* of S as the length of the shortest description of S: H(S) = min { | P | + | D | such that P is total and P(D) = S }.

For example, consider again the string S_0 mentioned above, which is completely patternless (random) except that each bit is doubled. Then, as we have seen, the shortest description of S_0 is DOUBLE (0010110…), i.e. a program which doubles bits operating on the input (data) string 0010110… Thus, $H(S_0)$ is the length of this description, i.e. | DOUBLE | + | S_0 |/2. This is so because the patternlessness of the pair prevents further compression of S and conversely any lesser compression, say PRINT (00001100111100…), is a longer description than this one and therefore not minimal.

As noted above, the meaningful part of the complexity (that is, the structure) of S_0 is this doubling feature and the *sophistication* of S_0 (and, more, generally, of the structure class of S_0) is | DOUBLE |. Note that the particular input string used with DOUBLE contributes nothing to the structure of S_0; it determines only the specific string within the structure class.

More generally, let P be the shortest program which is part of a shortest description of a string S. Then | P | is called the *sophistication* of S. If the structure of a string is very simple then the program which captures it will be short and thus the string's sophistication will be low; if the structure of a string is very subtle then the program which captures it will be long and thus the string's sophistication will be high. As expected, if R is a patternless random string, its sophistication is negligible (in contrast to its complexity which is maximal for strings of its length). Specifically, since its shortest description is the string itself, i.e. PRINT(R), its sophistication is merely | PRINT |.

SOPHISTICATION OF INFINITE STRINGS

In order to apply these concepts to the theory of self-organizing systems, we will first have to extend the definition of sophistication to include infinite strings.

Let α be an infinite string and let α^n be its initial segment of length n. Then a program P is a *compression program* for the infinite string α if there exists some constant c, such that for every initial segment α^n of α, there is a data string D_n such that $P(D_n) \supseteq \alpha^n$ and $| P | + | D_n | \leq H(\alpha^n) + c$. That is, a program is a compression program for α if it is part of an almost minimal description of every initial segment of α.

The sophistication of an infinite string is the length of the smallest compression progam for α.

EXAMPLE

Consider the following example which illustrates how the minimal description of initial segments of an infinite string change as more bits of the string are revealed.

Let π be characteristic string of the primes and let π^n be the initial segment of π of length n. Consider the following programs each of which generates the primes given appropriate data:

P_1 prints the data;

P_2 prints 0 for the even bits (greater than 2) and prints the data for the remaing bit;

P_3 prints 0 for the ith bit if i is a multiple of 3 (greater than 3) and prints the data for the remaining bits;

P_4 prints 1 for the ith if and only if i is prime;

Observe that P_4 does not require data at all (except for the mandatory single bit) and it generates an infinite string, namely π.

Now for j = 1,2,3,4 let $| P_j | = K_j$ and assume that $K_1 \leq K_2 \leq K_3 \leq K_4$. Assume further that if P' is any other program which could produce segments of π then $| P' | > K_4$.

Then using the definition of minimal programs for *finite* strings we have that the minimal program for π^n is:

P_1	if $n \leq 2(K_2\text{-}K_1)$
P_2	if $2(K_2\text{-}K_1) < n \leq 2(K_4\text{-}K_2+1)$
P_4	if $2(K_4\text{-}K_2+1) < n$

Any minimal program for π will produce exactly one string, π, given sufficient data. The same, of course, holds for any recursive string. Thus, using minimal descriptions as a basis for predicting continuations of π^n as n grew, we would successively narrow the class of expected continuations to the ranges of P_1, then P_2, and finally P_4, the range of which consists only of π itself.

Let us consider the progression of minimal descriptions we have just seen. As some pattern persists while n gets large, that pattern is incorporated into the minimal program. The precise value of n at which a particular pattern is incorporated into the minimal program depends on the relative efficiency of the description versus the next best one. For example, we never replace P_2 with P_3, since P_3 is larger and less efficient than P_2. We replace P_2 with P_4 precisely when n is large enough that the extra bits of program in P_4 are compensated for by the reduction (compared to P_2) in the number of bits of data used. Since for all sufficiently large n, $P_4(0)$ is a minimal description of π^n (and hence for all n, $P_4(0)$ is within a constant of being a minimal description of π^n), P_4 is a compression program for π.

TRANSCENDENCE AND SELF-ORGANIZATION

The important fact to observe is that not every infinite string actually has a minimal description. It may be the case that there is no fixed program which is part of a minimal (or near-minimal) description for all initial segments of some string. More precisely, it may be the case that longer and longer initial segments of a string are more and more sophisticated. We can think of such strings as having infinite sophistication and we call them *transcendent* string.

It should be further noted that although the precise sophistication of a (finite or infinite) string dependes on how the length of a program is measured (i.e., it depends on the choice of programming language), this is not the case with regard to infinite sophistication. A string is transcendent in one language if and only if it is transcendent in every other language [Koppel and Atlan 1991].

Recall the example above in which the consideration of longer and longer initial segments of the characteristic string of the primes revealed more and more sophistication until the process converged to a correct description. Now, for transcendent strings the process simply fails to converge; it continues to reveal more and more sophistication ad infinitum.

We say that an infinite α, can be programmed if there is some program which, if run indefinitely, outputs α. Such a string is commonly known as "computable" or "recursive". In our terminology, an infinite string can be *programmed* if it has a compression program which requires only a finite data string for all initial segusments of α. More generally, we say that a string α can be *simulated* if it has a compression program; that is, α is simulated if there is some finite description of its structure class (even if the specification of the specific string α from among its class requires an infinite amount of data). Thus, by definition, a transcendent string cannot be programmed or even simulated.

We suggest that sophistication as defined above is the appropriate formalization of "organization". A *self-organizing system* would thus be a system which, beginning from low sophistication, either trivial structure or randomness, evolves higher sophistication. In particular, self-organizing systems which evolve higher and higher sophistication *ad infinitum* are called *true* self-organizing systems.

Let us now return to the two defining properties of self-organizing systems given in the introduction, in light of the more precise concepts we introduced here. A system is *truly self-organizing* if:

a) beginning from a state of low sophistication the system converges to a state of infinite sophistication, and

b) the state towards which the system converges cannot be simulated

We have:

i. replaced the polysemic notion of "organization" with the well-defined concept of "sophistication" which captures the idea of meaningful complexity,

ii. insisted that the increase in sophistication is ad infinitum, and

iii. replaced the empirical condition "is not programmed" with the analytic condition "cannot be simulated".

Most importantly, however, it is now the case that the two conditions can actually be conflated to one, since *they are equivalent.* Both conditions simply assert that the system evolves towards transcendence.

BOUNDED SELF-ORGANIZATION

We have already seen that "simulating true self-organization" is an oxymoron, for by definition no such simulation is possible. We will show, however, that there is a very nice algorithm for simulating *bounded* self-organization.

Let us consider first how we might attempt to simulate true self-organization and undestand why the attempts fails. Begin by arbitrarily choosing some "seed" string which we call S_1. Now consider all extension of S_1, which are sufficiently long that at least one of them has strictly greater sophistication than S_1. From among these extensions of S_1, choose any one which has greater sophistication than S_1 and call it S_2. Similarly, for each i, choose from among the set of sufficiently long extensions of S_i, any one which has greater sophistication than S_i and call it S_{i+1}. Then $\lim_{i\to\infty} S_i$ is a transcendent string and any embodiment of this algorithm is a true self-organizing system. It has been suggested that this type of self-organization lies at the core of intentional behavior [Atlan 1992].

Where does does this scheme fail? That is, why can there not exist a simulation of a true intentional self-organizing systems? The answer is that the above scheme incorrectly assumes that there is some algorithm for determining the sophistication of a given string. In fact, there is no such algorithm since any method of testing various potential descriptions of a string (for the purpose of finding a minimal description) must sometimes result in entrapment in an infinite loop.

This is, of course, an entirely negative result. But it suggests how our problem can be limited so that the failure of the above scheme can be overcome. We need only replace the noncomputable task of finding a description of a string which is truly minimal with the computable task of finding a description of a string which is minimal among all descriptions which use bounded computational resources. Then we can avoid infinite loops in testing various potential descriptions, since no candidate description need be tested beyond the given bound. This observation leads us to a natural bounded version of self-organization which captures its essential features while allowing for the theoretical possibility of simulation.

First, we give a definition of a bounded version of sophistication. Let g be some increasing function from positive integers to positive integers. If P is a total program and D is data such that P(D) = S and P(D) can be computed in less than g(| S |) steps, then we say that (P,D) is a *g-description* of S. The *g-complexity* of S (written: $H^g(S)$) is the length of a minimal g-description of S. The *g-sophistication* of S (written: $SOPH^g(S)$) is the length of the shortest program which is part of a minimal g-description of S. Similarly, P is a *g-compression program* for an infinite string α if there exists some constant c, such that for every initial segment α^n of α, there is a data string

D_n such that $P(D_n)$ generates α^n within g(n) steps and $|P| + |D_n| \leq H^g(\alpha^n) + c$. In short, all of these definitions are the same as the original definitions except that now we consider only descriptions which generate the string with sufficient efficiency; n bits of the string must be generated in no more than g(n) steps.

We can now use this fact to define a bounded version of self-organization which can be simulated but otherwise captures the intuition behind the notion of self-organization. Recall that a function f is said to be o(g) if $f(n)/g(n) \to 0$ as $n \to \infty$.

A system is *g-self-organizing* if:

a) beginning from a state of low g-sophistication the system converges to a state of infinite g-sophistication, and

b) the state towards which the system converges cannot be simulated in o(g) steps.

As in the case of true self-sophistication, the two conditions are in fact equivalent.

CONSTRUCTING A G-SELF-ORGANIZING SYSTEM

In this section we will show that for appropriate functions g, g-self-organizing systems exist and we can in principle construct them.

We begin by identifying some appropriate function g.

Let α be a transcendent string and suppose that there exists some sufficiently fast-growing computable function f such that for every n, $SOPH(\alpha^{f(n)}) > n$. Then it is easy to see that there is some computable function f' such that for any finite string S, there exists an extension of S, S', of length $|S| + f'(n)$, such that SOPH(S') > n. We use this function f' to recursively define the function g. Let S(1) be some string of length 1 and let g(1) = c where c is large enough that some description can generate S(1) in c steps. Now, given g(i) and S(i), let $SOPH^{g(i)}(S(i)) = r$ and let k be the smallest number such that for some extension E of S(i) such that $|E| < i + f'(r)$, we have $SOPH^k(E) > r$. Finally, let $S(i+1) = \ldots = S(|E|) = E$ and let $g(i+1) = \ldots = g(|E|) = k$.

We can now construct a g-self-organizing string in the following way:

Let S_1 be, say, the string 1. Compute the 1-sophistication of each extension of S_1, of appropriately bounded length [specifically, of length less than $1 + f'(SOPH^{g(1)}(1))$]. If for one of these extensions E, $SOPH^1(E)$ exceeds $SOPH^{g(1)}(1)$, then choose $S_2 = E$. Otherwise, compute the 2-sophistication of each extension and then the 3-sophistication of each extension and so on until for some integer k1, some extensions E is found such that $SOPH^{k1}(E)$exceeds $SOPH^{g(1)}(1)$. Then choose $S_2 = E$. Now for any S_i chosen in this way, choose S_{i+1} by computing successively the 1-sophistication, 2-sophistication … of all appropriately bounded extensions of S_i until for some integer ki, some extension is found such that its ki-sophistication exceeds $SOPH^{k(i-1)}(S_i)$.

Then for the infinite string $\alpha = \lim_{i \to \infty} S_i$, $SOPH^g(\alpha)$ is infinite and hence the process which generates it is g-self-organizing.

SELF-ORGANIZATION AND NATURAL SYSTEMS

We have defined a process as simulatable if it is composed of computable and random components and have defined a system as self-organizing if it is not simulatable. The weakest version of the "physical Church-Turing thesis" is the claim that all processes which occur in nature are, in principle, simulatable. That is, it is the denial of the existence of true self-organizing systems. Thus, to assert the existence of true self-organizing systems is to deny a version of reductionism which claims that higher-order phenomena are reducible to lower-order phenomena which can be simulated (although the problem of finding the reduction might be intractable). The very reducibility, even if intractable, implies that the higher-order phenomena are also simulatable and hence are not self-organizing.

We have seen, though, that intentionality assumes the unlimited ability to create organization. Thus, the subjective experience of intentionality seem to argue for the view that people are indeed self-organizing systems, i.e., it argues against the physical Church-Turing thesis. Roughly speaking, this observation can be, and has been, dealt with in one of several ways. Dualists [Popper and Eccles 1977] simply deny the relevance of the physical Church-Turing thesis to cognition. Some non-dualists [Penrose 1989] deny the thesis' validity with respect to brain function (on the basis of certain not yet understood quantum effects in the brain). But the prevalent view might be summed up in our terminology as follows: the subjective experience of intentionality is driven by *bounded* self-organization, the existence of which has been shown to be consistent with the physical Church-Turing thesis.

It is interesting to contemplate what a self-organizing system would look like to an observer. Imagine that we were to witness a sequence of more and more sophisticated states, or more simply, longer and longer segments of an infinitely sophisticated string. How would we project the continuation of the observed segments? We have argued elsewhere [Koppel and Atlan 1991] that the very nature of induction is the projection of the most parsimonious description of the observed segment. That is, induction assumes that the sophistication of the whole string is precisely that of the observed segment. But since any finite segment has finite sophistication, it follows that we never project an infinitely sophisticated continuation. In other words, by the very nature of induction, *no set of observations can lead to the conclusion that an observed systems is self-organizing.* Rather, at each stage such a system would appear to us as partly computable and partly random, with perhaps, some of the randomness of an earlier stage having reemerged as complexity. Thus, we might observe that, at least in the short run, the systems satisfies a "complexity from noise" principle (already recognized as a necessary, though not sufficient condition for self-organization in [Atlan 1972 & 1987]), but we would never conclude that the system was indeed self-organizing.

CONCLUSION

We have placed the notion of self-organization on a firm formal foundation by defining it in terms of concepts from the theory of computable functions. This definition incorporates all the features generally associated with self-organization but presents some theoretical difficulties. The existence of self-organizing systems as we define them violates the physical Church-Turing thesis. Moreover, even if such a system exists, it would never be recognized as self-organizing by an outside observer. In order to soften the force of these difficulties, we have suggested a bounded version of self-organization which can be simulated using standard computational tools and which, in most contexts, might serve as a reasonable stand-in for true self-organization.

Prof. Moshe Koppel
Department of Mathematics
Bar- Ilan University
Ramat Gan 52900 Israel

Prof. Henri Atlan
Hopital de l'Hôtel Dieu
Service de Biophysique
1. Place des Parvis Notre-Dame
75181 Paris, France

REFERENCES

H. Atlan, L'Organisation Biologique et la Theorie de l'Information, Paris: Herman, 1972.

H. Atlan, "Self-Creation of Meaning", "Physica Scripta", 36, (1987) pp. 563-576.

H. Atlan, "Self-Organizing Networks", "La Nuova Critica", 19-20 (1992) pp. 51-70.

G. Chaitin, "A Theory of Program-size Formally Identical to Infomation Theory", "Journal of the ACM", 22 (1975) pp. 329-340.

M. Eigen, & P. Schuster, The Hypercycle: A Principle of Natural Self-Organization, Heidelberg, New York, Springer-Verlag, 1979.

H. von Foerster, (1960) "On Self-organizing Systems and their Environments", in Self-Organizing Systems, Yovitz and Cameron, eds., Pergamon Press, 31-50.

S.A., Kauffman, "Emergent Properties in Random Complex Automata", "Physica", 10 D (1984): 145-156.

M. Koppel and H. Atlan, "An Almost Machine-Independent Theory of Program-Length Complexity, Sophistication and Induction", "Infomation Sciences" 56 (1991) pp. 23-33.

G. Nicolis & I. Prigogine, From Dissipative Structures to Order through Fluctuations. New York, J. Wiley & Sons., 1977

R. Penrose, The Emperor's New Mind, New York: Oxford University Press, 1989.

K. Popper and J. Eccles, The Self and its Brain, Berlin: Springer-Verlag, 1977.

GIUSEPPE LONGO

THE DIFFERENCE BETWEEN CLOCKS AND TURING MACHINES

CONTENTS:

1. CLOCKS AND LOGIC

During several centuries the prevailing metaphor for the human brain was the "clock" This metaphor became precise with Descartes. For Descartes, there is a mechanical body and a brain, a << statue d'automate>>, similar to the fantastic automatic devices of his time and ruled by cog-wheels, gears and pulleys. Separated from it, but ruling it, is the << res cogitans >>. The physical connecting point of this dualism was provided by the pineal gland; the philosophical compatibility was given by the fact that the res cogitans too is governed by rules, logical/deductive ones. << Non evident knowledge may be known with certainty, provided that it is deduced from true and known principles, by a continually and never interrupted movement of thought which has a clear intuition of each individual step>>. Knowledge is a sequence or << a chain >> where << the last ring ... is connected to the first>> [Descartes,1619-1664; rule III]. In view of their role in the mathematical work of Descartes, it may be sound to consider these remarks as the beginning of Proof Theory; mathematics is no longer (or not only) the revelation or inspection of an existing reality, where "proofs" (mostly informal and incomplete, often wrong) were only meant to display "truth", but it is based on the manipulation of algebraic symbols and stepwise deductions from evident knowledge. Descartes' Analytic Geometry, an algebraic approach to Geometry, brought the entire realm of mathematics under the control of formal deductions and algebraic computations, as distinct from geometrical observation. In Algebra and, thus, in Algebraic or Analytic Geometry, proofs are sequences of equations, formally manipulated, independently of their (geometric)

A. Carsetti (ed.), Functional Models of Cognition, 211-232.

meaning. Three centuries later, Proof Theory will be the rigorous (mathematical) description of these deductions.

Consider now that mathematics is, for Descartes, the paradigm, the highest level, of human thinking. This is why, in spite of Descartes' dualism, there is a great unity in his understanding of the mind, from a modern perspective: both the res cogitans and its physical support obey "formal" or "mechanical" rules. Their description is derived from the concrete experience of the clocks and automata of the time as well as from Descartes' early steps towards modern metamathematical reflections and mathematical rigor, largely related to the nature of Analytic Geometry.

In the rationalist tradition, the connection between mechanical thinking and physical brain went much further. In Diderot and D'Alembert, clocks are more than a metaphor for the brain (and mind): they provide a precise scientific reductionist project, in a monist perspective. The philosopher and the harpsichord have the same degree of materiality; the same mechanical rules govern human thinking and the musical instrument. The logical syllogism works like a machine, and the machine works like a syllogism [Buffat, 1982]. The encyclopedists' radical mechanicalism aims at a *complete* mechanical theory of the world; perfect, "encyclopedic" knowledge is possible by the understanding of all the rules of deduction and all clocklike material mechanisms (<< les phénomènes sont nécessaires >> in mathematics and in physics). Vaucanson's fantastic "Canard" which could eat, drink and walk and his "Music Players" were the first steps towards the concrete realization of this project.

The reduction was not proposed in a shallow way; these authors hinted at a mechanical understanding of human functions and mind in the full awareness of the limits of the "clocks" they knew. They essentially meant to underline the need for a reduction of the complexity of human mental activity to scientifically describable facts, while being aware that a lot more chemistry and physics had yet to be understood. Moreover, << les machines sensibles >> had to be based on a great psycho-somatic unity; for Diderot, there is no intelligence without joy and suffering, without sexuality and feelings [Diderot, 1973]. But yet they thought that, in the end, the mechanics of all of this could be somehow fully described by a material machine whose physical structure contained all elements of this reduction. The difference between brain and clocks was only quantitative, a pure matter of complexity, we would say today.

This philosophical perspective cannot be distinguished from the progress in mathematics in the 18th century and from the growth of mathematical rigor. Leibniz' Mathesis Universalis, or the representation of the world in mathematical terms, is in the background. The philosopher is a << machine à reflexions >> exactly because he can inspect the world by the rules of mathematics and these rules can be implemented in clocks. Shortly later, with Laplace, the highest point of this view is reached: the Universe is similar to a perfect immense "clock", ruled by Newton's mathematically describable laws.

Laplace's philosophy largely dominated the 19th century's science when Boole, Babbage and Power worked. Their influence is clear: the complete descriptions by mathematical laws may cover the world and, in particular, human reasoning. Boole's "Laws of Thought" provide the first algebraic counterpart of Descartes' still informal metamathematics; mathematical thinking is being mathematized. Thus, mathematical, indeed human, reasoning can be implemented in machines better than ever. Babbage and Power enriched clocks and Pascal's arithmetic machine by multiple possible choices and combinations in the pieces of hardware; besides arithmetic operations, their machines could perform differentiation by a technique that we could call today a "finite elements method". In Power's machines, some parts of the hardware were as changeable as holes in strips of paper... The connection between the progress in Logic and machines was still informal and depended more on a common cultural frame, as almost a century was needed before boolean logic could be implemented in electronic circuits; yet the growth of mathematical rigor is simultaneous to the increasing interest in (thinking) machines. The background was given by the major progress of mathematics in the 19th century. Thus, once more, the mechanics of thinking machines followed or grew together with the understanding of (meta)mathematics. A dramatic change will be induced by further advances in Logic at the turn of the century.

2. TURING MACHINES AND THE DISTINCTION BETWEEN HARDWARE AND SOFTWARE

Frege is commonly considered as the father of modern Mathematical Logic. By his work, and that of Dedekind, Peano and Hilbert, metamathematics was finally described essentially as we know it today. At the base of this mathematization of mathematical deduction, there is the key distinction between language and metalanguage. Mathematics, the object of study, is carried on in an object language which is rigorously investigated in a metalanguage with the newly specified tools of Proof Theory. The aim is similar to Laplace's, at a "higher" level: according to Laplace, the language of mathematics (the partial differential equations of Analysis) can completely describe and predict the Universe. Similarly, with the formal tools of metamathematics one should be able to completely describe and decide mathematics (Hilbert's conjectures of completeness, decidability and provable consistency for formalized Analysis).

The mathematics of metamathematics though has a peculiarity, stressed by Hilbert: it is finitistic, as it must use finitely represented languages and finite procedures, such as deductions. The point then was to make precise what this finitistic approach could mean and, in particular, the exact extent of finite procedures. In 1930-36, Gödel, Herbrand, Kleene, Church and Curry proposed a precise mathematical notion of "effective or algorithmic function" within a Hilbertian perspective, (indeed, several ones, all proved equivalent by Kleene and Turing in 1936). In a sense, they formally described, as an algorithmic procedure, Descartes's early hints towards the stepwise process of human deduction; as a result, they obtained a mathematically sound theory of computable functions. All their different theories were original axiomatic approaches to brand new and deep mathematics, based primarily on functions more than on sets, but still in the frame of familiar mathematical languages and tools (Herbrand-Gödel's recursive functions, Church's lambda-calculus, Curry's Combinatory Logic ...). In 1936, Turing had an even more revolutionary idea: the invention of an abstract, mathematical, machine (TM), only vaguely inspired by previous mechanical devices, but absolutely unrelated, a priori, to mathematical theories of functions, sets or whatever. The core of his idea was the (abstract) distinction between "hardware" and "software": on one side the unchangeable material basis (a "read/write" head, moving on a tape), on the other, the changeable sets of instructions (the programs, as finite lists of instructions: move right/left, write/erase a symbol on the tape...). One can notice here the same distinction, as in metamathematics, between metalanguage and object language, as well as Tarski's precise partition between syntax (the stepwise computations of a formal TM) and semantics (the class of computable functions, in extenso). The equivalence of the Turing computable functions with the mathematical approaches proposed by the others, listed above, is still now a surprising, deep idea and a technically difficult theorem (though very boring). As a result, a finite machine, though abstract, could compute infinitely many functions by the invention of "software". Moreover, each of these formalisms possesses a "universal" function or machine; as a matter of fact, any finitely describable algorithm or program may be encoded by a number. Thus it may be fed as input to a universal algorithm or function that may use it on any other input number. In other words, a Universal TM, U say, when given two inputs, the coding of a TM, A, and a number p, may simulate the computation that A would carry on p. This was derived from Gödel's fundamental idea of encoding the metatheory in the theory (of numbers), a key step in the proof of Gödel's incompleteness theorem for first order Arithmetic. In conclusion, in Turing machines, the (mathematical fiction of) hardware, the "head and tape", is distinct from the software, the "programs", as well as from the inputs, possibly numbers; however, they all coincide by coding the entire machine description by numbers [Davis, 1958].

The conceptual distinction between "hardware" and "software", as well as Gödel's encoding, are at the basis of today's computers and computer programming. About ten years after Turing's mathematical description, early computers were designed by Turing himself and von Neumann (yet another great mathematician in this century, philosophically a formalist of Hilbert's school, who contributed to all main areas of Mathematical Logic and many of mathematics). Of course, the first thing to be said is that TM's, being abstract, are allowed to have an unbounded memory, the "tape", while actual computers are "finite state machines". However, this abstraction is one of the reasons of TM's up-to-dateness; any "a priori" limit set to memory in 1936, but even in ... 1986, would be ridiculous today. Most modern computers are still in the frame of TM's, conceptually; more precisely, they include an operating system and a compiler or an interpreter, i.e., a physical realization of a Universal TM. Their difference to clocks should be clear: it is due to the key distinction between hardware and software, as well as the auto-encoding possibility. In clocks (and Babbage machines) all feasible computations follow a predetermined algorithm, carved for ever in the material structure. The limited number of holes in Power's pieces of paper did not change essentially their nature as mechanical devices. Today's programming languages, by their formal, but "human-like" descriptions of the world, allow the simulation of all finitistically describable processes and realities on machines.

Or, at least it seemed so and still seems so to many that Turing and von Neumann immediately opened the way for the new metaphor of this century: the mind (indeed, the brain for a monist) as Turing Machine[1].

The "so called" Turing Test[2]. In his seminal 1950 paper, Turing proposes an "imitation game" in order to discuss the question <<can a machine think ?>>. Person C puts questions to a man A and a woman B, with the aim of discovering who is the woman. Then A is replaced by a machine. The claim is that, if rules fully govern human behaviour, we should be able to set up things in such a way that the machine may simulate A ([Turing, 1950; §.8]). However, is there a finite set of rules for predicting human behaviour? Turing understands that a positive answer could lead to views analogous to Laplace's. Yet he claims that the predictability he assumes <<is more effective>> than Laplace's, as Turing machines are discrete, while the Universe is not (§.5). And what about the brain? Turing acknowledges that neural machines need not be finite state machines. Yet in a simulation game the difference - finite/infinite - may be undetectable, similarly as a continuous differential analyser cannot be distinguished, by a *finite* number of numerical questions, from a digital computer (§.7). The point then boils down to the assumption that humans have a <<complete set of behavioral laws>>, i.e. a finite and fixed set of rules. Cautiously, Turing only concludes that there is no scientific argument against this assumption (§.8). More audaciously, by assuming that these rules may be actually given, there began the adventure of (strong) Artificial Intelligence.

3. SIMULATING TURING MACHINES BY CLOCKS

For a long time, very many great thinkers have viewed the mind as a clock; very many great thinkers still propose TM's as a paradigm for mental activities. But the time seems ripe for exploring a new reductionist project for rationality: the reduction of TM's to clocks. Or can we simulate a TM by a clock?

By such an investigation we may further elaborate on the distinction we hinted at between clocks and TM's; is this a purely quantitative (in the end, numerical) or a qualitative (conceptual) difference? In the first case, can we actually perform the tasks of a (universal) TM by using enough clocks? And what does the second alternative mean?

Of course, TM's are abstract mathematics (they have no limit on memory size), while clocks are concrete, finite machines. So let's be more realistic and try to simulate actual computers. It is not a potentially infinite tape that would make the difference for our intended comparison, since the mind has no infinite tape. Just take the largest, TM-like, real computer you need or may conceive. As a computer is a finite state machine, it may assume finitely many states, in a finitary language (i.e. based on a finite alphabet, possibly binary, made out of words of finite length). Thus, with enough clocks we may mimic all possible internal states (instantaneous descriptions, ID) of a large computer. It suffices to carve in steel all possible ID's and connect them along all possible paths, by wheels and pulleys, and... the reduction is performed.

In order to be more specific, consider a modern wonder, a Cray computer. Two years ago, the memory of a Cray was about 64 M words of 64 bits. This is inscribed in chips of about 2cm size, each containing up to 8 megabytes. Assume, optimistically, that the binary information in a bit may be encoded in a 1mm size binary mechanical device and then... compute. This is not the end though, as we have only discussed memory. Binary logic is implemented in chips containing up to 30 M ports. Moreover, the key point is that Crays are programmable. As they are finite state machines, though, Crays may compute only finitely many functions over a finite number of inputs, in contrast to TM's. Try now to mimic the previous memory and ports' structure, plus all possible, yet finitely many, computations as sequences of ID's, by a non-programmable mechanical device. Since the coding of a switch may require, at least, a 1mm square... then you will need more than the surface of our solar system.

There is a problem now. In Cray computers engineers have to optimize the internal design because of... the speed of light, or of electricity along the wires (about 1/3 of that of light or about 10 cm/nanosecond). The parallelism of these machines, due to a large number of concurrent and interacting processes, would be heavily affected by delays in communication. As a matter of fact, the interactions may be mostly sequentialized by interleaving; thus, it is very relevant that exchanges between processors and memory, at the frequency of 100 MHz (roughly 100 Mips or 10 ns), do not happen at a distance larger than a few centimeters. On the basis of Relativity Theory, it is easy to understand what it could mean if concurrent interacting computations are represented in different portions of the solar system....

Does this practical limit make a *qualitative* difference between clocks and Crays or TM's, their formal counterpart? I believe so. Yet let's put it in the following way. Assume that a "complexity issue" like this may still be considered only as a *quantitative* difference and that it does not affect our philosophical understanding of the formal reduction we have effected. In General Recursion Theory, for example, computability is unrelated to feasibility (see [Rogers,1967]), so that, by analogy, many may consider this reduction as "mathematically sound" and assume that the mechanicalism of rationalist reductionism may be equivalently based on clocks or TM's. However, even if the quantitative reduction were "in principle" possible (from todays' programmable finite automata to clocks), its practical unfeasibility because of the increase in complexity would be so high that a "qualitative jump" must be made in order to make it possible. In this case, the qualitative intellectual jump, is exactly what Turing proposed, by inventing programmable machines, based on a split between hardware and software and on the gödel-numbering of programs.

4. FINITE REPRESENTATIONS IN PHYSICS AND IN METAMATHEMATICS

In the previous section we tried to discuss how the "practical" issue of complexity in reductions and simulations may be so relevant as to blur the difference between "in principle / in practice". It makes no sense to say that, in principle, Crays (or the Macintosh I am writing on), may be reduced to clocks, since physical *principles* forbid it. However, the reader may still find a common basis for clocks and TM's: they were and are meant as paradigms for stepwise computations over finite representations. More precisely, given a few << true and known principles >> , for Descartes, or an initial state of affairs, ID, deductions or the system proceed by following a fixed and finite (decidable) set of rules. In particular, no further interference with the rest of the world is allowed. The other key assumption on which rationalist reductionism is based, is that any finitely representable process or piece of the world, including the brain, should be completely describable in either formalism, clocks or TM's. But what does "completely describable" mean? A way to understand it is given by the programs of Laplace and Hilbert. As already mentioned, they are strictly analogous: starting with a sufficiently accurate description of the initial conditions (a decidable set of axioms, for Hilbert) or a finitistically presented (decidable) set of partial differential equations (rules of inference) allows full predictability (complete, decidable knowledge), up to a sufficiently accurate approximation of the output (there is no need for approximation in Hilbert's finitistic metamathematics).

We know now that three physical bodies, only subject to gravitation, or a forced or double pendulum (also the top oscillates), escape Laplace's predictability because of the well-known sensitivity to initial or border conditions; there may be no continuous dependence of the evolution of the system on the initial conditions [Poincaré,1892]. And yet, three bodies or a double pendulum should be finitely representable: they occupy a finite portion of space, may assume finitely many states or positions (in view of quantum mechanics) and are subject to the simple laws of gravitation. Undecidability and incompleteness - see the Appendix - for systems including arithmetic (6 axiom schemata, instead of the initial conditions of three bodies or a double pendulum; three or four inference rules, instead of Newton's laws) are the corresponding negative results for the metamathematics of finitistic systems.

To summarize: in both cases, even closed "finite" systems are impredictable (undecidable, respectively), as the pendulum above or formal, first order Arithmetic. The common lesson is that a few and simple deterministic rules for finitary physical or logical structures may result in chaotic behaviors or very complex logical theories.

There is more, though, to be said about real mathematics. The existence of formally independent or undecidable sentences may be stated with no reference to meaning; it may be based on the construction of a formal sentence, provably underivable in a closed, finitistic logical system. However, the unprovability of a sentence wouldn't be so relevant if seen just as the purely formal independence from axioms of a "meaningless" proposition, as a property of the intended *structure*. For example, Gödel's diagonal formula in PA, the first order formal theory of *numbers*, is meaningless as a property of numbers [3]. It is *meaning* that steps in and gives the "essence" or significance of the negative results: what escapes the formalist, finitistic reduction is "truth" as provability in the broadest sense over the set of natural numbers, a very important structure in mathematics. This and other "structures" emerged in our natural and historical endeavor towards knowledge; they are determined and enriched by the unity of mathematics, as a growing and interconnected whole, where links and translations give understanding, suggest new methods of proof and give conceptual stability to invariants emerging from our relation to the world (such as the sequence of numbers, the group of symmetries in space etc.). In other words, the incompleteness theorems tell us that no fixed finitistic formal system can single out from the rest of mathematics structures as rich and relevant as the ones mentioned above. Numbers sit at the core of mathematics and reflect its unity and richness; because of this, we are willing to accept proofs based on them obtained by an ever growing variety of tools: Gödel's non-constructive proof of the independent sentence, transfinite induction, analytic methods, impredicative II order definitions, mechanical "follies" such as the proof of the four colour theorem (see the Appendix for more on proofs of unprovable sentences). The effectiveness and meaning of these tools may be given by different or new areas of mathematics (see [Kreisel&Macyntire,1982] for yet another understanding of constructivity in Algebra).

In conclusion: the understanding of incompleteness that we are trying to spell out here is the following. We do have a precise notion of truth over the numbers (Tarski's, for example), as we can define what a Unicorn is, but we do not *have a hold* of true properties, nor of Unicorns. We can characterize, in abstracto, the true sentences, i.e., what it means for a sentence to be true, but we do not know it, individually, unless we prove it in some *acceptable* way. We understand what truth (or a Unicorn) is but what we dynamically get is "only" provability, in an open sense. In spite of the claim of Platonists, the more or less precise definition of truth for mathematical propositions or for Unicorns and the failures of rationalist reductionism do not imply the existence of an Objective World where mathematical truths and Unicorns live; they only encourage us to look for further proof methods.

Even though it is not yet clear how much relevant mathematics lies outsides finitistically provable fragments, the independence results hint that mathematics is an essentially open system of proofs, open to meaning, over an ongrowing variety of structures and history. (See the Appendix)

5. MATHEMATICS AS AN OPEN SYSTEM

An actual deduction in mathematics, although a posteriori finitely representable, is not a formal game of rules in a fake laboratory of meaningless symbols, but lives by the ongoing interaction with other areas of mathematics and other forms of knowledge, not determined a priori. It is guided by the unity of mathematics as a growing interactive system, as well as by a variety of cognitive experiences. Historical, actual deductions are open to semantic analogies, which may suggest directions of work and conjectures. The a posteriori reduction of this interactive process to static formal systems devoid of meaning may be informative (meta)mathematics, but it is similar to the reduction of TM's to clocks, or even worse than that. Consider any relevant result in mathematics: it is rarely conjectured in a precise formal frame and then proved within it. Each step of the proof may require a truly relevant conjecture, a new general lemma. This may originate from an unexpected (possibly wrong) semantic analogy; it may use technical ideas from unrelated areas which require a full change in the formal frame, or even language, used up to that point. The sensitivity of the result, which may also change in nature or extension along the way, to "bordering" meanings and to the dynamic process of proof cumulates in the complexity issues related to the size of fully spelled-out formal proofs and which, by itself, resemble the TM/clock translation. Mathematics, indeed each real mathematical proof, proceeds as an open system. The state of knowledge rarely is a predetermined set of axioms and rules: theories are built, axioms are modified and rules amended.

Consider, for example, a proof by induction. Induction is one of the clearest inference rules, in a formal approach. It looks like a fully mechanizable mathematical rule, as, at the inductive step, only a finite amount of cases need to be checked: once P(0) is proven, check whether P(0), P(1).... P(n) entail P(n+1), and then deduce P(n), for all n. However, every non trivial theorem whose proof may be given by induction is based on a non obvious "inductive load". That is, in order to prove P(n) for all n, one generally needs to find a stronger property Q, which implies P, as a way to derive Q(n+1) from Q(0), Q(1).... Q(n). The possibly infinite choice of such a Q is a major difficulty for automatic theorem proving when dealing with relevant theorems [Dowek, 1991]. The mathematician's discrimination and judgement depend on a very broad context, still to be explored.

In short, proofs grow and change by interaction with the rest of mathematics and of the world, via language, common sense and space descriptions. <<The [mathematical] proof changes the grammar of our language, it changes our concepts. It establishes new connections and creates the notion of these connections >> [Wittgenstein, 1956].

An investigation of this process, both as an historical one and as a local, one-proof related process, is still largely missing. The a posteriori, formal study of a proof or of an entire mathematical discipline is a different matter.

For example, given a new deep theorem of number theory or analysis, it is possible that there exists a constructive or predicative system that proves it. Constructive versions of number theoretic results often shed light on relevant results. In "reverse mathematics", some outstanding colleagues of ours (Simpson, Feferman and others) have been able to reconstruct "backwards", predicative, indeed "least" axiomatic frames for a few fundamental theorems (see [JSL,1988]). This is extremely informative because of the generalizations it allows, and it is often very hard. It may even "justify" or found some branches of mathematics for those who still feel that there is a problem about the "logical foundation" or formal roots of mathematics - as if mathematical logic, a branch of mathematics, could found the tree of which it is a part (see [Wittgenstein, 1956]). The standard of rigor achieved today in definitions and proofs, also thanks to the work in logic, is *per se* a very robust "foundation" or basis. That approach says nothing, though, of the actual process of proof, the one that needs to be understood now or "found" by establishing its contextual dependencies from general knowledge and human experience.

The point is that each major result in mathematics, usually its very proof, led to brand new theories or even tools. Assumptions need to be added or formal frames to be invented. Lebesgues' measure theory required the full use of impredicative definitions, as early as 1902, when their status was still very obscure. Gödels' incompleteness theorem contributed to start recursion theory, originated the technique of Gödel-numbering and gave a first example of a provably non-constructive theorem. Gentzen established transfinite (eo-)induction in order to obtain a consistency proof for PA; cut-elimination also stemmed from his work as a new and fundamental technique. Through Girard's work it later gave rise to the non-constructive "candidates of reducibility" in a novel impredicative theory of proofs (II order lambda-calculus; see the Appendix).

Can all of this be generated in an initially fixed formal frame? In [Chaitin,1995], a very interesting information theoretic analysis of incompleteness is summarized: finite sets of axioms may not contain enough information to derive a sufficiently complex theorem. A theorem then may be independent from a given theory, simply because it essentially contains more or extra information. Chaitin's work may show that randomness, as he claims, or the sensitivity to contexts or border/initial conditions is part even of elementary branches of number theory.

In conclusion: the invention of new proof techniques, including proof techniques within broad contexts of knowledge, is part of the "theory building" and "tools building" so typical of mathematics.

It should be clear, though, that the formal analysis of proofs (current metamathematics) has had a major role: by examining results "in rigore mortis" or by the questions it raised, it gave us, in this century, a robust notion of mathematical rigor and many deep results (Lowenheim-Skolem, Gödel, Cohen's independence ... - by the way, mostly negative). Not to mention TM's and Crays, as we said above, since Computer Science originated from Logic. As a matter of fact, even the analysis of a frozen double pendulum, stopped two hours after it had started oscillating, may be of some use; by splines, one may interpolate as many intermediate states as quantum mechanics allows and obtain, a posteriori, a wonderful, computer assisted picture of the path in the bifurcation which lead it to move clockwise or counterclockwise.

This critique of the analysis of deduction as a closed system is not meant to claim the death of metamathematics or of mathematical representations; on the contrary, in today's Logic and Theory of Computation, we should try to mimic the revitalization that Poincaré's results, the KAM theorem or Ruelle's work brought to mathematical physics. Let's only hope not to be so slow: more than sixty years elapsed between Poincaré's three-body theorem and the rediscovery of the issue by the Russian school; probably a cultural blindness, due to the prevailing scientific paradigms. Indeed, it is a shame that as a student, in late sixties/early seventies, I was only taught Laplace's "rational mechanics", with not even a hint of what had happened ever since. This experience, common to many while students, is probably one of the cultural reasons for the still prevailing formalist and reductionist schools in Logic. I thus appreciated the apologies presented to the scientific community by Sir James Lightill, then chairman of the International Association for Mechanics : <<Here I have to pause and speak once again on the behalf of the broad global fraternity of practitioners of mechanics. We are deeply conscious today that the enthusiasm of the forebears for the marvellous achievements of Newtonian mechanics led them to make generalizations in this area of predictability which, indeed, we may have generally tended to believe before 1960, but which we now recognize to be false. We collectively whish to apologize for having mislead the general educated public

by spreading ideas about the determinism of systems satisfying satisfying Newton's laws of motion that, after 1960, were to be proved incorrect.>> [Lightill,1986] [4]

The point, of course, is not to transfer the tools of modern mechanics to logic or theoretical computing, but, while emerging from the Laplace/Hilbert cultural frame, to invent our own tools. This is being done, on a large scale, in the mathematics of computing, yet we need to understand better the dramatic change in perspective. For us, the logicians, working in the theory of computations, things have been simplified: we are being challenged by the new concrete achievements in computer design – but, so far, computers are simpler that Nature. The difficulty is that this requires an inversion of the historical paradigm which so far went from Logic to computing.

6. INTERACTIVE, ASYNCHRONOUS AND DISTRIBUTED COMPUTING

In the previous sections, we have tried to hint that the programs of Descartes, Laplace and Hilbert, namely, the paradigms of rationalist reductionism and of modern science, share a common aim: the understanding of the world by closed systems of axioms and rules. A few negative, very surprising, results set the internal limits of this approach. In Mechanics, Logic and Computing, these results stress that there is no relevant closed and complete system. In this section we argue that even the current artificial systems for computing are not completely represented by rationalist reductionism.

As a matter of fact, in the last twenty years or so, computer design started to go beyond the control of Proof Theory the which had originated it. By the progress of solid state physics and electronics, it became easy and cheap to have lots of processes going on at the same time and to let them interact with the rest of the world. This change dramatically accelarated in the last few years.

Interaction and on-line computations. Large computing systems, like airline reservations or banking systems, are based on a real time interaction with the environment. New information is inserted by autonomous agents; even rules may be modified by external agents or guarded commands, whose guards may depend on temporal as well as state sensitive firing conditions [Wegner, 1995(2)]. It is easy to prove that an "interactive Turing Machine", a machine which can read external inputs in real time, is strictly more expressive than an ordinary TM [Shœning,1995; Wegner,1995(1)]; an interactive machine is at least as powerful as an oracle TM (a machine that may ask during the computation whether a number belongs to an arbitrary set of numbers [Davis,1958; Rogers,1967]; for more on complexity of interactive systems, see [Goldwasser&al.,1985]). Further dynamicity is introduced by interactive interpreters. In LISP, for example (as in Universal TM's), the encoding of programs as data (the Gödel-numbering), allows the compilation of a program in LISP itself; or programs in LISP are "evaluated" (are given an operational meaning) within the language. More recently, in some object-oriented languages, this reflective property of interpreters has been extended by interaction. The interactive interpreters contain the specification not only of the system (axioms and rules), but also of an interactive device so that they can be modified by the environment. Engineers are (unconsciously?) approaching Wittgenstein's or Goodman's open understanding of rules.

Parallelism and concurrency. In modern computers, many processes are carried out in parallel. In some cases, these processes may be serialized, and their mathematical descriptions may take us back to the classical TM's. Often though, massively parallel computing relies on concurrent processes, whose output depends on the occurrence of many events in a precise time lap; only the simultaneous occurrence, for example, of a certain number of events may cause a given effect[5].

Distributed systems. Computing systems may be distributed in space. Namely, independent computing devices may concur in a way that interaction among components is more expensive than internal communication. Distributed systems are usually organized in layers: a physical layer, the data links, networks, sessions, presentation and application layers [Wegner,1995(2)]. Each layer has its own conceptual representation problem; moreover, they all affect each other while requiring different representations.

Asynchrony. One of the main problems with distributed computing is that of "time". Truly independent processors do not share the same clock. In a distributed system, possibly a large net of workstations, there is no notion of global or absolute time. Each processor has its own clock. There is no need to refer to relativistic effects to understand that their perfect synchronization is impossible; when individual operations are carried on in nanoseconds, slight differences in the measure of time heavily affect the ongoing computation. Not only in Einstein universes, but even in these artificial monsters we have created and use everyday, newtonian time has lost its meaning. In the best cases, one may define an "average time" by non-obvious techniques from meteorology.

A wide range of machines contains one or more of the previous facilities. A CM5, a modern Connection Machine, for example, may contain up to 1024 Sparc parallel processors (each with vectorial units) and reach 128 gigaflop (1 gigaflop = 1 billion operations per second). Clusters of workstations, with high speed network connections, may include many interacting computers, each open to human operators, who may modify data and programs, and each with its own clock. A large amount of practical and theoretical research is dedicated to the issues listed above. On the mathematical side, though, we are far away, for example, from the many deep results we have obtained in the triangular relation between Proof Theory, Category Theory and Type Theory. These key areas of mathematical logic are at the base of functional programming languages which are essentially sequential.

An analogy may help in understanding the new challenges we are facing in computing. Consider what is going on now in ... Paris. Millions of independent actions and events take place. Suddenly, M. Dupont is hit by a tile falling from a roof. The analysis of this event need not to be based on "chance". One may see the city as a distributed system of millions of parallel interacting asynchronous processes. The occurrence of a large number of them in a precise lap of time has caused the death of M. Dupont. Assume now that time is being reversed and that everything starts over again exactly as it began 24 hours earlier. Laplace would have imagined M. Dupont killed once more. By and by, Poincaré,

Arnold and Ruelle taught us that one may doubt the necessary death of M. Dupont: the sensitivity of “Paris” to the initial or border conditions (the throwing of dice the night before in a game by the carpenter on the roof; a message; a few electrons coming from outside Paris...). We may add to this the asynchrony of the ongoing processes. However, and also because of these reasons, time is irreversible and, up to now, no physical experiment could allow a double check on the fate of M. Dupont. Today we can. We can even decide to have or to avoid any (explicit) throwing of dice. In either case, it is a common experience of many computer scientists that, if a large distributed system of parallel interacting asynchronous processes is started over again with the same initial conditions, the same rules and programs, one generally obtains a different output. Asynchrony first may modify the result; but also (minor) interactions with the external world, on-line data and program changes and reflective interactive interpreters may act like resonance in gravitational systems and exponentially amplify indetectable variations of the initial or border conditions. It is worth noticing that Turing, in his 1950 paper, is concerned with the possible gap between formal and physical computability. The latter may be affected by the failure of Laplace’s predictibility for continuous systems, as <<one single electron ... could make the difference between a man being killed by an avalanche a year later, or escaping>>. As already mentioned, he believes that the difficulty does not show up, as it does in the case of <<differential analyzers>> (see above). Also von Neumann, while discussing his program towards an “electronic brain”, stresses the difficulties of formal descriptions, with reference to the complexities of measurement in physical theories, as well as the inadequacy of Turing Machines for distributive processes. As a matter of fact, a close analysis of the writings of these and other founding fathers, such as Descartes, Diderot, Laplace or Hilbert, often gives a much broader and critical view than that of their followers.

We are in a very stimulating moment for scientific knowledge: after sitting for centuries on the shoulders of giants, the fathers of modern science, and using the tools of rationalist reductionism, clearly understood today, but still difficult to handle, the very machines which are the topmost application of their developments force us to reconsider our methods. The grandchildren of clocks, the clusters of computers, are no longer representable by closed systems of axioms and rules. The key point is that in their design and performance there is at least as much emphasis nowadays on communicating, sharing, coordinating ... as on algorithms. (For more references, see [Conrad,1995] on “evolutionary” styles of programming; [Hasslacher,1995] for parallel and dynamic systems; [Makowsky,1995] for fascinating remarks on the role of “discrimination” in selecting data and rules).

In order to obtain linguistic or geometric representations of the world and of these physical devices, reductions will still be required. But a broader notion of rationality is needed, based on systems that, as a software engineer acknowledges, are opened to a <<world which is only partially known and progressively revealed by interaction or observation>> [Wegner,1995(2)]. This fruitful crisis became evident in mechanics first, but its somewhat different extension to everyday computing may more heavily affect Logic and common sense, and, thus, general scientific knowledge.

7. PLASTICITY AND THE WET BRAIN

We have observed above that there is a qualitative jump in moving from the early mechanical paradigms of closed systems, clocks, to the closed systems of Hilbert and Turing, the sequential computers. Moreover, the depth and rigor of their mathematics allowed us to prove internally the limits in expressive power of these very approaches. Asynchronous, concurrent distributed systems, open to interactions, seem to suggest a further qualitative jump, whose mathematics is not yet fully understood.

Nobody seems to doubt that our brain is a massively parallel, distributed, interactive device, even though a few still try to reduce it to TM's and claim that, "in principle", any finite piece of the world should be fully describable by symbolic manipulations. As CM's seem to be more expressive, can we now deduce that electronic connection machines are the new metaphor for the brain? Since further qualitative jumps may still be needed, any scientific mind should be suspicious of this nex metaphor. Let's list a few of these jumps.

Dynamic connections. Neurophysiologists tell us that connections in the brain change with time. Indeed, connections are stimulated by perception and psychological inputs; during the entire life span of an adult, not only in infants, stimuli affect changes in connections (a dramatic discovery, less than 10 years old). By comparing this fact to Turing's revolutionary distinction between hardware and software and the gödelization of programs, one may view it as a further qualitative jump that takes us away from actual computing where software does not concretely modify hardware, so far.

To be fair, this is exactly one of the features that our colleagues working on neural nets try to mimic. They do it mathematically, though, up to now. Namely, they do not have concrete machines whose hardware connections change according to software, but they can describe this process mathematically and simulate it numerically on large computers. Their mathematical merit is enormous, but it has a limit in complexity. Indeed, in CM's any node is potentially connected to any other node; thus, changing connections may surely be simulated. The point though is that the circulation of signals turns out as a major programming challenge, since traffic jams make this dynamic handling of computations extremely difficult. And this is so in computers whose complexity, as for the number of nodes, connections and operations per second is estimated to be between 1 ten-millionth to 1 millionth of the brain's complexity [Schwartz, 1989].

Informally, it is as if Turing, after inventing his mathematical abstraction, based on the distinction between hardware and software, hadn't been helped, 10 years later, by von Neumann and electronics and, instead, had been forced to simulate his TM's (or the 1950 Eniac computer) with clocks. At the scale of that early machine with tubes and valves, this simulation could still be done, perhaps, but the qualitative jump of his idea wouldn't have displayed all its revolutionary value. Real neural nets, whose physical connections change dynamically, are probably as far from Turing Machines as these are from clocks.

In view of the power of today's computers, the "equation", DynamicNets/TM's = TM's/clocks, may seem excessive. I believe, though, that we should not be lead astray by the relevant mathematics in abstract neural nets, nor by the power of TM's; we are probably only at the beginning of the adventure of true neural net machines[6].

Convection in fluids. The plasticity of the brain is also due to other fundamental phenomena, the complex blend, in vision, for example, between hologrammatic memorization of images and localization: certain brain lesions cause a general, but slight and not local, degradation of perception or image memory; others affect local reconstructions [Maffei&Meccacci,1979]. Moreover, the brain may often, though not always, substitute an area for another in carrying on a specific function; or it may use the same area for implementing a substitutive function. The latter case is studied, for example, in [Bach Y Rita,1994] where tactile experiences may replace vision in blind people, probably by using some areas of the cortex usually dedicated to vision. Finally, it is not underlined enough that a lot of communication in the brain goes by bio-chemical diffusion in liquids. Matters may be yet more complicated as the oscillations observed in exchanges of proteins between cells may be chaotic [Goldbeter,1995] - oscillatory interacting reactions may behave like a double pendulum. Pain, in particular, seems to be transmitted by diffusion via surrounding fluids. This clearly gives an essentially different form of transmission: slow, unreliable, but in all directions [Bach Y Rita, 1994].

More aspects of brain physiology are listed in [Schwartz,1989], each of which may represent a qualitative difference, as the one between clocks and TM's. All these features may be essential to brain functions; even if there seems to be redundancy in many individual functions, it is possible that the global functionality has no redundancy and that selection has been optimizing matters, in those regards. In this case, the only simulator of the brain and its functions would be an identical copy, embedded with its body into human history and society, the only way to acquire common sense and background knowledge, skills and practices. Yet the quest for knowledge is a very good reason to continue the investigation. Moreover, in the the search for a deeper understanding, also by formal/computational models of specific functions, we may happen to construct better and better machines, with indirect and fantastic fall-out. Man tried for long to simulate birds' flight. Human flight really began, stimulated by these attempts, when machines that parallel birds' flight were invented. We do not fly from branch to branch, today, but we can carry 50 tons at 1.000 km/h, a rather useful thing. Computers similarly parallel human reasoning. As for rigid wings, in computers (and human beings) rigid rules are not intelligence, but a substitute for it. The point now is to understand and scientifically describe the flexibility and multiplicity of intelligence, its "maze of background connections".

Both the analysis and the simulation of brain and its functions may require more mathematics, partly known, mostly yet to be invented. This may include to close, on a higher level, the gap between hardware and software which had been so useful in Turing's model, since the brain does not seem to implement it. As in the past, a change or a broader perspective in the philosophy of mathematics may possibly help.

8. IMPLICIT MATHEMATICS

We have briefly summarized the historical itinerary that brought us from the largely informal practice of Mathematics to the current clear understanding of axioms, rules and rigor. A crucial step was made when formal deductions were transformed into a new Bind of mathematics, the metamathematics of Frege and Hilbert; in other words, when the "explicit" part of mathematics was clearly defined and investigated. The final result of this itinerary has been such a clear display of mathematical rigor that logical deductions could be implemented in fantastic machines for playing chess, making flight reservations, changing our clerical work and everyday lives, except... for doing mathematics. Well, yes, there are several theorem provers; indeed, a few are even very effective, provided that they are not left alone, even in inductive proofs (see §.5 above). Only interactive programs achieve some results. Why is it that the linguistic, explicit rules of mathematics do not suffice to do mathematics? The formal, linguistic clarification has excluded, to say the least, the role of direct geometric manipulation, the open growth of mathematical practice, the overall unity of mathematical (and human) knowledge. In a word, it ignored what may be called "implicit" mathematics and its background knowledge.

In this century, many mathematicians tried to blend various forms of knowledge and historical or practical experiences into mathematics. It may suffice to recall the many remarks or writings by Poincaré, Hadamard, Polya, H. Weyl, Enriques.... Their perspective was little developed though, in view of the successes of the formalist approach, and because they did not fully succeed, except, in part, Poincare' or Weyl (see [Longo, 1989]), in turning their psychological remarks into a philosophy of mathematics.

The report of a personal experience has been often used in those reflections, and it may actually help, by the passion each researcher has for his own results. Let's immodestly dare to do so here. I recently collaborated on a theorem in my research area (by the way, a modern development of Church's formal language of functions [Longo et al, 1993] which see). Some observed that it is an elegant theorem. What does elegance mean? Is it a matter of "informal" symmetry, of similarities with properties of holomorphic functions which were totally unrelated? Or perhaps a clear intuitive understanding of computations on sets, the one-line statement, logically and "aesthetically" heavier on the left, lighter on the right... I cannot distinguish these judgements from the criteria by which we may appreciate a painting or a handicraft. The conjecture had been made possible by an improper reference to an informal model, a reference based on a contamination of languages where in an earlier version of the proof we had confused categories, sets and proofs. The "a posteriori" display of the "formal" steps we made is surely incomplete and their full TM formalization would be comparable to the... carving of steel clocks. Besides the (feasible?) a posteriori simulation, the *conjecture* is still missing. But also the proof technique and the frame language needed to be conjectured as well as the right lemmas to be singled out. Where do "implicit" judgements based on elegance, analogies with unrelated intellectual experiences stop (the musical/geometric role of a zig-zag we have in the proof has been clear to us) and formal matters begin? There is no magic here; or more exactly, we should stop investigating whether magic was applied or not applied; rather, we should investigate what is not captured by axiomatic methods and may require a broader analysis of mathematical knowledge. We should investigate its unity with other forms of knowledge or even, stress, with Diderot, that there is no knowledge without feelings, joy and sufferings.

Hardy insisted in several places on the purposeless esthethics of mathematics

[Hardy,1940]. His views, which are those endorsed in this paper, are still by no means widely accepted among mathematicians; they were meant to remove the mathematical work from rational investigation and practical aims. We should instead bring under the realm of scientific knowledge also the skills and differentiations from the practice of mathematics.

Some may say that we are confusing here the (formal, logical...) "justification" or "foundation" with the "context of discovery". Indeed, this is exactly what we need to reconcile, as that distinction is by now very well understood, in particular by the work in metamathematics which was meant to found the existing mathematics on certainty and rigor. The mistake instead is the continuing use of formal tools, invented to "justify" and set a standard of rigor, as tools for the analysis or even the progress of mathematical discovery (see, for example, [Michie,1995] on the Japanese V generation of computers or [Beeson, 1995] whose slogan is: "mathematics = logic+computation").

Since the founding fathers largely clarified for us explicit mathematics, we need now to go further: we need to reorganize knowledge and embed mathematics into the maze of back connections which relate it to perception, judgements and other forms of intelligence. This reorganization may encompass the understanding of a large varieties of individual, physical and historical experiences. In this century the way has been opened by H. Poincaré and H. Weyl, as already said, though from different perspectives. Further references and a few hints, concerning memory structuring, space and images and linguistic and geometric invariants, may be found in [Longo, 1992, 1995].

The difficulty is to turn scattered remarks into a foundation of mathematical *knowledge* (in contrast to the "internal" foundation of mathematics), by seeing mathematics as a discipline not transcending human experience but emerging from it. For example, we need to explore the cognitive process that makes a mathematician claim that he "sees" a many-dimensional differential variety (<<mathematics... whose main organ is "seeing">>, as Weyl suggests) and single out the mental invariants and images underlying this insight. Yet such an understanding should not be an isolated psychological remark but part of a philosophical proposal concerning knowledge.

To this aim, we should work at least in two directions. In short, we have, on one side, to use the rigor and generality of mathematics in order to design or describe the open systems of computations and life (mathematics as a tool); on the other, we should investigate actual mathematical reasoning as a broader paradigm: this has not yet been done (implicit mathematics as an object of study). In view of the relevance of the example set by mathematics for general reasoning and for the invention of metaphors and machines, this may help in computer design and cognition and their philosophy.

ACKNOWLEDGEMENTS

I would like to thank Peter Wegner for many stimulating discussions during the LICS'95 Conference in San Diego. Kim Bruce, Gilles Dowek, Jean Lassègue and Lorenzo Seno made several critical and insightful comments. Philippe Matherat and Roberto DiCosmo helped with comments and data on large computers.

APPENDIX: MORE ON THE PROOFS OF UNPROVABLE PROPOSITIONS

In a footnote we already hinted that the truth - over the standard model of first order Number Theory, PA - of Gödel's unprovable proposition, G, is proved by the very proof of its unprovability in PA. Since G says "this sentence is not provable in PA", then, if PA is consistent, i.e., non-contradictory, it cannot be proved in PA, and this proves that it is true. As this proof cannot be given in PA, it is non constructive, i.e., it cannot by derived by effective tools or by a stepwise construction or as a recursive function as these equivalent concepts are all expressible in PA (Gödel's key representation of the finitistic, effective metatheory in the theory). The second incompleteness theorem proves that the sentence "PA is consistent", also formalizable in PA, is unprovable as well. Indeed, the second theorem formalizes the previous proof of the truth of the improvable statement when assuming consistency, as it proves, in PA, that "PA is consistent" implies G.

We next discuss some other cases of proofs of constructively unprovable sentences. Let's first recall a common practice in mathematics concerning universal statements which are important both in everyday mathematics and in the theory of proofs. The discussion is slightly more technical.

Generic Proofs Given a mathematical theory which allows universal quantification, i.e., sentences such as $\forall x.P(x)$, how does one prove them? If the range of quantification is infinite, there is no way to explore and check each individual case. The practice of mathematics is usually the following: check what is the "intended range of quantification", i.e. over which set, domain or structure the universally quantified x is meant to be interpreted (the set of real numbers, for example). Then prove $P(a)$, where a is an arbitrary or *generic* element of that domain ("take a to be a real", typically), that is give a proof of $P(a)$ where *the only property of* a, *used in the proof*, is that a belongs to the intended domain. In Type Theory we would say: only the *type* of a is used in the proof. By this, the proof of $P(a)$ is a prototype or paradigm or pattern for the proof of $P(b)$, for any other b in that domain. Thus, one has a proof of $\forall x.P(x)$, i.e., a proof that $P(x)$ holds everywhere in the intended domain of interpretation of x.

Consider now the special case when the (intended) domain of interpretation is the set of natural numbers. How does this technique relate to induction? The implicit "regularity" of a generic proof - all proofs are given by the same reasons relative to a given formal frame - may not be present in the induction. Once we have proved $P(0)$, which may be ad hoc, one proves, in an inductive argument, the *implication* $P(n) \Rightarrow P(n+1)$. This is the proof that, sooner or later (in the case of nested induction), is generic in n, and it is based only on the *assumption* of $P(n)$, *not* on its proof (as for *derivable* vs *admissible* rules in logic) So, formally, $P(n)$ and $P(n+1)$ may be true for different reasons or their individual *proofs* may follow different patterns and yet, the proof of $P(n) \Rightarrow P(n+1)$ may be generic in n.

From generic to specific proofs. In the same period as the consistency of PA was shown, by Gödel, to be unprovable in PA, Gentzen gave a proof of consistency for PA. Clearly this proof cannot be given in PA itself. A key technique in Gentzen's approach is called "cut-elimination". The idea is the following. A proof of a proposition Q is *direct* when it contains no sentence "more complex" than Q, where more complex refers to the logical/syntactic structure of Q, e.g., the number of quantifiers. However, in mathematics and its theory of proofs, there are plenty of indirect proofs; typically, one may use a general lemma, e.g., a proof of $\forall x.P(x)$, in order to prove a special case, P(a), which may happen to be part of Q. As well known, these kinds of general lemmas are widely used, since they may be given by a generic proof, as above. Cuts are eliminated when one gets rid of these detours and directly proves P(a). By this, a proof becomes direct or is turned into a fully known and unique form, a "normal form". To show that one can eliminate all cuts is hard and not formalizable in PA. Indeed, "cut-elimination" implies, *within PA*, the consistency of PA, which is unprovable. This is the formal reason for the unprovability of "cut-elimination"; the actual reasons are spelled out in [Girard&al.1989] where Tait-Girard's technique for cut-elimination, called "the candidates of reducibility", is used. In short, given a set of formulae A in a theory T (e.g., Gödel's T), and a set Red(A) of candidates of reducibility relative to A, the crucial statement "t is in Red(A)" can't be proved uniformly for all t and A or by using the same formal paradigm for all of them, in the language of T. As a matter of fact, its very formalization in T has a different logical complexity (the quantifiers alternate differently) depending on A. Thus, in particular, no generic proof in t and A, in the sense above, applies, within T. This is why, even though each individual statement is provable in T, the intended universal quantification is not.

Paris-Harrington theorem and the type of a generic proof. A recent and interesting unprovable sentence of PA is given in [Paris&Harrington,1978]. In contrast to Gödel' diagonal sentence, the proposition is interesting as a "concrete" property of numbers, since it is a "minor" variant of a well-known and relevant property due to Ramsey (the finite Ramsey Theorem, see [Drake,1974]). The finite Ramsey Theorem is provable in PA, while Paris and Harrington showed that their sentence, call it $\forall x.PH(x)$, implies, within PA, the consistency of PA. There is an alternative way to prove the unprovability of $\forall x.PH(x)$, based on work on models of arithmetic by Paris and Kirby: one may give, for any non-standard model of PA that realizes $\forall x.PH(x)$, another non-standard one that does not realize it, see [Berestovoy&Longo,1981]. The point though is that $\forall x.PH(x)$ is true over the numbers, i.e., in the standard model. It is much easier to show the truth of $\forall x.PH(x)$ than its unprovability, but it is truth that interests us here, as we want to look at its non-constructive proof and at the role of generic proofs. We just mention the key steps.

For each number n one can show that $PA \vdash PH(n)$, i.e. that PA proves PH(n). The argument is "per absurdum": fix an arbitrary (generic) number n and assume that there is a model M which does not realize PH(n). Then, by using König's Lemma and an infinite version of the Ramsey Theorem, one gets a contradiction. Both the lemma and the theorem are non-constructive; the first says "in a finitely branching infinite tree there is an infinite branch". This seems obvious, but there is no way to obtain uniformly and effectively the infinite branch. The second is a more complex infinitary statement, see [Paris&Harrington, 1978; Berestovoy&Longo,1981]. In the end, though, we have, for all n, $PA \vdash PH(n)$. That this, per absurdum and via a non-constructive argument, one shows that there must exist, for each n, a constructive proof of PH(n), without providing it explicitly.

Note now that in the proof the number n is used generically. Why can't we go from it to a proof of $\forall$x.PH(x)? The point is that n is used in the proof as a natural number, i.e., the proof refers to its intended "type" as that of the standard model (its property of being a number is used both in the finite branching of König's tree and in the application of the Ramsey Theorem). But first order arithmetic, PA, has also non-standard models: thus, x in $\forall$x.PH(x) is meant to range over the domain of all possible models. In contrast, then, to the previous case (the candidates for reducibility), we do have a general paradigm or a generic proof, within PA, but this paradigm is restricted to a specific domain of interpretation of x. The unprovability of $\forall$x.PH(x) proves that this restriction, in the generic proof, is unavoidable. (Moreover, there is not even the possibility to express this restriction by a bounded quantification and prove $\forall$x.(N(x) $\Rightarrow$ PH(x)), as there is no way to define a formal type or predicate N of the natural numbers in PA.)

In conclusion, we briefly reviewed two relevant cases of true but unprovable theorems. There is no metaphysics involved, though, and no platonistic reality; their proofs simply escape the limited frame of finitism, as they cannot be given in PA (and this is provable), yet their truth is proved. In both cases, a typical mathematical technique is used (a technique not sufficientey discussed in logic), that of generic proofs. In either case, the generic argument cannot be extended to a proof of the universally quantified assertion in PA, but for different reasons: a lack of logical uniformity, as for "candidates of reducibility", the specific type of the generic variable n, as for the Paris-Harrington proposition. Clearly, both proofs can be fully formalized "a posteriori", in a non finitistic fashion: take (II order, impredicative) ZF or a similar formal set theory, extended exactly as to handle transfinite or ε_0-induction (i.e.; assuming the existence of a large enough cardinal), and that is all you need, once you know the proof.

Prof. Giuseppe Longo
Département de Mathématiques et d'Informatique
École Normale Supérieure
45 Rue d'Ulm
75230 Paris, France

NOTES

[1]See [Turing, 1950] and [Davis, 1995] for more on Turing, von Neumann, Logic and the brain; [McCorduck,1986] for a 30 years survey about this view.

[2]Turing makes little use of the word "test" and not in the usual sense of A.I., see [Lassègue,1994].

[3]Gödel shows that deductions, as constructive inferences, are representable in PA and encodes, as a formula of PA, the metatheoretic assertion "this sentence is not provable in PA"; thus, by the very proof of its unprovability in PA, i.e., by finitistic or constructive methods, he gives a non-constructive proof of its truth (see the Appendix).

[4]This remark about the little of attention to contemporary mathematical physics paid by logicians, who so long insisted on symbolic or functionalist reductions, is suggested by note 7, p.270 in [Putnam, 1989].

[5]The relevance of the issue and the amount of literature is enormous, see [Lynch et al.,1994] for a recent text.

[6]Berry, Vuillemin and many others currently work at the design of machines whose electronic connections

change with time and programming. They can only handle, so far, machines with a few thousands nodes, in contrast to the 10 billion neurons of our brain.

[7] See [Lakatos, 1976] for a description of the dynamic notion of rigor in mathematics; besides, it is amazing to browse in history and see how many wrong proofs were given in the past, by topmost mathematicians, from Fermat to Cauchy - often though these people had such a deep insight into mathematics that their theorems were true....

REFERENCES

Bach Y Rita P., Memosynaptic diffusion neurotransmission and late brain organization, 1995.

Buffat M., "Diderot, le corps de la machine", "Revue des Sciences Humaines", n. 186-187, 1982.

Beeson M. J., "Computerizing Mathematics: Logic and Computation" in [Herken,1995].

Berestovoy B., Longo G., "Il risultato di indipendenza di Paris-Harrington per l'Aritmetica di Peano", GNSAGA, Roma, 1981

Chaitin G. J.., "An Algebraic Equation for the Halting Problem" in [Herken,1995].

Conrad M., "The Price of Programmability" in [Herken,1995].

Davis M., Computability and Unsolvabilty, MacGraw-Hill, 1958

Davis M., "Mathematical Logic and the Origin of Modern Computing" and "Influences of Mathematical Logic in Computer Science" in [Herken,1995].

Descartes R., Regulae ad directionem ingenii, 1619 - 1664.

Diderot D., Le rêve de d'Alembert, Flammarion, 1973.

Dowek G., "Démonstration automatique dans le Calcul des Constructions", Thèse, Univ. Paris VII, 1991.

Drake F., Set Theory, North Holland, 1974

Enriquez F., Problemi della Scienza, Torino, 1909.

Enriquez F., Natura, Ragione e Storia, Antologia di scritti, Einaudi, 1958.

Girard J.Y., Lafont Y., Taylor P. Proofs and Types, Cambridge U.P., 1989.

Goldbeter A., "Non-linearity and chaos in biochemical reactions", lecture delivered at the Academia Europaea, Cracow, 1995.

Goldwasser S., Micali S., Rackoff C., "The knowledge complexity of interactive proof systems", "17th ACM Symp. on Theory of Computing", New York, 1985.

Goodman N., Fact, Fiction and Forecast, Harvard U.P., 1983.

Graubard S. (ed.), The Artificial Intelligence Debate, M.I.T. Press, 1989.

Hadamard J., The psychology of invention in the mathematical field, Princeton U.P., 1945.

Hardy L., A mathematician's apology, 1940.

Hasslacher L., "Beyond the Turing Machine" in [Herken,1995].

Herken R., The Universal Turing Machine, Springer-Verlag, 1995.

Hoffmann P., "Modèle mécaniste et modèle animiste", "Revue des Sciences Humaines", 186-187, 1982.

Kreisel G., A. Macintyre "Constructive Logic versus Algebraization I" in Brouwer Centenary Symposium, North-Holland, 1982.

JSL, Feferman S., S. Sieg, S. Simpson, three papers dedicated to the partial realization of Hilbert's program, "Journal of Symbolic Logic", 53, 2, 1988.

Lakatos I., Proofs and Refutations, Cambridge U. Press, 1976.

Lassègue J., "L'intelligence artificielle et la question du continu", Thèse, Univ. Paris X - Nanterre, 1994.

Lightill J., "The recently recognized failure of predictability in Newtonian dynamics" "Proc. R. Soc. Lond.", 407 (35-50), 1986.

Longo G., "Some aspects of impredicativity: notes on Weyl's philosophy of Mathematics and on todays' Type Theory" in Logic Colloquium 87, Studies in Logic (Ebbinghaus et al. eds), North-Holland, 1989.

Longo G., "Reflections on Mathematics and Computer Science", "European Congress of Mathematics", Paris 1992; (Ebbinghaus ed.), 1994.

Longo G., "Sur l'émérgence de l'objectivité des Mathématiques", "Intellectica", 1995.

Longo G., Misted K., Soloviev S., "The Genericity Theorem and Parametricity in the Polymorphic λ-calculus", "Theoretical Computer Science", 1993, (323-349).

Lynch N. et al., Atomic Transactions, Kaufmann, 1994.

McCorduck P., The Universal Machine: Confessions of a Technological Optimist, McGraw Hill, 1986.

Makowsky Y.A., "Mental Images and the Architecture of Concepts" in [Herken, 1995].

Maffei L., Mecacci L., La Visione, Mondadori, 1979.
Paris J., Harrington L., "A mathematical incompleteness in Peano Arithmetic", in Handbook of Mathematical Logic, 1978.
Poincaré H., Les Méthodes Nouvelles de la Mécanique Celeste, Paris, 1892.
Poincaré H., La Science et l'Hypothese, Flammarion, Paris, 1902.
Poincaré H., La valeur de la Science, Flammarion, Paris, 1905.
Putnam H., "Much Ado About Not Very Much", in [Graubard,1989].
Rogers H., Theory of Recursive Functions and Effective Computability, 1967.
Turing A, "Computing Machines and Intelligence", "Mind", LIX, 1950.
Schwartz J., "The New Connectionism", in [Graubard,1989].
Shœning U., "Complexity Theory and Interaction" in [Herken,1995]
Wegner P., "The expressive power of interaction", Brown Univ. Techn. Rep., 1995.
Wegner P., "A framework for empirical Computer Science", Brown Univ. Techn. Rep. CS-95-18, 1995.
Weyl H., Das Kontinuum, 1918.
Weyl H., Symmetry, Princeton University Press, 1952.
Wittgenstein L., Remarks on the foundation of Mathematics, 1956.

JEAN PETITOT

SHEAF MEREOLOGY AND SPACE COGNITION

CONTENTS

INTRODUCTION

Contemporary research concerning the cognitive links between perception, language, and action - see for instance Talmy's works - have revolutionized the dominating traditions of formal semantics. They have led to what Husserl already called in *Erfahrung und Urteil* a *genealogy of logic,* a perceptive genealogy of cognitive symbolic structures. Such a Gestalt-like perspective on symbolic structures raises a lot of new problems. It makes traditional simple - and even trivial - problems appear as complex, non trivial, ones.

It is one of these problems I want to address here, namely that of the most elementary and primitive forms of *predication in perceptive judgements*: "the snow is white", "the sky is blue", "the ball is red", etc.

For classical elementary logic these atomic formulae "S is p" stay at the lowest level of complexity. Both their syntax and their semantic are trivial. But it is no longer the case if one takes into account the perceptive situations to which these statements refer. Indeed, these situations compel us to introduce in the formalization of symbolic logical structures, *topological and morphological* structures of a completely different kind. If one wants to do semantics that way, one is therefore committed to elaborate a *mathematically integrated theory of topology and logic.*

Let us see briefly why the problem of taking into account the perceptive roots of predication is a non trivial one. On the perceptive side we find *a geometrical descriptive eidetics* of the morphological (gestaltic) organization of perception. I use here the term "morphology" in Thom's sense: a morphology is a set of qualita-

A. Carsetti (ed.), Functional Models of Cognition, 233-252.

tive discontinuities on a substrate space. But on the logical side we find a logical eidetics of judgement: a formal syntax and a formal semantics (e.g. a verconditional theory of denotation). But between these two eidetics there exists a dramatic gap, what Thom (1988:248) called an "insuperable break". Our challenge here is to fill this gap using some sort of synthesis between logic and topology.

The gap is linked with deep traditional philosophical issues, for instance that of the relations existing between *analytic and synthetic* rules. Morphological eidetics depends on synthetic laws of perception. On the contrary, logical eidetics depends on analytic laws.

I. THE MORPHODYNAMICAL SCHEMATIZATION OF PREDICATION

Let us consider how René Thom has morphologically schematized a perceptual attributive judgement "*S* is *p*" such as "the sky is blue". The truth conditions of such a statement depend on the way, as Husserl would say, the "intentions of signification" are *intuitively filled* (in the sense of an *Erfüllung*) at the *ante-predicative and pre-judicative* level. Without an adequate description of this intuitive filling-in, one can't elaborate a correct theory of predication.

Let W be the spatial extension of the substrate S. W is filled by a quality p belonging to a genus G (e.g. colours). The filling-in is then described by a map

$$f : W \to G,\ w \to f(w).$$

Husserl has given a deep analysis - lacking mathematical modeling - of this fact in his third Logical Investigation. I have shown that to model adequately *the unilateral dependence relation* linking the extension W with its dependent part - its *moment* - p, one has to interpret the map f as a *section* of the (trivial) *fibration*

$$\pi : E = W \times G \to W.$$

It is exactly the point of view mathematized by Thom.

Let me recall briefly what are fibrations and sections of fibrations.

Mathematically, a fibration is a differentiable manifold E endowed with a *canonical projection* (a differentiable map) $\pi : E \to M$ over another manifold M. M is called the *base* of the fibration, and E its *total space*. The inverse images $E_x = \pi^{-1}(x)$ of the points $x \in W$ by π are called the *fibers* of the fibration. They are the subspaces of E which are projected to points.

In general a fibration is required to be *locally trivial*, that is to satisfy the two following axioms:

(F_1) All the fibers E_x are diffeomorphic with a typical fiber F.

(F_2) $\forall x \in M$, $\exists U$ a neighborhood of x such that the inverse image $E_U = \pi^{-1}(U)$ of U is diffeomorphic with the direct product $U \times F$ endowed with the canonical projection $U \times F \to U$, $(x, q) \to x$. (See figures 1, 2).

In our case, we have $M = W$ and $F = G$.

Let $\pi : E \to M$ be a fibration and let $U \subset M$ be an open subset of M. A *section s* of

π over U is a *lift* of U to E which is compatible with π. More precisely, it is a map $s : U \to E, x \in U \to s(x) \in E_x$, i.e. such that $\pi \circ s = \mathrm{Id}_U$. In general s is supposed to be continuous, differentiable or analytic. It can present discontinuities along a singular locus. (See figures 3, 4).

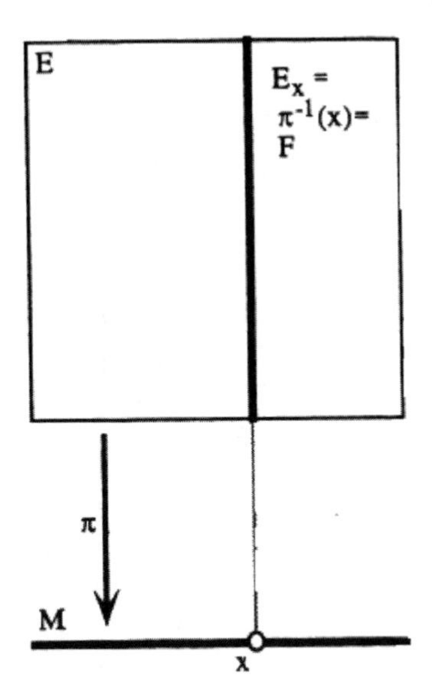

Figure 1

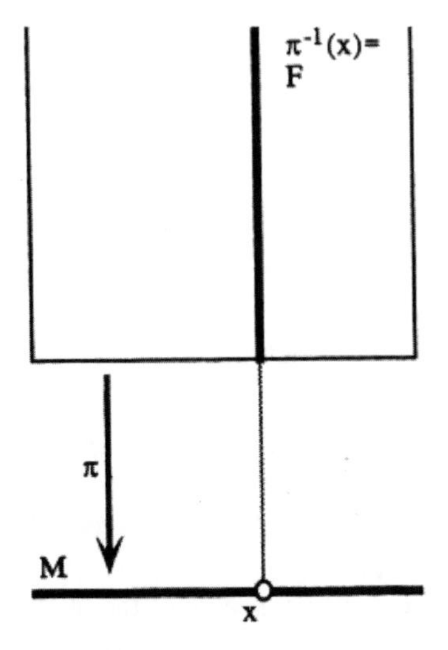

Figure 2

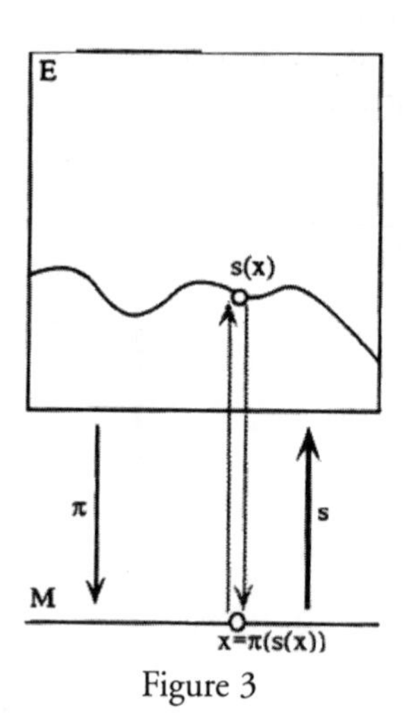

Figure 3

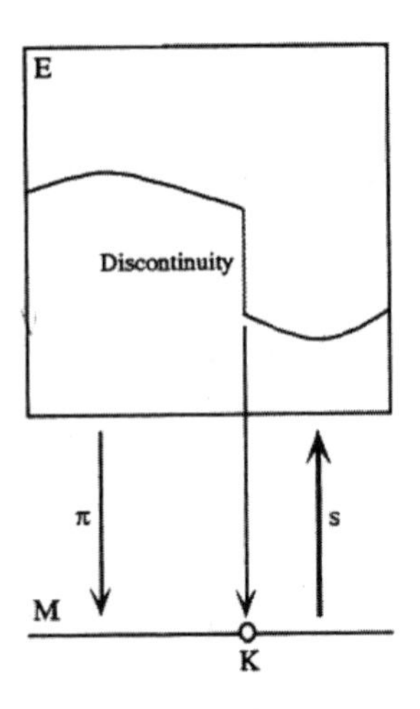

Figure 4

It is conventional to write $\Gamma(U)$ for the set of sections of π over U. If there exists a local trivialization of π over U (i.e. $E_U \to U \approx U \times F \to U$), it transforms every section $s : U \to E$ in a map $x \to s(x) = (x, f(x))$, that is in a map $f : U \to F$. Therefore, the concept of section generalizes the classical concept of map.

The main reason why it is of interest to conceive of the qualitative filling-in of spatial regions as sections of fibrations is that one can easily retrieve in that way the dependence relations (i.e. the ontological hierarchy) between "essences". The fact that spatial extensions are ontologically prior to qualities such as colours is modeled by the fact that the extensions W are the *base spaces* of the fibrations, and can therefore exist in an *independent* way, while the qualitative genera G are the *fibers* of the fibrations, and can therefore only exist in a *dependent* way.

In general the genus G will be *categorized* and decomposed in sorts. Let p be such a category (e.g. "blue") and let ∂p be its boundary in G. The perceptive statement "S is p" is morphologically interpreted *by the following geometrical fact:*

the image $f(W)$ of the section $f: W \to (W \times G)$ is encapsulated in the cylinder $W \times \partial p$.

These morphological descriptions can be generalized. If for instance there exists an order relation $<$ on G (e.g. p is lighter than p'), the statement $S(W) < S'(W')$ means that $\rho(f(w)) < \rho(f'(w'))$ where ρ is the second projection $\rho : W \times G \to G$ (that is $\rho(f(w))$ is the codomain of values of f).

The problem I want to address is very simple. In the framework of a semantic theory, what can be the link between two definitions of truth (for atomic formulae):

(i) the tautological Tarskian one: "S is p" is true iff S is p;

(ii) the morphological Thomian one: "S is p" is true iff $f(w)$ is encapsulated in the cylinder $W \times \partial p$?

II. THE INDEXICAL AND MODAL STATUS OF SPATIAL EXTENSION

In the categorial syntactic structure of a statement such as "S is p", the *specificity* of the spatial extension W of S is in some way *bracketed,* "implicitated". Of course, one can focus on the extension W itself and consider its regions, its points, etc. But in doing so one changes the level of description. At the level at which S is processed as an individual entity sharing some properties - that is at the level at which the symbolic description "S is p" is relevant - the specificity of W is bracketed. This strange sort of "epoche" concerns also the *relations* between qualities, as when we say "S is lighter than R".

This problem has been well pointed out by Wittgenstein in his *Remarks on Colour* (see Mulligan 1992).

"A language-game: Report whether a certain body is lighter or darker than another. But now there's a related one: State the relationship between the lightnesses of certain shades of colour. The form of the propositions in both language games is the same "X is lighter than Y". But in the first it is an external relation and the proposition is temporal. In the second it is an internal relation and the proposition is timeless".

The tautological aspect of Tarski's definition of truth is a symptom of this "epoche". It is what remains when syntax and semantics uncouple themselves from their perceptive rooting.

As a matter of fact, one can say that *every* perceptive statement is in some sense analog *to an indexical one*. It refers to its perceptive situation in much the same way as an indexical statement refers to its pragmatic context. The intuitive *Erfüllung* of the intentions of signification operates in a *counterfactual* fashion and one needs therefore some sort of Kripkean semantics to take account of this fact.

If one wants to understand how classical logical structures can have a non classical *space-semantics*, one has to understand how classical variables can semantically denote *sections of fibrations* in such a way that *extensions can condition truth.*

III. COMPUTATIONAL VISION AND RECEPTIVE FIELDS JET CALCULUS

Before tackling this point, I want to emphasize the *neuro-physiological* relevance of the geometrical concept of fibration.

The most influential specialists applying differential geometry to computational vision have hypothesized that cortical columns implement a *multi-scale jet calculus* and that it is essentially that way the brain does geometry. Let us consider for instance Jan Koenderink's perspective.

David Marr already showed in the late seventies that the retinal ganglion cells perform what is now called a *wavelet analysis* of the signal $I(x,y)$, that is a *multiscale and spatially localized* analysis. They act as *filters*, that is they convolute the signal by the receptive profile of their receptive fields. Now, these receptive profiles are approximatively Laplacians of Gaussians ΔG. According to the fact that ($\ast$ = convolution):

$$\Delta G \ast I = \Delta(G \ast I)$$

one can conclude that such cells smooth locally the signal at a certain scale and compute its second partial derivatives.

Using such smoothed partial derivatives one can for instance detect qualitative *salient discontinuities:* it is Marr's celebrated *zero-crossing* criterium.

Now, zero-crossings are differential geometrical invariants which belong to what is called the *2-jet* of the smoothed signal. Jets constitute an *intrinsic* version of the old idea of Taylor expansions. Let us take for instance the Taylor expansion of an $\mathbb{R}$-valued differentiable map $I : \mathbb{R}^2 \to \mathbb{R}$ up to order 2 at a point (x_0, y_0).

$$T_2I(x,y) = I(x_0,y_0)+(x-x_0)\frac{\partial I}{\partial x}(x_0,y_0)+(y-y_0)\frac{\partial I}{\partial y}(x_0,y_0)+$$

$$\frac{(x-x_0)^2}{2}\frac{\partial^2 I}{\partial x^2}(x_0,y_0)+(x-x_0)(y-y_0)\frac{\partial^2 I}{\partial x \partial y}(x_0,y_0)+\frac{(y-y_0)^2}{2}\frac{\partial^2 I}{\partial y^2}(x_0,y_0).$$

It is constructed via simple algebraic operations from the 8-uple:

$$\{(x,y); I(x,y); (\partial_x I,\partial_y I); (\partial^2_{x^2} I,\partial^2_{xy} I,\partial^2_{y^2} I)\}.$$

Now we introduce a space $J^2 = \mathbb{R}^8$ with coordinates

$$\{(x,y); z; (\alpha,\beta); (\lambda, \mu,\gamma)\}$$

and we define the *2-jet* of I as the map $j^2 I$: $\mathbb{R}^2 \to J^2$ given by

$$j^2 I(x,y) = \{(x,y); I(x,y); (\partial_x I,\partial_y I); (\partial^2_{x^2} I,\partial^2_{xy} I,\partial^2_{y^2} I)\}.$$

This map can be defined in a coordinate-free (geometrically intrinsic) manner. Actually J^2 is a *fibration* of base $\mathbb{R}^2_{(x,y)}$ and fiber $\mathbb{R}^6_{(z; (\alpha,\beta); (\lambda,\mu,\gamma))}$ and the 2-jet $j^2 I$ is a *section* of J^2.

Jets are therefore spatial fields of differential data, and using them one can compute *geometrically relevant differential invariants and morphological features* (fold points, cusps, end points, curvature, etc.) using only *arrays of point-processors.*

If one introduces the hypothesis that the receptive fields of certain classes of cortical cells approximate partial derivatives of Gaussians of order > 2 (up to order 3 or 4) then one can give a *multi-scale* version of this jet calculus. Koenderink has strongly advocated the thesis that such a receptive fields jet calculus is implemented in retinotopic arrays of cortical columns.

Let us recall briefly the *columnar structure* of the primary visual cortex (area 17 or area V1) extensively investigated since the pioneering works of Hubel, Wiesel and Mountcastle.

The basic functional module is a "cube" organized in a *retinotopic, columnar and layered* manner.

(i) The *retinotopic structure* preserves the topographic connexions of the retinian ganglion cells (the signal processing of which is transmitted and amplified by the lateral geniculate body). We get that way *a retinotopic fibration*, the base-space of which is constituted *by glued local receptive fields.*

(ii) The *columnar structure* (which is orthogonal to the cortex surface) is essentially constituted by *orientation columns,* columns of *ocular dominance,* and colour plugs. Orientation columns (the diameter of which is about 50μ) are set out orthogonally to the columns of ocular dominance. Their preferential orientation varies from 0° to 180° by 10° steps. They constitute *hypercolumns* the size of which is about 800μ-1mm.

(iii) The *layered structure* is composed of six layers. Its depth is about 1,8mm. The geniculate fibers project on layer IV. Layer VI is a feedback one. Layer V projects on the colliculus. Layers II and III receive the axons of layer IV and project their efferent fibers in other cortical regions where different attributes (shape, colour, movement, stereopsis, etc.) are further processed in a *non* retinotopic way.

The orientation columns yield a beautiful example of a *neurally implemented fibration.* Actually, they implement the fibration $\pi: E \rightarrow W$ the base of which is the visual field and the fiber of which is the *projective line* $F = \mathbb{P}^1$ of plane directions. William Hoffman (1989) has explained how the cortex implements what is called the *contact structure* of E.

Let M be a n-manifold. A *contact element* $c = (x, H)$ of M is constituted by a point $x \in M$ and by a $(n\text{-}1)$-hyperplane H of the tangent space T_xM of M at x. (See figure 5).

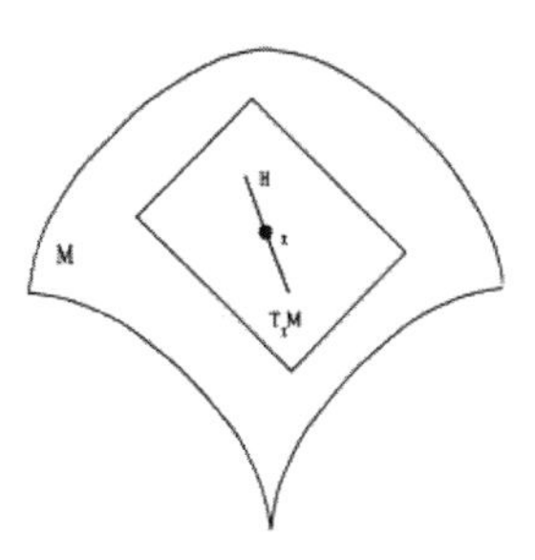

Figure 5

The *contact bundle* CM of M is the $(2n\text{-}1)$-manifold of these contact elements. CM is a fibration - a *fiber bundle* - $\pi : CM \rightarrow M$, the fiber over x being the projective space $C_xM = \mathbb{P}(T_xM)$ of hyperplanes of the vector space T_xM. Here we have $n=2$, the Hs are the directions and therefore $C_xM \cong \mathbb{P}^1$. (See figure 6: M is represented as one-dimensional and in the second figure the directions are represented as points of $\mathbb{P}^1$).

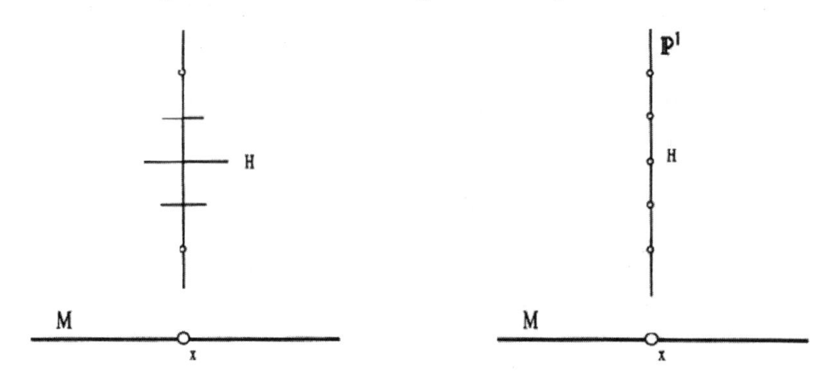

Figure 6

The contact bundle CM is isomorphic to the *projectivized* bundle of the *cotangent bundle* T^*M. Indeed a contact element (x, H) can be defined by a 1-form (a linear form) α_x on T_xM satisfying $\text{Ker}(\alpha_x) = H$. But $\lambda\alpha_x$ $(\lambda \neq 0)$ defines the same H and therefore $CM \cong \mathbb{P}(T^*M)$.

There exists on the $(2n\text{-}1)$ contact bundle CM what is called *a canonical contact structure*. It reflects the *canonical symplectic structure* of T^*M. In general a *contact structure* on an N-manifold $\mathcal{M}$ is defined by a *non degenerate* field of $(N\text{-}1)$-hyperplanes $\mathcal{K}_c$ which are called *contact hyperplanes* (non degeneracy forces N to be odd). Let $c = (x, H) \in \mathcal{M} = CM$. A tangent vector

$$(\xi, \Theta) \in T_c\mathcal{M} = T_{(x,H)}CM$$

belongs to the contact $(2n\text{-}2)$-hyperplane $\mathcal{K}_c$ iff its projection ξ on T_xM belongs to H (i.e. $c = (x, H)$ moves in such a way that the x-velocity ξ belongs to H). (See figure 7).

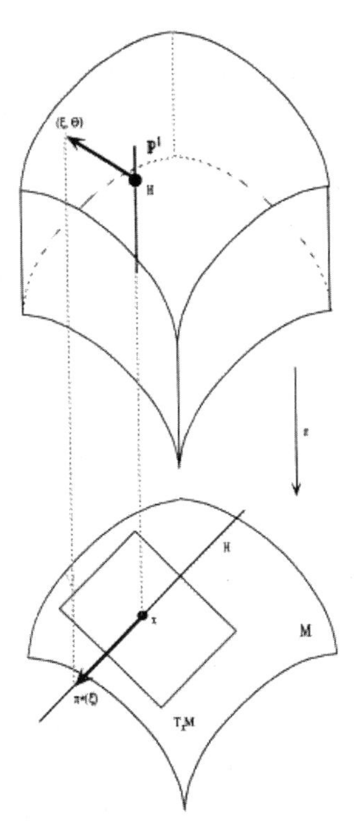

Figure 7

Let Γ be a sub-manifold of M and let $C_M\,\Gamma$ be the manifold of the $c \in CM$ *tangent* to Γ. $C_M\,\Gamma$ is always of dimension n-1 and is tangent to the field $\mathcal{K}_c$. $C_M\,\Gamma$ is therefore *an* $(n$-$1)$*-integral manifold* of the canonical contact structure of CM. In the visual case n = 2. $\mathfrak{M} = CM$ is of dim = 3 (coordinates, x, y, φ = direction of H). If $c = (x, H) \in CM$ and $(\xi,\Theta) \in T_{(x,H)}CM$, we have:

$$(\xi,\Theta) \in \mathcal{K}_c \leftrightarrow \xi \in H.$$

If Γ is *a curve* in M and if we consider che contact elements $c = (x, H)$ which are *tangent* to Γ (i.e. $T_x\Gamma = H$) we get a *curve* $C\,\Gamma$ (a (n-1)-submanifold) in CM which is *an integral curve* of the canonical contact structure. Indeed, when *x moves along* Γ, (ξ,Θ) is such that $\xi \in T_x\Gamma$ and $T_x\Gamma = H$. Therefore $\xi \in H$ and therefore $(\xi,\Theta) \in \mathcal{K}_c$. This shows that a boundary curve (a contour curve) is a section of a fibration and that it can therefore be processed by arrays of point processors.

W. Hoffman explains very well how:

(i) the concept of receptive field corresponds to the concept of local chart in differential geometry;

(ii) the neurally implemented canonical contact structure eexplains the processing of visual contours as *integral curves*;

(iii) the perceptual constancies are invariant structures describable by means of actions of symmetry Lie groups on the contact structure of the projective bundle of directions. He concludes claiming that "fibrations (...) are certainly present and operative in the posterior perceptual system if one takes account of the presence of "orientation" micro-response fields and the columnar arrangement of cortex" (p. 645).

For colours, the situation is a little bit more complex. There exist parvocellular layers in the lateral geniculate body constituted of ganglion cells sharing a *spectral antagonism:* (R+/G-), (G+/R-), (Y+/B-), (B+/Y-); there exist cortical plugs specialized in colour processing; their exist colour *blobs* in the area V1; and, according to Semir Zeki, the area V4 is specialized in colour processing. But it seems nevertheless plausible to hypothesize that colour is also processed through fibrations (with a dynamical antagonist spectral structure in the fibers).

This shed some light on the *morphological* aspects of perceptive statements.

What now, of the *logical* ones? How the morphological eidetics can be unified with a logic of judgement?

IV. HUSSERL'S ERFAHRUNG UND URTEIL

In *Erfahrung und Urteil, Untersuchungen zur Genealogie der Logik,* Husserl tries to clarify phenomenologically the *origin* of predication. He elaborates a theory of predicative statements such as "*S* is *p*", in the framework of formal apophantics (a syntactic theory) and formal ontology (a semantic theory such as set theory). But his perspective is "genealogic". According to him, classical logic *conceals* the fundamental problem of *evidence* in the logical concept of truth. By "evidence", Husserl means here *perceptive presentation as immediate acquaintance* with the world. Predicative statements such as "*S* is *p*" are rooted in ante-predicative and pre-judicative experience and it is the nature of such a founding relation which constitutes for Husserl the main problem. His "basic theme" is "categorical judgement founded in perception" (p. 70). Now, according to him, formal logic neglects perceptual evidence.

"The formal character of logical Analytics consists in the fact that it does not consider the material quality of what is given prior to statements, and that it looks to substrates only in what concerns the categorial form they take in the statement" (p. 18).

He wants therefore to shaw that logic is originally based on the experience of the world, that the foundation of logic in perception has a universal scope and concerns its essence.

In *Erfahrung und Urteil* Husserl summarizes his analysis of perception previously done in the third *Logical Investigation, Ding und Raum und Ideen I.*

(i) The morphological analysis of the filling-in of spatial regions by qualities and the constitutive role of *qualitative discontinuities* in segmenting visual scenes (see the opposition between *Verschmelzung* and *Sonderung*).

(ii) The essentially *adumbrative* status of perception (*Abschattungslehre*). The object is a noematic pole unifying in a coherent way a continuous flux of adumbrations. Its noematic content expresses the *synthetic* relations of foundation between spatial extensions and their dependent moments such as contours or colours. It rule governs perceptive *anticipations* which, being filled-in by a "fluent variability"of adumbrations, are "indeterminated and general", but which are nevertheless also *prescriptively ruled by typed relations.* For instance, there exists a "typical genericity" of the filling-in relations between extensions and qualities.

He analyzes then very precisely the origin of the "primitive logical categories" from the ante-predicative synthesis. The main problem is to understand the "categorial genesis" of the *substrate/determination categories* which transforms the *perceptive synthesis,* where a substrate S possesses a dependent moment p, in the *predication* "S is p". Husserl's main thesis is that this categorial transformation arises from a reflective *thematization* which enables consciousness to grasp the dependent moments and typifies the substrates in "subjects" and the dependent moments in "predicates". According to him:

"We find there the origin of the first "logical" categories" (p. 127).

Husserl points out therefore a fundamental conversion from the *synthetic* perceptive unity unifying an extension S with its dependent moments p to the *analytic* syntactic unity of the statement "S is p".

"In the most simple predicative judgement a *double information* is processed" (Husserl 1954, p. 247).

Indeed, underlying the syntactic "subject/predicate" information concerning the "functional forms" of the terms of the proposition, there exists another information concerning the "kernel forms" "substrate=independence" and "adjectivity=dependence". This underlying information is presupposed by the syntactic one. Predication is a process based on "the covering of the kernel forms as syntactic material by the functional forms" (p. 248).

It is this logical typification of the synthetic dependence relations in syntactic categories I want now to model. As far as the dependence relations between extensions and filling-in qualities can be schematized by sections of fibrations, we see that, if we want to model Husserl's "genealogical" analysis of logic, we need to understand *how variables which denote sections of fibrations can be syntactically typified in such a way that the resulting semantics coud be that sort of Kripkean semantics we need to explain the "indexical" status of perceptive statements.* Moreover, this must be done in the framework of formal apophantics and formal ontology.

Now, we will see that it is possible to achieve this goal using the tools of the geometrical logic yielded by *Topos theory.*

V. SHEAVES, TOPOÏ AND LOGIC

Some remarks to begin with.

1. Topos theory has been invented by Grothendieck for solving very sophisticated problems of algebraic geometry (he wanted to construct generalized cohomological theories). It can therefore seem strange to use it for solving elementary and non mathematical problems. But actually, when they are treated as *neuro-cognitive* ones, these "naïve" problems are not at all elementary and require sophisticated tools to be modeled adequately.

2. Category theory is certainly the best formal ontology we have now at our disposal.

3. Some specialists of topos theory have already applied these tools to clarify some basic semantic problems. For instance Gonzalo Reyes (who works, with Eduardo Dubuc, Anders Koch, Ieke Moerdijk and Marta Bunge on the applications of topos theory to *Synthetic Differential Geometry*) has recently used these techniques to formalize the Kripkean theory of proper names as rigid designators.

4. In general, topos theory is used in formal logic as a tool for *typed* λ*-calculus* (see e.g. the works of John Mitchell, Philip Scott, Giuseppe Longo, etc.). As we will see, when variables are typed by objects in a topos, the categorical properties of the topos (to be a cartesian closed category, to possess a subobject classifier, etc.) lead to an intuitionistic "internal logic" which is a typed λ-calculus. As far as the Curry-Howard correspondance shows that (in intuitionistic logic) formulae can be treated as types, proofs as λ-terms, and the reduction of a proof by cut-elimination as the reduction of a λ-term to its normal form, topos theory has become fundamental for understanding the semantic of formal - and in particular of programming - languages.

In these applications, the geometrical origin of the sheaf and topos concepts is concealed. Here, we aim at a completely different sort of applications. We want to use this geometrical origin to explain the symbolic ascent of logical forms from morphological ones.

As we already saw, in the "receptive field" perspective one needs to conceive of sections of fibrations as resulting from a gluing of local sections. This leads naturally to the concept of *sheaf* which, even if it can seem a very abstract one, is actually very deeply rooted in spatial intuition.

1. THE CONCEPT OF SHEAF

For the concept of sheaf, the primitive topological concept is no longer that of a point but that of an open set (what Peter Johnstone calls "pointless topology").

At an abstract level, a fibration is characterized by the sets of its sections $\Gamma(U)$ over the open sets $U \subset M$. If $s \in \Gamma(U)$ and $V \subset U$, we can consider the *restriction* $s|_V$ of s to V. The restriction is a map $\Gamma(U) \to \Gamma(V)$. It is clear that if $V=U$, then $s|_V = s$ and that if $W \subset V \subset U$ and $s \in \Gamma(U)$, then $(s|_V)|_W = s|_W$ (transitivity of the restriction). We get therefore what is called a *contravariant functor*

$\Gamma : \mathcal{O}^*(M) \to$ **Sets** from the category of the open sets of M in the category **Sets** of sets. (The objects of $\mathcal{O}(M)$ are the open sets of M, and its morphisms are the inclusions of open sets).

Conversely, let Γ be such a functor - what is called a *presheaf* on M. To have a chance of being the functor of the sections of a fibration, Γ must clearly satisfy the two following axioms.

(S_1) Two sections which are locally equal must be globally equal. Let $\mathcal{U} = (U_i)_{i \in I}$ be an open covering of M. Let $s,s' \in \Gamma(M)$. If $s|_{U_i} = s'|_{U_i}\ \forall i \in I$, then $s=s'$.

(S_2) Compatible local sections can be collated in a global one. Let $s_i \in \Gamma(U_i)$ be a family over $\mathcal{U} = (U_i)_{i \in I}$. If the $s_i|_{U_i \cap U_j} = s_i|_{U_i \cap U_j}$ when $U_i \cap U_j \neq \varnothing$, then they can be glued together: $\exists s \in \Gamma(M)$ such that $s|_{U_i} = s_i\ \forall i \in I$.

(S_1) and (S_2) can be expressed in a purely categorical manner. For instance (S_2) says that the arrow

$$e : s \to \{s|_{U_i}\}_{i \in I}$$

is the *equalizer*

$$\Gamma(U) \xrightarrow{e} \prod_i \Gamma(U_i) \underset{q}{\overset{p}{\rightrightarrows}} \prod_{i,j} \Gamma(U_i \cap U_j)$$

of the two projections p, q corresponding to the inclusions $U_i \cap U_j \subset U_i$ and $U_i \cap U_j \subset U_j$.
In fact these axioms characterize a more general structure - and even more pervasive in contemporary mathematics - than the structure of fibration, namely the structure of *sheaf*. It can be shown that if the axioms (S_1) and (S_2) are satisfied, then one can represent the functor Γ by a general fibered structure $\pi : E \to M$ (called un "étale" space and which is not necessarily a locally trivial fibration) in such a way that $\Gamma(U)$ becomes the set of sections of π over U. In a nutshell, the fiber E_x - called in that case the *stalk* of the sheaf Γ at x - is the inductive limit (the colimit):

$$E_x = \lim_{V \subset \in \mathcal{U}_x} \{(\Gamma(U), \Gamma(V \subset U)\}$$

(where $\mathcal{U}_x$ is the filter of the open neighborhoods of x). The stalk E_x is the set of *germs* s_x of sections at x. E is the sum of the E_x. If $s \in \Gamma(U)$, it can be interpreted as the map $x \in U \to s(x) \in E_x$. The topology of E is then defined as the finest one making all these sections continuous.

2. GENERALIZATIONS.

There are many generalizations of this basic situation. The best known is that of Grothendieck's topologies: open coverings are defined in terms of "sieves" and the concept of sheaf is generalized to this structure of "site".

Another generalization is that of *frames* and *locales*. One consider frames, that is lattices which share the properties of the lattices of open sets $\mathcal{O}(X)$. They are complete distributive lattices with finite meets and general joins. Locales are the objects of the dual category (in the case of topological spaces the correspondance is

$$f : X \to Y \text{ continuous} \Rightarrow f^* : \mathcal{O}(Y) \to \mathcal{O}(X)).$$

Points are then defined as morphisms $p : A \to 2 = \mathcal{O}(1)$ (in the case of topological spaces, if $A = \mathcal{O}(X)$, points correspond to true points $p : 1 \to X$). Let $Pt(A)$ be the set of points of A. A topology is defined on $Pt(A)$ by the subsets

$\varphi(a) = \{p \in Pt(A) \mid p(a) = 1\}$. $\mathcal{O}(Pt(A))$ is the best approximation of A by a "spatial" locale.

3. THE CONCEPT OF TOPOS.

Now, the main point, is that the category of sheaves on a space M shares the categorical properties of a topos and that (intuitionistic) logical languages can therefore be *inter-*

preted in it. According to Lawvere's perspective, such a topos $\mathcal{E}$ can therefore be conceived of as a universe of discourse *the objects of which are variable entities depending on the elements U of* $\mathcal{O}(M)$. $\mathcal{O}(M)$ operates therefore as a set of "possible worlds". The "possible worlds" are here *spatial extensions.* This dependence on spatial extensions shares the two main properties we need:

(i) it is constitutive of the concept of logical truth and therefore of semantic relevance;

(ii) but it is not directly "visible" in the syntax of the internal logic of $\mathcal{E}$ and it is therefore syntactically irrelevant.

Actually, as long as the formal constructions in such a universe are constructive (in the intuitionistic sense) one can completely bracket and forget the spatial dependence at the syntactic level. We get therefore a very elegant explanation of the puzzling fact that spatial extension operates in perceptive statements in a rather "indexical" and "pragmatic" manner.

Ieke Moerdijk and Gonzalo Reyes (1991) have proposed an interesting philosophical commentary to Lawvere's main ideas.

"Topos theory has brought to light and given the means to exploit a complementarity (or duality) principle between *logic* and *structure*" (p. 10).

In the framework of classical model theory, a theory T is of the general form $T = S + I$ where S is the axiomatics of a type of structures and I a set-theoretical interpretation satisfying the "trivial" and "tautological" Tarskian semantics. But one can also interpret S in a topos $\mathcal{E} = \mathbf{Sh}(\mathcal{C})$ by means of a *sheaf semantics.* One can then work in $\mathcal{E}$ as one would work in a standard universe of sets provided one does it with a non classical (intuitionistic) logic reflecting the new interpretation.

Let us now recall briefly the structure of a topos. If A is a sheaf on M, $A(U)$ will denote the set $\Gamma_A(U)$ of sections of A over U. It is easy to show that the sheaves on a base space M constitute a category $\mathbf{Sh}(M)$.

3.1. EXPONENTIALS

$\mathbf{Sh}(M)$ is a *cartesian closed* category. This means that it has products and fibered products or pullbacks, a terminal objet - classically denoted by 1 - and *exponentials* B^A. An exponential object is an object which "internalizes" in the objects of $\mathbf{Sh}(M)$ the morphisms $f: A \rightarrow B$. Such "internalization" of functorial structures are called *representable* functors. Technically, the functor $(\bullet)^A$ is the *right adjoint* of the functor $A \times (\bullet)$. This means that we have for every object C of $\mathbf{Sh}(M)$ a functorial isomorphism $\mathrm{Hom}(C,B^A) \cong \mathrm{Hom}(A \times C,B)$. E.g. for $C=1$, we get $\mathrm{Hom}(1,B^A) \cong \mathrm{Hom}\ (A,B)$. But an arrow $f: 1 \rightarrow B^A$ is like an "element" of B^A. In fact, if A is a sheaf, an arrow $s: 1 \rightarrow A$ is a *global section* of A, that is an element $s \in A(M)$.

If we take $C=B^A$ and Id_{B^A}, the right adjunction defines what is called a *counit* $\varepsilon: A \times B^A \rightarrow B$ such that for every $f: C \rightarrow B^A$ the associated $h: A \times C \rightarrow B$ is given by $h = \varepsilon \circ (1 \times f)$. The counit generalizes the map $(x,f) \rightarrow f(x)$ in set theory and is therefore called the *evaluation map.*

The sheaf B^A is defined using the evident restrictions $A|_U$, $B|_U$ to open sets: $B^A(U) = \text{Hom}(A|_U, B|_U)$. It is called the "internal Hom" or the sheaf of germs of morphisms from A to B.

3.2. SUBOBJECT CLASSIFIER

Sh(M) possesses also what is called a *subobject classifier* Ω, that is an object which "internalizes" the sets of subobjects, making the subobject functor representable. A subobject $m : S \rightarrow A$ is a monomorphism (an injective map in the case of the category of sets). This means that if f and g are two morphisms from an object R to S, then $m \circ f = m \circ g$ implies $f = g$. It is equivalent to say that the fibered product $S \times_A S$ defined by m is isomorphic with S. A subobject classifier is a monomorphism $True : 1 \rightarrow \Omega$ such that every subobject $m : S \rightarrow A$ can be retrieved from $True$ by a pull-back:

$$\begin{array}{ccc} S & \rightarrow & 1 \\ \downarrow & & \downarrow \text{True} \\ A & \underset{\varphi_S}{\rightarrow} & \Omega \end{array}$$

We get therefore a functorial isomorphism $\text{Sub}(A) \cong \text{Hom}(A, \Omega)$. φ_S is called the *characteristic map* of the subobject S.

In the category **Sets** of sets, $\Omega = \{0,1\}$ is the classical set of boolean truth-values. Here - and it is perhaps the main difference between a topos like **Sh**(M) and the classical topos **Sets** -, $\Omega(U)$ *depends essentially on the topological structure. It expresses the localization of truth in a sheaf topos.*

By definition $\Omega(U) : \{W \subset U\}$. It is trivial to verify that Ω is a sheaf. The *True* map $True : 1 \rightarrow \Omega$ is defined by $True(U) : 1 \rightarrow U \in \Omega(U)$ that is by the *maximal* element of $\Omega(U)$: to be true over U is to be true "everywhere" over U. The global section $T \in \Omega(M)$ selected by *True* is nothing else than the whole base space M itself and $True(U)$ is its localization to U. If S is a subsheaf of the sheaf A, its characteristic map $\varphi_S : A \rightarrow \Omega$ is given by the maps $\varphi_S(U) : A(U) \rightarrow \Omega(U)$ which map $s \in A(U)$ to the largest $W \subset U$ s.t. $s|_W \in S$. It is easy to verify that the monic map $S \rightarrow A$ is effectively the pull-back of *True* by φ_S. As Mac Lane says: φ_S gives "the shortest path to truth" $W \subset U$ for $S \rightarrow A$.

3.3. ELEMENTS, PROPERTIES AND PARTS

In a topos, the morphisms $a : B \rightarrow A$ are called *generalized elements* of A, or elements *defined on* B (this denomination comes from algebraic geometric and, more precisely, from Grothendieck's theory of schemes). Among the elements, the most important are those *defined on open sets* U, that is precisely the sections $s \in A(U)$. If $U \in \mathcal{O}(M)$, we can consider the Yoneda sheaf $y(U) = \{W \subset U\}$ and verify that $A(U) \cong \text{Hom}(y(U), A)$. The elements defined on the terminal object 1 are "global". We will see that an arrow $\theta : A \rightarrow \Omega$ is a "predicate" for A, that is a "property" of its generalized elements. Among all predicates, there is the predicate $True_A : A \to 1 \to \Omega$. It is easy to verify that an element $a : B \to A$ factorizes through a subobject $S \to A$ iff char$S(a) = \varphi_S \circ a = True_B$. φ_S is therefore the predicate of A which is true exactly for those elements of A which are in S. The *unicity* of φ_S expresses the *extensionality principle.*

Using the exponentials and the subobject classifier we can define the *parts* of an object A as another object $P(A) = \Omega^A$. We get the functorial isomorphisms:

$$\mathrm{Sub}(A) \cong \mathrm{Hom}(A, \Omega) \cong \mathrm{Hom}(A\times 1, \Omega) \cong \mathrm{Hom}(1, \Omega^A) = \mathrm{Hom}(1,P(A)).$$

We have therefore $\Omega = P(1)$.

This shows that there are 3 equivalent descriptions of a subobject $m : S \to A$.

(i) its "extension" S: we will see that it can be symbolized as in **Sets** by $\{a|\varphi_S(a)\}$;

(ii) its characteristic map $\varphi_S : A \to \Omega$ which is a "predicate" of A;

(iii) the global section $s : 1 \to P(A)$ which is its "*name*".

The evaluation map $\varepsilon_A : A\times\Omega^A = A\times P(A) \to \Omega$ is a "membership" predicate: if $a : B \to A$ is an element of A, and if $s : 1 \to P(A)$ is (the name of) a subobject of A, then $\varepsilon_A \circ (a\times s) = True_{B\times 1}$ iff $\varphi_S \circ a = True_B$, that is iff a is an element of S.

3.4. TOWARDS LOGIC

The existence of an intuitionistic "internal logic" in a topos depends essentially on the fact that, being the set of the open sets W of the topological space U, $\Omega(U)$ is (functorially) an *Heyting algebra.* Ω is therefore a sheaf of Heyting algebras (an Heyting algebra object in **Sh**(M)). The consequence is that the "external" set of subobjects Sub(A) and the "internal" one $P(A)$ are also Heyting algebras, the canonical isomorphism Sub(A) $\cong$ Hom(1, $P(A)$) being an isomorphism of Heyting algebras.

4. TOPOI AND LOGIC

Now, the central fact is that a topos is exactly the categorical structure which is needed for doing logic. *But this logic is spatially localized.* For details see e.g. Saunders Mac Lane, Ieke Moerdijk 1992.

4.1. TYPES AND LOCALIZATION

We can associate with each topos **Sh**(M) a formal language $\mathcal{L}_M$ called its Mitchell-Bénabou language, and a forcing semantics called its Kripke-Joyal semantics. The crucial point is that a sheaf X can be considered as a *type* for variables x, which are interpreted as sections $s \in X(U)$ of X. *We get therefore at the same time a typification and a localization of the variables.* This achievement fits perfectly well with Husserl's description and explains Wittgenstein's remark. It provides them with a correct mathematical status.

(i) Sections are tokens denoted by variables belonging to types (species, essences). They are "concretely" particularized by the specification of their localization U and by their specific values. But as an element of type X, s particularizes an abstract unilateral relation of dependence, the relation "quality→extension" which is constitutive of X.

(ii) The relations between particular sections $s \in X(U)$, $t \in Y(V)$ are *external.* The relations between X and Y are *internal.* Nevertheless the linguistic expressions which express them are formulae in the formal language $\mathcal{L}_M$ associated to **Sh**(M), and are *the same.*

4.2. SYNTAX

How are the terms and the formulae of $\mathcal{L}_M$ syntactically constructed? Here is a summary of their inductive construction.

A term σ of type X constructed using variables y, z of respective types Y, Z has a *source* $Y\times Z$ and is interpreted by a morphism $\sigma{:}Y\times Z\rightarrow X$ which expresses its structure.

(i) To each $X \in$ **Sh**(M) considered as a type are associated variables, x, x', ... They are interpreted by the identity map $1_X : X\rightarrow X$.

(ii) Terms $\sigma : U\rightarrow X$, $\tau : V\rightarrow Y$ of respective types X and Y yield a term $<\sigma,\tau>$ interpreted by $<\sigma,\tau> : W = U\times V\rightarrow X\times Y$ of type $X\times Y$.

(iii)Terms $\sigma : U\rightarrow X$, $\tau : V\rightarrow X$ of the same type X yield the term $\sigma{=}\tau$ of type Ω interpreted by

$$(\sigma{=}\tau) : W = U\times V\rightarrow X\times X\overset{\delta_X}{\rightarrow}\Omega$$

where δ_X is the characteristic function of the diagonal subobject $\Delta : X\rightarrow X\times X$.

(iv) A term $\sigma : U\rightarrow X$ of type X and a morphism $f : X\rightarrow Y$ yield by composition a term $f\circ\sigma$ of type Y.

(v) Terms $\theta : V\rightarrow Y^X$ and $\sigma : U\rightarrow X$ of respective types Y^X and X yield a term $\theta(\sigma)$ of type Y interpreted by

$$\theta(\sigma) : W = V\times U\rightarrow Y^X\times X\overset{e}{\rightarrow}Y$$

where e is the evaluation map.

(vi) In particular, terms $\sigma : U\rightarrow X$ and $\tau : V\rightarrow\Omega^X$ yield a term $\sigma{\in}\tau$ of type Ω interpreted by

$$\sigma{\in}\tau : W = V\times U\rightarrow X\times\Omega^X\overset{e}{\rightarrow}\Omega.$$

(vii) A variable x of type X and a term $\sigma : X\times U\rightarrow Z$ of type Z and of source $X\times U$ yield a term of type Z^X interpreted by $\lambda x\sigma : U\rightarrow Z^X$.

(viii) Ω is the type of the *formulae* of $\mathcal{L}_M$. As it is an Heyting algebra, we get the logical operations of propositional calculus: $\varphi\wedge\psi$, $\varphi\vee\psi$, $\varphi\Rightarrow\psi$, $\neg\varphi$. It is easy to verify that if $\varphi(x,y) : X\times Y\rightarrow\Omega$ is a formula, we can write the subobject of $X\times Y$ classified by its interpretation in the "set theoretic" manner: $\{(x,y) \in X\times Y \mid \varphi(x,y)\}$.

(ix) One of the most remarkable facts of topoi theory is that it is possible to define *quantification* in a purely categorical manner. Let $f : A\rightarrow B$ be a morphism of **Sh**(M) and consider the "inverse image" functor $f^* : \mathrm{Sub}(B)\rightarrow\mathrm{Sub}(A)$ defined by composition with f. Its internal version is the morphism $P(f) : P(B) \to P(A)$. The fact is that $P(f)$ has *two adjoint functors*: a left adjoint one $\exists_f : P(A) \to P(B)$ and a right adjoint one $\forall_f : P(A) \to P(B)$. They generalize the two adjunctions in **Sets**. If $f : A \to B$, $S \subset A$ and $T \subset B$,

$$\exists_f(S) = \{b \in B \mid \exists a \in S \ (f(a) = b)\} = f(S).$$
$$\forall_f(S) = \{b \in B \mid \forall a \in A \ (f(a) = b \Rightarrow a \in S)\} =$$
$$= \{b \in B \mid f^{-1}(b) \subset S\}.$$

and $\begin{cases} f^*(T) \subset S \Leftrightarrow T \subset \forall_f(S) \\ S \subset f^*(T) \Leftrightarrow \exists_f(S) \subset T \end{cases}$

We have for sections s

$$s \in \exists_f(S)(U) \text{ iff } \{V \mid \exists t \in S(V), f(t) = s|_V\} \text{ covers } U,$$
$$s \in \forall_f(S)(U) \text{ iff } \forall V \subseteq U, f^{-1}(s|_V) \subseteq S(V).$$

Now, let $\varphi(x,y) : X \times Y \to \Omega$ be a formula of two variables in $\mathbf{Sh}(M)$. Let $p : X \to 1$ be the canonical projection and $P(p) : P(1) \to P(X)$. We get the adjunctions:

$$\Omega^X = P(X) \underset{\overset{\exists p}{\to}}{\overset{\overset{\forall p}{\to}}{\underset{P(p)}{\leftarrow}}} P(1) = \Omega.$$

We consider $\lambda x \varphi(x,y) : Y \to \Omega^X = P(X)$ and we get the formula of source Y:

$$\forall x \varphi(x,y) = \forall_p \circ \lambda x \varphi(x,y) : Y \to \Omega.$$

These categorical constructs show that the formal language $\mathcal{L}_M$ is, at the "linguistic" level, exactly of the same nature as the classical formal language of set theory. The main difference is that we have introduced a subtle dialectics *between the type of the variables and their localization.*

4.3. SEMANTICS

The Kripke-Joyal semantics of topoi is a *forcing* semantics generalizing Cohen's one. A variable x of type X denotes a *section* $s \in X(U)$, that is a morphism $s : U \to X$ where U is now the sheaf defined by U. The semantic rules are rules $U \Vdash \varphi(s)$ (U forces $\varphi(s)$). Let $s : U \to X$ and let $\mathrm{Im}(s) \in \mathrm{Sub}(X)$ be the image of s. One defines

$$U \Vdash \varphi(s) := \mathrm{Im}(s) \subseteq \{x \mid \varphi(x)\},$$

that is $U \Vdash \varphi(s)$ iff $U \xrightarrow{s} X \xrightarrow{\varphi} \Omega$ factorizes through $\{x \mid \varphi(x)\}$:

$$U \Rightarrow \{x \mid \varphi(x)\} \to 1 \xrightarrow{\text{True}} \Omega.$$

The semantic rules are:

(i) $U \Vdash \varphi(s) \wedge \psi(s)$ iff $U \Vdash \varphi(s)$ and $U \Vdash \psi(s)$.

(ii) $U \Vdash \varphi(s) \vee \psi(s)$ iff there exists an open covering $(U_i)_{i \in I}$ of U s.t. for every i, $U_i \Vdash \varphi(s|_{U_i})$ or $U_i \Vdash \psi(s|_{U_i})$ (intuitionistic rule for disjunction)

(iii) $U \Vdash \varphi(s) \Rightarrow \psi(s)$ iff, for all $V \subseteq U$, $V \Vdash \varphi(s|_V)$ implies $V \Vdash \psi(s|_V)$.

(iv) $U \Vdash \neg \varphi(s)$ iff there does not exist $V \subseteq U$, $V \neq \emptyset$ s.t. $V \Vdash \varphi(s|_V)$ (the negation is intuitionistic because Ω is a Heyting algebra and not a Boolean one).

(v) $U \Vdash \exists y\, \varphi(s,y)$ (y being of type Y) iff there exist an open covering $(U_i)_{i \in I}$ of U and sections $\beta_i \in Y(U_i)$ s.t. for every $i \in I$ $U_i \Vdash \varphi(s|_{U_i}, \beta_i)$.

(vi) $U \Vdash \forall y\, \varphi(s,y)$ iff for every $V \subseteq U$ and $\beta \in Y(V)$ we have $V \Vdash \varphi(s|_V, \beta)$.

The subtilities of this form of semantics result from the following problem. In a topos, one can interpret set-like expressions such as

$$\left\{ (x_i)_{i \in I} \in \prod_{i \in I} F_i \mid \varphi(x_i) \right\}$$

defined by $(a_i) \in \{(x_i) \mid \varphi(x_i)\}(U)$ iff $U \Vdash \varphi(a_i)$. But one has to be sure that such expressions define *subsheaves*. For this, one has to "sheafify" the sub-functors appearing in the set-like constructions.

5. SHEAF MEREOLOGY AND BOUNDARIES.

Two last points to conclude.

1. There is a natural *mereology* applying to sections of fibrations and sheaves. It satisfies all the axioms of mereo-topology (Smith 1993) except the one asserting the *unicity* of the universe of objects under consideration. Actually, an union of sections is a merging, a fusion, a gluing of compatible sections. The union of the sections s satisfying a predicate φ is therefore in general not a single section but only a set of *maximal* ones. (See Petitot 1994).

2. In perceptive statements, *boundaries* play a fundamental role. And one knows since Brentano that boundaries are somehow paradoxial entities. To tackle this point, we can use Lawvere's idea of *co-Heyting* algebras. In a Heyting algebra of open sets, the negation $\neg U$ of U is the *interior* of its complementary set, that is the largest open set V such that $U \cap V = \emptyset$. In a co-Heyting algebra of closed sets, the negation $\neg F$ of F is dually the smallest element H such that $F \cup H = 1$ (that is the closure of its complementary open set). One has

$$\neg(F \cap H) = \neg F \cup \neg H, \text{ but only } \neg(F \cup H) \subseteq \neg F \cap \neg G.$$

One defines then the *boundary* ∂F of F as the intersection $F \cap \neg F = \partial F$. ∂F is therefore defined by logical "contradiction". The boundary operator satisfies the Leibniz rule:

$$\partial(F \cap H) = (\partial F \cap H) \cup (F \cap \partial H)$$

Boundaries are characterized by $\partial B = B$ that is by $\neg B=1$ or $\neg\neg B=0$. In general, the double negation $\neg\neg F \subset F$ is the "regular core" of F (the closure of its interior). One has of course $F = (\neg\neg F)\cup\partial F$.

CONCLUSION

1. There exists a topological geometrical eidetics of the morphological structures of perception. This eidetics is neurologically relevant.
2. There exists an indexical status of spatial extensions in perceptive statements. This fact requires some sort of Kripkean modal semantics.
3. It is relevant to try to model Husserl's ante-predicative "genealogy" of logic.
4. The geometric logic yielded by topos theory can do the job. It has therefore a deep cognitive relevance.

Prof. Jean Petitot
École des Hautes Études en Sciences Sociales
54, bd Raspail
75270 Paris, France

BIBLIOGRAPHY

Asperti, A., Longo G., 1991.Categories, Types, and Stuctures, Cambridge, MIT Press.

Hoffman, W., 1989. "The visual cortex is a contact bundle", "Applied Mathematics and Computation", 32.

Husserl, E., 1954. Erfahrung und Urteil, Untersuchungen zur Genealogie der Logik., Hamburg, Claassen&Goverts.

Mac Lane, S., Moerdijk, I., 1992. Sheaves in Geometry and Logic, New York, Berlin, Springer.

Moerdijk, J., Reyes, G., 1991. Models for Smooth Infinitesimal Analysis, New York, Berlin, Springer.

Mulligan, K., 1992. "Internal Relations", "Australian National University Metaphysics Conference".

Petitot, J., 1990. "Le Physique, le Morphologique, le Symbolique. remarques sur la Vision", "Revue de Synthèse", 1-2, 139-183.

Petitot, J., 1992. Physique du Sens, Paris, Editions du CNRS.

Petitot, J., 1993. "Phénoménologie naturalisée et Morphodynamique", "Intellectica", Paris.

Petitot, J., 1994. "Phenomenology of perception, Qualitative physics, and Sheaf mereology", "16th. International Wittgenstein Symposium, Philosophy and the Cognitive Sciences", Kirchberg/Wechsel, (R. Casati, B. Smith, G. White eds), Vienna, Verlag Hölder - Pichler - Tempsky, 387-408.

Petitot, J., Smith, B., 1991. "New Foundations for Qualitative Physics", Evolving Knowledge in Natural Science and Artifical Intelligence, (J.E. Tiles, G.J. McKee, G.C. Dean eds.), 231-249, Pitman, London.

Petitot J., 1998, "Modépes morphodynamiques de segmentation spartiale" "Cahiers de Géographie du Qué bec", 42, pp. 335-47.

Smith, B., 1993. "Ontology and the logistic analysis of reality", Preprint.

Thom, R., 1988. Esquisse d'une Sémiophysique, Paris, InterEditions.

White, G., 1993. "Mereology, Combinatorics, and Categories", Preprint.

ARTURO CARSETTI

LINGUISTIC STRUCTURES, COGNITIVE FUNCTIONS AND ALGEBRAIC SEMANTICS

A *model* for a *first-order language* L is, from an extensional point of view, a structure $M = < U, I >$ where U is a non-empty set, called the *domain* (or *universe*) of M and I is an *interpretation function* that assigns appropriate items constructed from U to the non-logical terminology of the language L.

L is equipped with the usual variables, a countable set of predicate constants, connectives and quantifiers, punctuation marks, a countable set of function constants, a distinguished 2-ary predicate constant: "=" (in the case that L is a first-order language with equality) and a special predicate T (the truth predicate). 0-ary predicate constants are called proposition constants; 0-ary function constants are called individual constants.

In this sense, I assigns to each n-ary ($n \geq 0$) relation (or predicate) constant from L, different from "=", an n-ary relation on U, to the constant "=" (if present in L) the identity relation on U and to each n-ary ($n \geq 0$) function constant an n-ary function from U^n into U. 0-ary relations on U and 0-ary functions from U° into U are identified with truth-values and elements of U respectively. Thus, I assigns to each individual constant in L an element from U.

Given a structure M we write $| M |$ to denote the domain of M. For a first-order structure M we denote by A (M) the set of *assignments* over M: i.e., the set of all functions from the set of variables into $| M |$. For each structure and assignment there is, as usual, a *denotation function* that assigns a member of U to each term of the language.

The relation of *satisfaction* between models, assignments and formulas is, then, defined in the customary manner.

A formula S is said to be *satisfied* in a structure M by an assignment $a \in A$ (M) iff $V_{M,a}$ $(S) = 1$. Instead of writing $V_{M,a}(S) = 1$ we write also $M,a \models S$.

A formula S is *satisfiable* iff $M, a \models S$ for some structure M and some assignment a over M. A formula S is *true* in M iff $M, a \models S$ for any assignment a. If S is true in M, then M is called a *model* of S. S is (logically) *valid* iff S is true in every structure M.

The notion of model that we have just introduced is essentially an extensional one and we well know that extensional models are inherently inadequate to represent the way things are. At the level of extensional semantics we are in front of a coarse-grained representation of properties and relations and we are not able to examine closely the intrinsic modalities of this representation, modalities which appear, in any case, strictly linked to a specific range of possible continuous variations of the properties and relations.

So, in order to obtain a more adequate form of representation, instead of working exclusively with models as defined before, we shall take into account, from the beginning of our analysis, also a set W of new primitives: the *possible worlds*. The objects $w \in W$ are considered as primitive insofar as we assume (in the initial stage of our considerations) that they do not have any set-theoretic structure of their own. For each $w \in W$ there is a domain U^w of individuals that are actual in w. One of the members of W is considered as representing the actual world.

A. Carsetti (ed.), Functional Models of Cognition, 253-286.

Let us see how, within this conceptual framework, it is possible to get a finer-grained representation of properties and relations. In the old models the traditional interpretation function represents a given property by a set of objects-individuals. This set represents the extension of the given unary predicate in the model. At the level of *possible worlds semantics* we must, on the contrary represent a property P with a function f_P that assigns to each $w \in W$ a subset of U^w, thinking of $f_P(w)$ as the set of things that have property P in world w. In other words, an interpretation function appears as a two-argument function. In the old models, as we have seen, I assigns to each individual constant in L an element from U. Now we have to take into consideration the fact that, at the level of possible worlds semantics, the meaning of an individual constant is given not only by its extension, but also by the way in which this same extension is determined.

According to Frege, actually, the intension of a name must include more than just its reference. It must also include the way in which the reference is given. So, in place of a one-argument function $I(a)$, we must construct a two-argument function $I(a,w)$.

In this sense, a model for a *modal first-order language* L_M is a structure $M = < U, W, R, I >$ where U is a non-empty set, called the *domain* of M, W is a non-empty set, the set of *possible worlds*, R is a binary relation of "*accessibility*" on W, I is a function which assigns to each pair consisting of an n-ary relation constant ($n \geq 0$), different from "=", and an element w of W, an n-ary relation on U, to the constant "=", if present in L_M, the identity relation on U and to each n-ary function constant ($n \geq 0$) a function from U^n into U.

An *assignment* over a structure M is specified in the usual way. The *denotation function* for this type of semantics is a simple extension of the denotation function from first-order semantics.

A formula S is said to be true in M iff $V_{M,a,w}(S) = 1$ for any $a \in A$ (M) and any $w \in W$. If S is true in M, then we say that M is a *model* of S. We can write, instead of $V_{M,a,w}(S) = 1$: M, a, $w \models S$. A structure M is said to be a *T - system* iff R is reflexive. If, in addition, R is transitive, M is said to be an *S4-system*. If R is an equivalence relation, then M is said to be an *S5-system*. A formula S is said to be *X-valid* (where $X \in \{T, S4, S5\}$) iff S is true in every *X-system*.

Despite the fact, however, that the introduction of possible worlds makes the description much more adequate, this same description still remains a coarse-grained one. For example, the classical problems concerning the logical equivalence that are present in classical extensional models, arise again also in the theory of possible worlds. Many sentences that are true in the same possible worlds do not possess, in fact, the same semantical content or the same sense. How will it be possible to define further parameters in order to characterize these particular sentences in an univocal way? From a more general point of view, the real problem concerns the ultimate nature of possible worlds. Is there "life" on possible worlds? Is it possible to go into the inner structure (if existing) of these worlds? Is it possible to consider possible worlds as complete wholes, as maximal consistent sets or, on the contrary, do we have to see them as partial situations or short "stories"? What link exists between possible worlds and the sample-space points of the usual measure-theoretical treatment of probability calculus? As we said, in the beginning, at the moment of their introduction, the possible worlds have been considered as primitives.

But very soon, after the logical analyses performed by J. Hintikka, K. Thomason, K. Fine, P. Tichy etc. the *inner modal structure* of these worlds has become the object of specific logical investigations. A precise distinction between protoworlds and worlds has been introduced. The link existing between the different worlds and between individuals and attributes has constituted the object of meticulous studies. In this way a precise hermeneutic work started. The effects of this work play a central role within the present philosophical and epistemological debate concerning the ultimate "reasons" of first-order intensional semantics, and the link, for instance, existing between Model theory, Topology and Category theory.

One of the most important facets of this debate is represented by the analysis of the character of "partiality" of possible worlds, of their inner logical articulation and of the conditions of their inner coherence. At present, in addition to these synchronous aspects the diachronical aspects, concerning the "life" existing on the possible worlds have also become an object of specific discussion. Actually, the inner coherence of possible worlds appears to be strictly linked to the relation of accessibility existing between them, precisely to the action of the constraints imposed by the successive articulation of this specific relation.

Moreover, from a semantical point of view, when we enter with the tools of our analysis into the inner modal structure of possible worlds, we necessarily have to take into consideration the peculiar role played by the observer, by his capacity of determining, in a partial way, the conditions of their revelation and consistent expression. The information content is not, in other words, only a property of the source (in our case of the inner structure of possible worlds), but it also depends on the ability of the observer to construct sequential tests, particular logical investigations, to find adequate "distance" functions at the level, for instance, of the accessibility relation. In particular, the analysis of the distance functions and of specific measure functions defined with reference to higher-level languages and multilayered temporal frames, can be exploited by the observer in order to identify new patterns of data, new regions of the world under investigation and to "guide" the process of progressive revelation of these same data. In this way, new primitives will appear. Specific protoworlds will reveal themselves. The conditions of atomicity of possible worlds will eventually change. Some new constraints, in turn, can express their hidden presence. A strange life lying in the deep dimension will progressively manifest itself.

In order to take into account and to face such a complex reality, i.e., to go into the paths of the inner structure of possible worlds and of their ultimate connections, it appears suitable to resort, for instance, to the introduction of non-classical logics, to "creative" logics in particular, capable of elaborating in a finer-grained way the problems concerning the logical equivalence as well as the relationships existing between logical inference and rational inference. It is according to these theoretical tools that the "life" existing on possible worlds seems, finally, to have the possibility of entering the stage of our awareness. If we want to give a consistent explication of the meaning of linguistic espressions, of the deep information canalized by them, we have to situate these expressions within a general theory of meaning capable of giving an adequate explanation of the actual and global flow of real information. For the theorists of *Situation Semantics*, for example, the information flow concerns real things, living entities which interact with their environment.[1] Meaning lies in the systematic relationships existing between different kinds of real situations. These crossed relations or constraints permit a given situation to contain information concerning other different situations.

In this theoretical context the logic proper for a given natural activity (as, for instance, language), finally, reveals itself as nothing but the set of constraints that govern this same activity. This kind of logic contains, however, many more constraints than those which we are aware of as human beings.

Within the existing Reality a deep information exists that partially escapes us. The rational inference is precisely that particular type of cognitive activity that aims to explore facts in order to extract additional information implicitely contained in them. This certainly does not exclude the validity of the utilization of the rules of classical extensional semantics. These rules, however, concern only a particular sort of constraints, only some of the modalities that are necessary to pick out deep information. So, in order to collect additional information we have, to explore further constraints, to utilize ever new rules. But, we may wonder, which paths do we have to follow for it to be able to carry out a conservative extension of the inferential activity? Do we have to introduce further parameters or indices, contextual ones for instance, in the field of standard semantics? Or would we rather extend the logical tissue of first-order logic by following the paths of stronger and richer logics? Or do we still have to enter the mysterious kingdom of non-standard models?

In the second half of the sixties such questions as well as other problems concerning, in general, the extensional approach to that particular kind of natural (self-organizing) activity represented by natural language induced R. Montague to propose a conservative extension of classical tarskian semantics capable of making the information hidden in the deep structures of natural language more visible. This extension was realized by Montague by grafting, first of all, the theory of types onto the old trunk of tarskian semantics, by taking into consideration an articulated set of intensional indices and, finally, by resorting to an innovative utilization of the λ-calculus. As a result, as is well known, Montague succeeded in outlining a system of *Intensional Logic* capable of representing in an adequate way significant portions of natural language.[2]

The starting point of Montague's analysis possesses a precise model-theoretic character and consists of associating with each *category* C of natural language expressions a set of *possible denotations* called T_C or the *type* for C, defined in terms of the semantic primitives of the model. In standard first-order logic the primitives of a model are given by a non-empty universe of discourse U and by the set of the two truth values. Models may be different insofar as the chosen U will be different. But, once a universe is chosen, the types for categories are fixed. In this way, the type for Sentence is given by B (we indicate by B the set of the two truth values), and the type for individual constants and variables is given by U. The type for monadic predicates is the power set of U. In general, T_{P_n}, the type for n-place predicates, is the power set of n-tuples over U: U^n. (We have to remark, from the beginning, that the set B is a set with a boolean structure, in other words, the type for sentences is a boolean algebra).

Extending the primitive intuitions outlined by Russell concerning the theory of types and the successive sophisticated formalizaton of such a theory offered by Church, Montague discovered the articulation of types directly at the level of the inner structures proper to natural language and utilizes the analysis of such an articulation as a sort of picklock in order to enlighten the more secret aspects of this particular kind of cognitive structure. Once the types for the primitive categories are constructed, it is possible, then, to build the types for complex categories in terms of the types of the more simple categories. So, at the level of the classical theory of types, we may, starting from B and U, define the set B^U, that is the set of all possible functions from U into B: $(U \rightarrow B)$. Each function f in B^U characterizes a particular set of elements of U: the set of elements of U for which f assumes value 1.

Such a function, called *characteristic function,* can represent the extension of a predicate, i.e., the set of entities that are true with respect to a given predicate. The process can be naturally iterated. We can, for example, define all the functions from U into B^U : $(B^U)^U$. Each function h in $(B^U)^U$ individuates a relation: the relation existing between x and y iff $h(x)(y) = 1$. The logical mechanism which is at the basis of such an iterative process becomes completely clear once we define the set *Type* of logical types.

Let *Type* be a particular set that satisfies the following conditions:

1) $e \in Type$;

2) $t \in Type$;

3) if $a \in Type$ and $b \in Type$, $<a, b> \in Type$.

According to these conditions *Type* is a set of types; e and t represent, respectively, "entity" and "truth". $<a, b>$ represents the type of formulas from objects of type a to objects of type b.

If α is an element of a given type a, then the semantic value of α (called the *denotation* of α) belongs to a well determined set which is called D_a and is referred to as the set of possible denotations of type a. The sets D_a are recursively defined by the following rules:

1) $D_e = U$;

2) $D_t = B$;

3) $D_{<a, b>} = D_b^{D_a}$.

(The notation W^V represents the set of mappings from a set V into a set W).

The conceptual scaffolding represented by type theory allows Montague to establish two precise homomorphisms at the level of the simulation of natural language: the homomorphism between natural syntax and formal syntax and the homomorphism between formal syntax and semantics. From a formal point of view, stating these two homomorphisms allows, in turn, for the introduction of the λ-calculus in the landscape represented by the theoretical frame of intensional logic.[3] With regard to this, we have to underline, once again, that at the origin of Montague's system there is the precise association of a (unique) logical type with each syntactic category of natural language. In this way, at the level of this type of denotational semantics we can present, as a guide, the following correspondence table:

Natural expression	Set of possible denotations	Usual terminology
	U	Individual
Sentence	B = {1, 0}	Truth value
Intransitive verb	$((W \to U) \to B)$	Set of individual concepts
Noun phrase	$((W \to ((W \to U) \to B)) \to B)$	Set of properties of individual concepts

Figure 1

The correspondence between formal syntax and semantics is also well illustrated by the following table:

Type	Denotation
e	Individual
t	Truth value
$< s, e>$	Individual concept
$<<s, e>, t>$	Set of individual concepts

Figure 2

Within the framework of this theoretical construction, a model **M** is given by a fourtuple $<U, W, \boldsymbol{R}, V>$ where:

1) U is a set of individual objects;

2) W is a set of possible worlds w;

3) $\boldsymbol{R}$ is a set of accessibility relations between the possible worlds;

4) V is an interpretation function: it assigns elements of U to the individual constants and it assigns sets of n-tuples of elements of U to the n-place predicate constants for each possible world w.

The *assignment function* g assigns elements of U to the individual variables of the language: for each type a, the function g assigns elements of a well defined set D_a to the variables of type a.

The *semantic value* of any expression α of the formal language is obtained recursively from the semantic values of the basic elements of the language and from the rules that define the semantics of the logical connectives, the quantifiers, the modal operators and the intension and extension operators.

The semantics of the basic elements is defined as follows:

a) if b is a non-logical constant, then

$$[b]^{M, w, g} =_{\text{def}} V(w, b);$$

b) if x is a variable, then

$$[x]^{M, w, g} =_{\text{def}} g(x).$$

If we refer this kind of theoretical construction to the examples given earlier concerning the possible denotations associated with the different categories of natural language expressions, we can immediately understand how the multiplicity of types present in Montague's system of intensional logic may be utilized in order to give a broad representation of the meanings of the different linguistic expressions.

As is well known, at the linguistic level fundamental syntactic categories are represented by common nouns, proper nouns, determiners, transitive verbs, etc. If we indicate, according to the notations adopted by Keenan and Faltz[4], by N the category of *common nouns*, by $\overline{N}$ the category of *noun phrases* and by $\overline{N}_{\text{prop}}$ the category of *proper nouns*, it is easy to realize that common nouns, insofar as they are not distinguished in first-order logic from P_1's (the monadic predicates), are normally thought of as subsets of U. In other words, the set of possible denotations of N's (= P_1's) is U^*: the collection of all the subsets of U. So, in first-order logic $T_N = U^*$. But, taking T_N to be the power set of U implies that T_N must be considered as a power set algebra, hence as a boolean algebra.

Even easier to define is the type for $\overline{N}_{\text{prop}}$. At the level of first-order logic, actually, the type for $\overline{N}_{\text{prop}}$ is directly taken to be U. Things are different and more problematic, however, when we try to introduce a definition in first-order logic for the type for $\overline{N}$. This kind of logic, in fact, has no such category although it does have the equivalent of a subcategory of it: that is $\overline{N}_{\text{prop}}$. The problem really concerns the same definition of that particular semantic primitive represented by U (the type, as we have just seen, for $\overline{N}_{\text{prop}}$). Actually, U is simply not versatile enough to provide denotations for complex $\overline{N}$'s such as, for instance, "every man".

"Imagine, for example" as Keenan and Faltz remark "a U with just two members, say John and Bill, and let *man* be interpreted as U itself (so everything which exists in that universe is a man). Then if all complex $\overline{N}$'s were to take their denotations in U, *every man* would have to denote either John or Bill, say John. Then *every man is John* would be true, which it obviously isn't since Bill is a man and he is not John".[5]

In this sense the standard first-order logic reveals itself to be a theoretical instrument which is not strong enough to mirror, within its general frame, the subtleties proper to the syntax of natural language. We can easily recognize these particular subtleties as present when we try to cope with the action of generalized quantifiers, complex noun phrases, transitive verbs, etc. As Barwise and Cooper showed in 1981 with a wealth of examples, when we try to translate the expressive power of a fragment of natural language in a logical and mathematical setting like that one that characterizes the extensional first-order logic, we are immediately faced with precise losses of meaning, misunderstandings and so on.

A solution to these particular problems was, precisely, proposed in 1970 by R. Montague. The new conceptual scaffolding outlined by the great American logician was successively improved, criticized and enlarged by Barwise, Cooper, Kamp, Partee, Keenan, Faltz, etc. In the light of Montague's first intuitions, proper nouns must not be considered as denoting elements of U but rather as denoting the collection of subsets of U which contain a fixed element. More precisely, we can refer to the following definition:

Definition 1. For all $x \in U$, $I_x =_{\text{def.}} \{K \subseteq U : x \in K\}$.

In other words, the entity x is represented by the set of all the sets to which x belongs. More precisely, I_x will be called the *individual* generated by x. $T_{\bar{N}\text{prop}}$ will be, then, given by: $\{I_x : x \in U\}$.

If we simply consider the elements of T_N as extensional properties, the above definition permits us to think of the denotation of "John" as the set of properties which hold of some fixed element of U. In this sense, the elements of U will be in a one-to-one correspondence with the set of individuals. So, the proper noun "John" may be considered, according to the terminology introduced by Barwise and Cooper, as a *generalized quantifier*. In order to see, by means of an example, how it is possible to interpret proper nouns as "John" and complex N's in the same type let us simply consider that "every man" formally denotes: $\cap \{I_x : p \in I_x\}$. But, conversely, a property p is "a member of an individual I_x iff $x \in p$, since p is a subset of U and I_x is the set of all the subsets of U which contain x as a member. Thus, we might also write the denotation of *every man* above $\cap \{I_x : x \in p\}$, and analogously for *a man*".[6]

What we have just said, in a very concise way, allows us to understand that the system of intensional logic outlined by Montague constitutes effectively an innovative extension of traditional first-order logic. In this last kind of logical system proper nouns are considered as individuals and are submitted to a formal processing completely different with respect to the sort of processing normally utilized in the case of generalized quantifiers. When we assume, however (as is the case of Montague's logical system), an entity or individual $x \in U$ and define a set $I_x =_{\text{def}} \{K \subseteq U : x \in K\}$, this particular individual, for instance the particular simple *NP* "John", can ben taken as denoting the family of sets which contain "John". In other words, the *NP* containing just the word "John", represented by $[\text{John}]_{NP}$ will denote the family of sets containing "John". The *NP* in question represents, thus, a quantifier.

Within the conceptual framework of Intensional Logic as outlined by Montague, the aforesaid family or collection of sets can be represented, for each expression α of type *e*, in the following way: $\lambda P[P\ (\alpha)]$. The generalized quantifier $\lambda P[P\ (\alpha)]$ codes, in other words, the same information that the individual α codes. This specific expression describes-translates, so to speak, at the deep level, the role played by a proper noun such as "John", allowing us to understand the function performed by this kind of linguistic expression within the "landscape" constituted by natural language and with respect to the semantic system concerning the links existing between the observer and the surrounding world. So, if $<e, t>$ is the type of simple predicates, $<<e, t>, t>$ will be the type of generalized quantifiers.

If we indicate by *P* the set of properties ($T_N =_{def} P : P$ is a complete and atomic boolean algebra or *ca*-algebra) we can directly affirm that those particular functions called generalized quantifiers (or quantifiers of type <1>) denote in $[P \rightarrow \mathrm{B}]$ the set of functions with domain *P* and range included in B.

The conceptual revolution underlying these intuitions may be understood in a very simple way, if we refer to an example introduced by Barwise and Cooper in 1981.[7] When we try to translate a sentence containing *NP*'s like: (a) $[\text{Some person}]_{NP}$ $[\text{sneezed}]_{VP}$ into predicate calculus, this same sentence is, normally, represented, ignoring tense, as: (b) Ex $[\text{person}\ (x) \wedge \text{sneeze}\ (x)]$. In effect, (b) is not really a translation of (a), but of a logically equivalent, but linguistically different, sentence: "something was a person and sneezed". On the contrary, what we aim to obtain is a sentence that will be true just in case the set of sneezers (represented by $\hat{x}$ [sneeze (x)]) contains some person, that is: (c) (Some person)$\hat{x}$ [sneeze (x)].

In this sense, there is no mismatch between the syntax of noun phrases in a natural language like English and their usual representation in traditional predicate logic.

From a general point of view, the noun phrases act, semantically, like the logician's generalized quantifiers and even proper nouns can be treated as quantifiers. Hence the appearance, at the surface level, at the level of scientific awareness, of a linguistic universal can be expressed by the following definition, as indicated by Barwise and Cooper: every natural language has syntactic constituents (noun phrases) whose semantic function is to express generalized quantifiers over the domain of discourse. According to this definition we can recognize the universal character of noun phrases. Moreover, the quantifier universal serves also to distinguish natural language from other languages like, for instance, the formulation of first-order predicate calculus.[8]

The rediscovery in our century of the problematics concerning linguistic universals realized under the guidance of Montague, Bach, Barwise, Cooper, Keenan etc. has led many scholars in the eighties to develop an analysis of the relationships existing between logic and linguistics in a completely new way. As a consequence of this kind of analytical work the laws of logic no longer appear as parts of a wider kingdom represented by abstract Mathematics, but as specific and articulated properties of natural language as well as of the use that we normally make of this kind of language. Hence the equation: laws of thought = laws of language. It is exactly logic that, step by step, comes to reveal itself as a part of linguistics.

The theoretical scaffolding supported by the aforesaid general considerations constituted, in particular, the object of the careful analysis of the formal structures of natural language as developed in the eighties by Keenan and Faltz. The final result of this analytical work was the individuation of a new "area" of research: Keenan-Faltz *Boolean Semantics.* In 1985 these scholars came to present a nearly complete revision, in boolean terms, of the system of intensional logic outilined by Montague. This revision brought to light many of Montague's hidden assumptions. It also extended its formal applicability. Keenan and Faltz start by noting that, from an ontological point of view, U is no longer the type for any category of natural language (or for L_C: the *core language* outlined by Keenan and Faltz in order to represent the basic predicate-argument structure of (first-order) English). There are no expressions which denote elements of U, even if in Montague's system denotations for common nouns, proper nouns, etc. are, generally, defined in terms of U.

As Keenan and Faltz remark: "U rather appears now as a kind of noumenal world of entities which 'supports' the phenomenological world of individuals. U then is an essential *mystery* on this approach and the ontology which comprises it is essentially mysterious. In the extension of this approach we propose below the mysterious U is eliminated completely, though proper nouns, full noun phrases and common nouns otherwise preserve exactly the character they have on the mysterious approach".[9]

The existence of such mysterious dialectics is well testified by the fact, for instance, that a simple linguistic expression like "John" appears, on the one hand, to be linked to entities living in U, and, on the other, it can be translated into a complex formula like: $(\lambda P)\ (P(j))$ itself interpreted as the set of properties which hold of the denotations of j.

It is precisely in order to cope with such difficulties that Keenan and Faltz propose the elimination of U from the ontology of the extended version of first-order logic as outlined by Montague. From a conceptual point of view, the elimination of U can be articulated in three steps: 1) it is, first of all, necessary to characterize in an alternative way T_N so that T_N can be taken as the real primitive instead of U; 2) individuals must be defined as subsets of T_N which satisfy certain conditions; 3) $T_{\overline{N}}$ must be defined as the power set of T_N and regarded, thus, as a boolean algebra.

In the light of these precise guidelines, the interpretation of the proper name "John", finally, simply appears as an arbitrary element of the type for $\overline{N}_{\text{prop}}$. In extended first-order logic T_N was considered, as we have seen, as a power set boolean algebra. Since power set boolean algebras are a special case of boolean algebras, namely, they are complete and atomic algebras, T_N will be taken, from the beginning, as an arbitrary complete and atomic algebra. It possesses the boolean structure of a power set algebra, but need not be the power set of some set. In this way, Keenan and Faltz obtain, by means of simple conceptual tools, a theoretical result of great importance: the type for N is now defined solely in terms of its boolean properties. We have, in other words, the possibility of defining individuals directly in terms of the boolean structure of T_N, eliminating all reference to U.

Let us now examine in which way, from an algebraic point of view, this set of assumptions can arrive at consistent formulations. In Montague's system the individual generated by $x \in U$ was characterized, as we have just seen, as $\{K \subseteq U : x \in K\}$. But, as our authors remark, $x \in K$ iff $\{x\} \subseteq K$. So, we can define, in an equivalent way, I_x to be $\{K \subseteq U: \{x\} \subseteq K\}$. We have, however, to note that $\{x\}$ is an element of $U^* = 2^U$. In fact $\{x\}$ is an atom. Taking T_N as an arbitrary *ca*-algebra it is, then, possible to present an equivalent definition of individuals as follows:

Definition 2. For B any boolean algebra and b any atom of B, $I_b =_{\text{def}} \{p \in B: b \leq p\}$ We call I_b the *individual* generated by b. We denote by I_B the set whose elements are exactly the individuals I_b.

The elimination of U leads Keean and Faltz to outline directly a new form of ontology for L_C, an ontology that presents itself essentially as a pair $<P, \mathrm{B}>$ where P is an arbitrary complete and atomic boolean algebra called "the (extensional) properties" and B is the boolean algebra 2 called "truth values". With respect to such an ontology it is, then, possible to define the types for P_0, N, $\overline{N}$, $\overline{N}_{\text{prop}}$ in the following way:

1) $T_{P_o} =_{\text{def}} \mathrm{B}$;
2) $T_N =_{\text{def}} P$;
3) $T_{\overline{N}} =_{\text{def}} P^*$;
4) $T_{\overline{N}_{\text{prop}}} =_{\text{def}} I_P$.

We have just defined P as an algebra of properties since its elements are effectively represented by properties. In this sense, T_N will turn out to be isomorphic to the type for P_1. So, what characterizes, at the deep level, the type for N is its structure, a structure that possesses a specific boolean character. In particular, we have to remark that in first-order logic T_N is the power set of U which is itself the set of denotations for proper nouns and, thus, constitutes the set of primitive individuals. In a corresponding way, despite the existence of the differences that we have reported above, also in the logical system outlined by Keenan and Faltz, T_N is indistinguishable from the power set of the individuals in L_C. Actually, P $(=T_N)$ is isomorphic to $(I_P)^* = 2^{I_P}$, the power set of the individuals on P. However, in L_C individuals are not considered as primitives: they are defined in terms of properties, namely, as sets of properties. Taking individuals to be sets of properties provides a natural way to represent the idea that individuals possess or fail to possess certain properties according to whether these sets contain or fail to contain these properties as members. Precisely with respect to this frame of reference, properties can be considered, as we have said, either as the interpretations of *N's* or as objects determined by the interpretations of P_1*'s*.

As a consequence of Definition 2 above, we realize immediately that each individual must contain, by construction, exactly an atom. From this it is easy, in turn, to show that an atom is a property which is contained in exactly one individual. Conversely, if p is a property which is contained in exactly one individual, then it is possible to show that p must be an atom, provided we assume that T_N is atomic. Actually, we have just seen that $I_b =_{\text{def}} \{p \in B: b \leq p\}$. But we know that in a boolean algebra B a *principal filter* is a filter generated by a single element $b \in B$, and we also know that there is a precise correspondence between the atoms of B and the principal *ultrafilters*. This fact involves that, at the level of T_N, the individuals should be precisely those subsets of T_N which satisfy the Meets, Joins and Complements conditions. In other words, the individuals must be just the I_b's defined earlier.

It is in this sense, that, with reference to the logical system outlined by the American linguists, L_C appears to satisfy a further adequacy criterion, the criterion of *extensionality:*

Definition 3. For all properties p and q, $p = q$ iff the set of individuals which contain p is the same as the set which contain q.

Let us now recall briefly some of the simple formulas that we have utilized until now. We have just seen that the set of the individuals on P, $I_P = \overline{T_N}_{\text{prop}}$. On the other hand, $\overline{T_N} = P^* = 2^P$. P^* is a power set algebra. But $T_N = P = (I_P)^* = 2^{I_P}$. So $\overline{T_N} = P^*$ is the power set of T_N. This means that the full set of possible N denotations is uniquely determined by the boolean operations, once the individuals are given.

$\overline{T_N}$ is the set of all sets of properties and we have just seen that the individuals are, in fact, ultrafilters. Let us recall, however, that also in Montague's approach the type for common nouns is the collection of subsets of U, and constitutes, thus, a power set boolean algebra. Individuals are precisely particular subsets of this algebra.

In this context we have to remember, first of all, that, from an algebraic point of view, 1) every finite algebra is complete; 2) the power sets are examples of boolean algebras par excellence. With reference to 2) we have seen earlier that a power set boolean algebra is a special case of boolean algebra characterized by the presence of two additional properties: atomicity and completeness. There is, however, another important property of complete boolean algebras that is particularly useful at the level of the generation of linguistic structures. We can prove the following theorem as stated by Keenan and Faltz:

Theorem 1. A complete boolean algebra B has a set of (complete and atomic) *ca* - free generators iff B is isomorphic to the power set of a power set.

For example, $\overline{T_N}$ is extensionally P^*, the power set of T_N, itself taken to be a *ca*-algebra and, thus, isomorphic, as we have just said, to the power set of its atoms. With respect to this conceptual frame, the set of individuals must then be considered as a set of *c*-generators for $\overline{T_N}$. Individuals, however, are not only a set of *c*-generators for $\overline{T_N}$, they are also free. $\overline{T_N}$ is, precisely, a *ca*-algebra which has a set of free generators. In fact, the only algebras that have a set of *ca*-free generators are just those which are isomorphic to power sets of power sets and have cardinality $2^{(2^n)}$ for some cardinal n. In other words, if we have a model of n individuals there will be, at the level of the model, $2^{(2^n)}$ sets of properties. We immediately have to underline, however, that most boolean algebras do not have a set of free generators. For instance, the types of Modifier categories are not algebras with individuals. The linguistic structures, in fact, are, in general, well differentiated and polymorphic.

From a mathematical point of view we can express the properties just mentioned, in a synthetic way, as follows. Let G be a set of generators of a boolean ring B; then the following properties are equivalent:

a) G is independent;

b) B is free on G

If B is n-free, card $B = 2^{2^n}$ and B possesses 2^n atoms (or 2^n ultrafilters) which are the simple polynomials on G. So, when we have a boolean algebra that is isomorphic to the power set of a power set, we are simply faced with a particular composition of power sets according to which the cardinality of the first power set refers to the number of atoms and the cardinality of the second power set refers to the number of elements of the entire complete boolean algebra. In the case of P^*, for example, we have: $P^* = 2^P = 2^{2^{I_P}}$. Let us recall that by I_p we indicate the elements of the set $I_P =_{\text{def}} \{I_p: p \text{ an atom of } P\}$. P is isomorphic to $(I_P)^*$, the power set of the individuals on P and there is a one-to-one correspondence between atoms and individuals (kinds of individuals). When we pass from P to $T_{\bar{N}} = P^*$ we have, as a result, that the individuals now play the role of generators with respect to the entire free algebra P^*. P^* is the power set of P, itself taken to be a c-algebra and, thus, as isomorphic to the power set of its atoms. P^* has as its atoms, at the surface level, the old elements of $P = 2^{I_P}$. This fact can be interpreted also as the result of the application of a particular kind of inner transformation concerning the specific interplay existing between atoms and generators. In effect, starting from generators considered as predicates we obtain atoms interpreted as collections of attributes; but if we start from generators considered as closed individuals (i.e. as complex constructions characterized by the presence of a closure process which determines the identification of a predicative "shape") we obtain atoms considered as worlds-urns. Once certain specific condition are given, to compose a power set with another power set can involve a precise transformation of the role played by the individuals. As a consequence of this kind of change, the corresponding filters at the algebraic level will change as well.

* * *

Once given an adequate interpretation for common nouns and proper nouns, the next step is represented by the introduction of a procedure of interpretation also for extensional determiners. The elements of Det ($=\bar{N}/N$) can be considered, from a general point of view, as functions from P into P^*. But it is also necessary to impose specific constraints on the meanings of extensional English Dets.

For example, the expression "every student" will be interpreted as the value of the "every" function at the student property. "Every student" denotes the set of properties which the individual students have in common. In the same way, the expression "some student" will denote the set of properties which at least one individual student possesses. From a general point of view, the function "every student" concerns collections of properties. Such a function can be directly considered as a predicate concerning not single entities but sets of entities.

If we limit our attention to the more simple (and common) determiners, i.e. one-place determiners like, for instance, "most" etc., we can immediately realize, therefore, that the essential difference existing between the generalized quantifiers or quantifiers of type <1> and quantifiers of type <1,1> lies in the fact that *NP*'s denote in $[P \rightarrow \mathrm{B}]$ (a *NP* as, for instance, "John" combines with property denoting expressions to form sentences) while Det_1's combine with property denoting expressions to form *NP*'s, denoting, thus, in $[P \rightarrow [P \rightarrow \mathrm{B}]]$.

These short hints show us some of the further directions of the analysis and allow us to realize, in some way, that until now we have only managed to examine the first steps of a very long path of research. The aim of our investigation does not consist, however, in presenting a general survey of the results so far obtained in the field of natural language (boolean) semantics: on the contrary, we want, by means of a critical examination of the conceptual bases of some of the more powerful systems of Intensional Logic and the more adequate theories of reference, to present some new ideas and suggestions, at the methodological level, for the possible outlining of a new kind of approach to the problems concerning the link between semantic structures, on the one hand, and cognitive functions, on the other. In this regard, let us remark that, from a theoretical point of view, what we have said until now permits us to understand how the construction of generalized quantifiers works. In this sense, we can simply affirm that two fundamental operations are at the basis of the definition of proper names (that is of that particular grid through which we normally refer to the events in the world):

1) the construction of a complex attribute that acts as a sieve;

2) the statement that a specific class defined by partitioning the family of sets provided by the model is not void.

These specific considerations permit us, in turn, to realize, in a very clear way, that the conceptual scaffolding built by Barwise, Cooper, Keenan, Faltz etc. is, from an epistemological point of view, essentially characterized in an antinominalistic vein. In effect, for a nominalist, as well as for a nominalist realist, proper nouns constitute the real and primary vehicle of reference. Their ontological land is populated by empirically distinguishable individuals; this is exactly the opposite point of view with respect to the conceptual approach advocated by Keenan and Faltz. In fact, as we have just seen, in the opinion of these scholars reference may be expressed only within the realm constituted by a sophisticated sequence of linguistic and conceptual constructions. Therefore, they deny any significative role for the noumenal world of primitive entities *U*. Individuals are no longer held to be encounterable and primitive objects. Let us also remark, in particular, that this specific kind of theoretical approach which possesses a very "radical" character and which, in some way, parts from the customary paths followed by the traditional first-order logic, partially tunes Keenan and Faltz into the theory concerning the roots of reference as outlined by the great American philosopher W. V. Quine.

Quine also, in fact, ends up assimilating proper nouns into predicates drawing, thus, a precise boundary line with respect to his initial nominalistic strategy. In his opinion, the words which do the referring in the sentence "Socrates is a philosopher" do not occur in the surface structure of the sentence. They can be made explicit only by the use of quantifiers and variables. So, we have: "There is something that Socratizes and is a philosopher". In this sense the equation "$x = a$" must be reparsed in effect: "as a predication '$x = a$' where '$= a$' is the verb, the 'F' of 'Fx'. Or look at it as follows. What was in words 'x is Socrates' and in symbols 'x = Socrates' is now in words still 'x is Socrates', but the 'is' ceases to be treated as a separate relative term '='. The 'is' is now treated as a copula which, as in 'is mortal' and 'is a man', serves merely to give a general term the form of a verb and so suit it to predicative position. 'Socrates' becomes a general term that is true of just one object, but general in being treated henceforward as grammatically admissible in predicative position and not in positions suitable for variables. It comes to play the role of the 'F' of 'Fa' and ceases to play that of the 'a' ".[10]

So, "x is Socrates" now has the form of "x is round". In other words, when we write in predicative terms "x Socratizes" we have, in effect, defined the attribute which works as a sieve in order to make a specific partition, in order, that is, to individuate a class. The successive and necessary step consists in giving a particular form of existence to this class, in affirming that it is not void. Here we find the peculiar role of quantifiers: (Ex) (x is Socrates) or (Ex) (x = Socrates).

Quine works essentially in the landscape represented by monadic and polyadic predicates, whereas Montague makes essential reference to T_N and $T_{\overline{N}}$. But we have remarked before that a precise equivalence exists between T_N and P_1 in Montague's system. Let us also note that in the same way that T_{P_1} is isomorphic to the power set of I_P, T_{P_n} is isomorphic to the power set of $(I_P)^n = ((I_P)^n)^*$.

In this theoretical perspective, at the objectual level, the variable becomes, thus, for Quine, the essence of ontological discourse. Once a theory is formulated in quantification terms, its referential objects can be said to be the values of its quantified variables. The relative clause and the quantifiers, in Quine's view, constitute the real roots of reference.

Followirg this line of thought, we now understand why it is perfectly admissible in accordance with Quine's opinion to infer $(Ex)Fx$ from $F(a)$, and why it is impossible to construct $(EF)\ Fx$. Actually, he admits existence assumptions concerning the individuals: the objects, that is, of the domain, but he doesn't admit existence assumptions concerning properties or relations. Hence, a specific distrust with respect to second-order predicate logic, a distrust that is perfectly coherent with the psychogenetic reconstruction of the referential apparatus outlined by him in his work.

In this way it is possible to affirm that, in Quine's opinion, what is recognized as being is exactly, in first instance, what we assume as true for monadic predicates (complements included). We really know only insofar as we regiment our system of the world in a consistent and adequate way. At the level of proper nouns and existence functions we have already given some clues about the standard form of a regimented language whose complementary apparatus consists of predicates, variables, quantifiers and truth functions. However, if the discoveries in the field of Quantum Mechanics should oblige us, in the future, to abandon the traditional logic of truth functions, it will be our received notion of existence which will be challenged. New forms of theoretical regimentation and new processing procedures will, thus, appear nearly at the same time.

Our ontology as well as our grammar, as Quine affirms, are ineliminable parts of our conceptual contribution to our theory of the world. And, in effect, we have seen how the passage, for instance, from the ontology characterizing Montague's system to the ontology outlined by Keenan and Faltz, marks the stages of a progressive unfolding of our knowledge of the modalities proper to human beings of understanding both the surrounding world and themselves. So, it doesn't appear possible to think of entities, individuals and events without specifying and constructing, in advance, a specific language that would be used in order to speak about these same entities, in accordance with Wittgenstein's aphorism "Meaning is use". We are faced, in general, with a conceptual perspective like the *"internal realist"* perspective advocated by Putnam, whose principal aim is, for certain aspects, to link together the philosophical approaches developed respectively by Quine and Wittgenstein. Actually, Putnam conservatively extends the approach to the problem of reference outlined by Quine. In his opinion, to talk of "facts" without specifying the language to be used is to talk of nothing. "Object" itself has many uses and as we creatively invent new uses of words, "we find that we can speak of 'objects' that were not 'values of any variable' in any language we previously spoke."[11] The notion of object becomes, then, a sort of open land, an unknown territory as well as, on the other hand, the notion of reference. The exploration of this land appears to be constrained by use and invention. But, we may wonder, is it possible to guide invention and to control use? Are there any rules (inductive rules, for instance) at the level of this strange, but completely familiar land?

A possible answer to these difficult and ancient questions appears implicitly contained in the original formulation of Putnam's program. In fact, it appears quite clear that the invention of new words necessarily coincides with an extension and an inner reorganization of language in use. This extension will concern, however, not only the extensional aspects of the first-order structures proper to natural language, but also the intensional aspects and the second-order aspects. We have to stress that, as a matter of fact, language can go beyond any limit as determined by a particular abstract conception of linguistic formal rules. As Putnam remarks: "Reason can go beyond whatever reason can formalize."[12] These considerations can be paraphrased in a different way by saying that, as we precisely and consistently specify the methods of reasoning permitted at the level of a given language, we determine an upper bound to the complexity of our results. We cannot prove, in other words, according to Goedel's intuitions, the consistency of a particular system by methods of proof that are restricted enough to be represented within the system itself.

In this way, according to Goedel's conception, as revisited for instance by Chaitin, we have to step outside the methods of reasoning accepted by Hilbert; in particular, we have to extend Hilbert's finitary standpoint (and in our case, as a consequence, the conditions of predicative activity as defined by Quine) by admitting the necessary utilization of certain abstract concepts in addition to the merely combinatorial concepts referring to symbols. For this purpose we must count as abstract those concepts that do not comprise properties or relations of concrete objects (the inspectable evidence) but which are concerned with thought-constructions, with the articulation of the intellectual tools of invention and control proper to the human mind.

If we take the mathematical proofs as examples of this kind of construction, the concepts just mentioned appear, then, to concern not only the simple physical combinations of signs-symbols but the ultimate sense of these same symbols. The utilization, at the semantic level, of abstract concepts, the possibility of referring to the sense of symbols and not only to their combinatorial properties, the possibility, finally, of picking out the deep information existing in things (as well as in the language considered as an expression of a precise cognitive activity) open, as is easy to understand, new horizons at the level of the problems of reference.

When, for example, we affirm "the apples are red" we realize perfectly, as speakers of a natural language, that the predication constitutes a link which is stronger than the simple conjunction; the predication, in fact, requires the "immersion" of the apples into the red. From a perceptual, linguistic and conceptual point of view, we are really in front of an object considered as a whole, an object, that is, that possesses a precise character of global unity, a character which anticipates even more complex forms of organicity. A book, for instance, is composed of pages which themselves, insofar as they are parts depending on an organic whole, do not constitute an object. These same pages, however, become simple objects once they are torn out of the book. Husserl was the first to introduce in our century a sort of logical calculus of parts and wholes. He thought that only certain "organic" wholes constitute real objects and, in this sense, he introduced the notion of *"substratum"*. The dialectics existing between part and whole really "homes" in the same act of predication, namely in the language considered as action and intervention. We have to discover the secret functional aspects of such dialectics if we want to enter the mysterious kingdom of primary linguistic expressions. But to face this kind of dialectics means also to face the problems concerning the *symbolic dynamics* that governs the dependency links, the relationships between parts and wholes. If, as we have just seen, at the ontological level we can recognize the existence of an entire sequence of expression levels of increasing complexity, the dialectics part-whole must be compared to the successive constitution of the different levels as well as to the net of generative constraints through which this kind of process expresses itself. It is only within the secret meanders of this symbolic dynamics that we can, finally, find the theoretical instruments capable of showing us the correct paths that we need to pursue in order to obtain a more adequate characterization of the Fregean *Sinn*.

At the moment, the aim that many scientists working in the field of natural language semantics want to attain consists, precisely, in individuating the logical forms subtended to the dynamics just mentioned as well as the relationships existing between this very dynamics and the processes of meaning construction. If, for instance, we insert at the level of the ontology proposed by Keenan and Faltz, the dialectics of the dependency links and, at the same time, decide to consider the possible worlds not as external indices or pseudotypes but as types in every respect, the static landscape outlined by Keenan and Faltz suddenly becomes alive.

At the level of the arising new horizons, precise algebraic structures have to be grafted onto the previous structures and complex functional spaces will unfold with reference to a specific frame of geometric interactions. In consequence of the dynamic transformations of the dependency links, the same "shape" of real objects, as it is normally recognized by us, will change. The same modalities through which we regiment the universe of the events as well as the modalities through which we induce, in a cooperative way, by coagulum and selection a specific coherent unfolding of the deep information content hidden in the external Source will be subject to a specific transformation. So, the ontological discourse connects itself with a precise genealogical and dynamic dimension, a dimension that must be linked, however, to a specific continuous procedure of conceptual recovery if we want this same dimension to continue unfolding freely, performing, so, its specific generative function.

In this sense, the true cognition appears as a recovery and as a guide, as a form of realized simulation. With respect to this particular framework, the simulation activity, the construction, for instance, of an adequate semantics for natural language, presents itself as a form of interactive knowledge of the complex chain of biological realizations through which Nature reveals itself to our minds in a consistent way. To simulate is not, however, only a form of self-reflection or a kind of simple recovery performed by a complex cognitive net in order to represent itself at the surface level. The simulation work offers, in effect, real cognitive instruments to the semantic net in order to perform a self-description process and to outline specific procedures of control as well as a possible map of an entire series of imagination paths. The progressive exploration of these paths will allow, then, external information to canalize in an emergent way and to exploit new and even more complex patterns of interactive expression and action. It is exactly the framing of this particular kind of possible emergence that will assure the successive revelation of ever new portions of the deep information. We have to underline, however, that this particular irruption of the Other (not yet known) can express itself only within those specific schemes of the imagination and within that variant geometrical tissue of the Ideas which characterize, in an ultimate way, the cognitive activity of the Subject.

* * *

Let L be a classical (two-valued) first-order language which contains a family π of predicates. From a general point of view, the set F of all formulas at the level of the two-valued predicate calculus becomes, as is well known, a boolean algebra after identification of equivalent formulas. In particular, from a narrower point of view, let π be a finite family of monadic predicates $P_1, P_2 \ldots P_m$. By $\varepsilon(\pi)$ we denote the smallest set of compound predicates including π and closed with respect to the application of the rules concerning the connectives $\neg, \wedge, \vee$. Then we define a *congruence* (or equivalence) relation ζ between the elements of ε (π). The boolean ring $\bar{\varepsilon}$ (π) associated to the preboolean set ε (π), that is, the quotient of ε (π) modulo the congruence relation ζ, will be called the (monadic) *predicate algebra* generated by π. In our monadic case $\bar{\varepsilon}$ (π) is *atomic* and the atoms are "represented" by the *Q*-predicates (i.e., the elements of the algebra are equivalence classes and not simple predicates).

These partial results can be generalized by resorting to the theory of *polyadic boolean algebras* outlined by Halmos, a theory which constitutes an algebraization of the lower predicate calculus without any reference to the notion of formula. Finally, we must underline that, in the frame of our small logical construction, every element of ε (π) admits a *disjunctive* (or *conjunctive) normal form.*

Starting from *k generators* there are, from a general point of view, 2^k possible minimal polynomials, each of which can either be present or absent in a given canonical form. There are therefore 2^{2^k} possible distinct canonical forms and consequently a *free boolean algebra* with *k* generators has exactly 2^{2^k} different elements. If, however, a boolean algebra with *k* generators is not free, then it will have less than 2^{2^k} elements. In other words, if, for instance, we have only 2 generators, *a* and *b*, and if they satisfy the additional relation or constraint $a \geq b$', the total lattice telescopes. In this sense, the real problem is to individuate the nature of (possibly hidden) relations existing between the generators. By resorting to generators and relations it is, then, possible to outline a model, at the monadic level, partially different from the traditional urn model, a model that is based on a field of generators and a set of relations rather than on a domain of individuals with given properties.We can outline, in other words, a *frame* based on: 1) the specification of some subbasic elements (the *generators*); 2) the derivation from these elements of all possible joins of meets of subbasics; 3) the specification of certain *relations* to hold between expressions of step two.

Such a framework represents a very general method of Universal Algebra. Within it, it is possible to outline a particular frame with generators only and no relations. This frame (or better the *presentation* relative to it) will be called *free.* By extension, the algebra it presents will be called free. The algebra $\bar{\varepsilon}$ (π), for instance, that we have considered before, is a ring free on *G*. *G* is an independent generator and card (*G*) = card (*H*). (We write *H* to denote the set of the elementary predicates).

When we construct, at the level of monadic predicates, the normal forms, we are again dealing with a series of possible worlds. With reference to these worlds, it is, then, possible to introduce the individuals by resorting to particular logical "insertion" techniques which may constitute a bit of (relative) novelty with respect to the traditional methods of logic. In this way, we can finally recover the logical form proper to the carnapian *state-descriptions* and to the classical (monadic) constituents. So, also the (monadic) urn models will appear as a particular case within these more general logical structures.

The theoretical design that we have just outlined possesses a sequential and compositional character. We start with primitive predicates and we go along the derivation from these predicates of all possible joins of meets of these same elements. In this way, we trace the design of different kinds of worlds and their composition. This design remains, however, as we have seen, within the boundaries of a unique algebra generated by *G*.

The perspective changes, suddenly, if we take into consideration, generalizing the initial carnapian frame of reference, models that are not relative to the design, at a compositional level, of a unique algebra, but result as the product (not in the strict boolean sense) of the interaction between a multiplicity of nuclei of generators. In this way a net of algebras can arise, a net that can span patterns of connectivity of a very high level. If we construct such algebras as *heterogeneous algebras,* the net will reveal itself as a *net of automata* (once certain specific conditions are satisfied). We shall be in the presence of a boolean net of connected automata whose primitive inputs are represented by monadic predicates. This type of structure can express itself as a sequential structure, but it can also express itself as a *recurrent structure*. If we decide to follow the paths of a recurrent structure, some radical changes arise with respect to the traditional urn models and structure-descriptions. In particular, we must now take into account the birth of new concepts: the concept of *cycle*, the concept of *attractor*, the concept of *trajectory*. The resulting net appears to be a *self-organizing* net. With respect to the predicates we shall not deal any longer with the traditional forms of compositionality, throughout the successive development of the layers of the complexity. On the contrary, we shall be obliged to consider specific forms of constrained modulation, linked to the successive constitution of specific basins of attractors. We will be able, in this way, to inspect the progressive unfolding of complex forms. It is within the frame relative to these abstract complex forms, linked to a *generative dynamics* which appears determined according to temporal connection schemes, that we must now insert the individuals and, consequently, define the different types of worlds. The concept of accessibility, the traditional alternativity relation, must also be adapted to these complex forms.[13]

So, when we assume the existence of a circular net of binary-valued elements where each element is connected to the others but not back to itself, we necessarily put ourselves within the boundaries of a landscape outlined according to a temporal dynamics scanned by the successive determination of specific constraints, also in consequence of the choices of the branches relative to the occurring bifurcation that have been made.

As is well known, the *schemes* from a kantian point of view are, essentially, *time determinations*. In a self-organizing net the successive bifurcations, the recurrent delimitations imposed on the primitive predicates-inputs, appear, actually, as temporal and connected determinations of information fluxes. In this sense such determinations appear to concern (differently from Hintikka's appraisal of Kant's primitive intuitions), not the (direct) successive presentation-construction of individuals, but the (previous) construction of patterns of constraints, of clusters of selective choices. Hence the essential link with the traditional contemporary definitions of complexity at the propositional (and monadic) level.

So, insofar as these determinations of time articulate modulating in a recurrent way the action of the generators, the self-organizing nets present themselves progressively as frozen surface images of the originary informational Source, as a tool for the further construction-revelation of its inner creativity.

The process of self-organization that characterizes the models appears, in other words, to be strictly connected to the mysterious paths of the expression-construction of the Source.

It appears as a sort of arch and gridiron for the construction (and the recovery) of the "Other" (the Source) through the constraints of its proper "sacrifice".

The modulation of the articulated structure of the predicate-inputs changes progressively with time. It will be "fixed" when the process has attained the attractor. It is the attractor that individuates the real "form" of the attribute.

In this sense the deep meaning appears as relative to precise semantic *fixed-points*, to a manifold of underlying functions. The fixed-points of the dynamics represent the "true" revelation of the specific tuning that characterizes the predicates. As we have just said, such a revelation can realize itself, however, only through successive determinations of time, through the action of specific temporal schemes. According to these lines of thought we have, thus, the possibility to outline, at a monadic level, a new particular kind of model: a self-organizing structure not bound to sets and individuals, (with relative attributes) but to generators and fluxes of tuned information. In this new theoretical framework the simple reference to possible worlds (as in Frege or Hintikka, for instance), in order to take into account the structure of intensionality is no longer sufficient. One has also to resort to the relation of alternativity and, in particular, to the dynamics of the constraints, to the recurrent paths of the informational flow as well as to the role played by the observer and the very interplay existing between intervention and change.

At the level of this type of structure, we can individuate the existence, as a matter of fact, of an essential plot between the successive "presentation" of the constraints and the action of the schemes, on the one hand, and the articulated design of mutations, cancellations and contractions of the predicates-inputs that characterize the higher layers of formal constructions, on the other.

The internal tuning which marks the articulation of this generative dynamics is very sophisticated and requires careful analysis. To explore this kind of tuning it is, in particular, necessary to exploit specific control procedures capable of being expressed in the framework of orders or spaces of higher level. In order to identify at least some of the basic roots of this kind of emergent development, it appears particularly suitable to submit now the conceptual bases of predicate algebra introduced earlier to close examination, so that it will be, finally, possible to introduce some new conceptual instruments such as the concept of constituent, the concept of depth, the concept of compatibility, etc., all concepts that, as we shall see, will reveal themselves to be of great importance in the course of our analysis.

* * *

Let L be a very simple applied first-order language which contains n different individual constants which stand for n different individuals and m primitive monadic predicates which designate primitive properties of the individuals. In the frame of this *monadic language* a *Q-predicate* is a conjunction of predicates in which every primitive predicate occurs either negated or unnegated (but not both) and no other predicate occurs at all. The Q-predicates specify all the possible kinds of individual; they represent the *atoms* of the algebra of predicates. The number of Q-predicates is $2^m = K$. A full-sentence of a Q-predicate is a *Q-sentence*. A *state-description* is a conjunction of n Q-sentences. Thus, a state-description completely describes a possible state of the world.

Naturally, we can also take into consideration m primitive binary predicates. In this way, we shall have the *Q-relations* of a binary language which specify all the possible relations existing between two individuals. They constitute an immediate generalization of Carnap's primitive Q-predicates.

Generalizing this kind of procedure, Hintikka introduces, first of all for monadic first-order logic without identity, the concept of constituent and distinguishes attributive constituents from constituents. An *attributive constituent* (or *individual constituent* in our terminology) specifies the different *kinds* of individuals that can be defined by means of the linguistic resources and, precisely, by means of the predicate symbols $P_i(x)$. An arbitrary individual constituent will be called:

$$(1) \qquad Ct_v(x) = \prod_{i=1}^{v} P_i(x) \qquad (i = 1, 2, \ldots\ldots m) \quad (v = 1, 2, \ldots\ldots 2^m)$$

Each such conjunction is associated with an attribute and the set of such attributes induces not a partition of the logical space relative to the possible worlds, but a partition, in each world, of the individuals.

A monadic *constituent*, then, simply specifies for each individual constituent whether or not it is instantiated:

$$(2) \qquad \Pi \ (E\,x)\ Ct_v\,(x)$$

In a pure monadic language each formula is equivalent to one which uses only the first variable of the language. Each formula with only one variable is expressible as a disjunction of the above constituents. (But remember that all this logical construction is essentially linked to a precise assumption: the existence of empty domains of individuals).

Thus, statement (2) specifies the different *kinds* of possible worlds that can be defined by means of the linguistic resources. The constituents will be abbreviated: C_1, $C_2 \ldots C_{2^K}$ $(K = 2^m)$.

Starting from the monadic case, Hintikka generalizes the obtained results to the general (polyadic) first-order languages, attaining, thus, the general definition of *distributive normal form* (but with an important distinction: in monadic first-order logic, like in propositional logic, distributive normal forms yield a *decision method.* In view of Church's undecidability result they cannot do this in the full first-order logic (with or without identity)).[14]

In order to perform such an extension it is necessary to define, first of all, the concept of *depth* with reference to a given sentence S of L.

A sentence S of L is of depth d if the number of layers of nested and connected quantifiers in S is d. We indicate with $\mathbf{C}_d$ the set of all constituents of depth d that can be formed in L. Each $C^{(d)} \in \mathbf{C}_d$ describes a kind of possible world which can be described by means of the resources of the language L when this description is restricted so that it is possible to consider only the sequences of d individuals in their relation to each other. So if $C^{(d)}$ is true in a model M, it gives us a ramified list of all the different kinds of sequences of individuals of length d that one can find in M.

In this sense, if we call the range of predicates the "intensional base", a possible world shows itself simply as one conceivable way in which the predicates forming the intensional base are distributed through the universe of discourse.

That constituents really describe the basic alternatives is well illustrated by the fact that each sentence S with depth d (or less) can be effectively transformed into a disjunction of a number of constituents with depth d (and with the same predicates as S). Each sentence S of depth d has, thus, a distributive normal form. So, for example, the content of each of the C_i's is given by a set of individual constituents, those claimed by the constituent to exist. Moreover, since a formula which uses just one variable is said to be of depth 1, each depth-1 proposition is expressible as a disjunction of the C_i's.

In order to understand in what sense the constituents are pictures, we can imagine the world as an urn from which individuals may be drawn and inspected for their primary properties.

A constituent specifies what kinds of balls-individuals one can draw from this urn after one another. In this sense, a constituent $C^{(d+e)}$ of depth $d+e$ greater than d, allows us to examine a world by means of the increased number of steps-experiments. It specifies, in other words, what kinds of sequences of individuals of lengths $d+e$ one can find in its models. We can, evidently, have constituents that are *inconsistent*. In particular we have to distinguish between constituents that are trivially inconsistent (when the inconsistency can be inspected from the same structure of $C^{(d)}$) and constituents whose inconsistency results hidden in such a way that it can be brought out only by expanding $C^{(d)}$ into deeper normal forms.

As far as the analogy between constituents and urn models is concerned, we must point out, however, that there is a deep difference between the use of urn models in classical probability theory and the urn interpretation of first-order logic. At the level of polyadic languages, actually, we cannot establish the attributes of the balls-individuals by examining them alone. A single individual, drawn from the urn, bears different combinations of relations to the so far unexamined individuals present in the urn. So, two sequences of individuals may be identical as far as the revealed properties are concerned and yet differ from each other in the kinds of relations they bear to the unknown individuals lying in the urn. Moreover, the practice concerning urn models in probability calculus puts us, from a logical point of view, in front of a further radical novelty.

In the ordinary interpretation of a constituent it is assumed that the model is *invariant* in every respect between the draws. But, as is evident, the world can change. The problem is: in which way can a world change? When we think of a change in a model we can think, in the first instance, of a change in the attributes of individuals of the model or of a change in its domain. But as the practice of models in the probability theory shows, a model can change also *with* draws. If we make draws without replacement, the domain evidently changes at each draw. This is a simple example of the intervention of an observer on the ultimate structure of the model.

Following the indications given by Rantala, we can outline an urn with an internal change mechanism. The observers are informed in advance of the functioning of this mechanism. This can happen, for instance, insofar as the global functioning of the mechanism is characterized by set-theoretical terms.

If the mechanism of the urn does not impose any constraint at all on the observers, we have an analogue of a completely invariant model for a first-order language. In other words, if the observer knows the initial state of the universe and the set-theoretical structure of the change mechanism, he can predict the evolution of the states of the model. But if in the model some changes happen and if these changes are not entirely independent of the intervention of the observer who is inspecting the world step by step, we may see a complete metamorphosis of the game. The choice by the observer of the balls, the rational decisions and calculations relative to these choices, the constraints that these choices can impose on the change mechanism, necessarily prime a strange play between the mutations internal to the world and the selective activities coming from the outside.

In this way the metamorphosis of the model appears as dependent not only on the internal change mechanism and on the external perturbations, but also on the abilities of the observer and his intelligent behavior. We are dealing with a partially holistic reality with a peculiar character of circularity.

When we are faced with a model which changes, the old definitions of truth and consistency proper to the standard first-order semantics will change as well. A new type of semantics necessarily arises, a semantics which relates the truth of a proposition not only to the world but also to the results of the observational process and to the successive changes that "happen" at the deep levels of the world and its ultimate connections. In this sense the non-standard semantics reveals itself in perspective also as a *semantics of processes.*

An individual constituent, at the monadic level, is, substantially, as we have just said, an atom of boolean algebra (the algebra of predicates) and identifies the different kinds of individuals which can be defined by means of the expressive resources of the language in use, namely by means of the primitive predicate symbols $P_i(x)$ and of the customary boolean operations. In this sense, as is obvious, the conjunction represented by a monadic individual constituent is associated with a (complex) attribute. In turn, the set of such attributes induces not a partition of the primitive logical space relative to the possible worlds living at the propositional level, but a partition in each world of the individuals.

When we pass from individual constituents to constituents, we simply specify whether a given individual constituent is present or not in the different worlds. (From an effective point of view things are more complex. At the predicative level we have, in effect, to distinguish between individual constituents (in our terminology) and closed individuals or closed *points of reference* which work as a support for the full expression of the individual constituents). But it is precisely the task of quantifiers to affirm that a particular entity possessing a specific attribute is instantiated. The quantifier, in effect, acts, as we have seen, as a sieve in order to realize a partition within the family of sets provided by the model.

Once the partition in each world of the individuals has been effected, we can make reference only to the first variable of the language and, as a result, each formula becomes expressible as a disjunction of the constituents. But we can also utilize a multiplicity of variables. Thus, new and more complex languages will arise. At the level of such languages the algebra will be no longer free. If, for example, we assume a language $L_{n=3}^{\pi=3} = (2^3)^3$ where $\pi = 3$ are the predicates and $n = 3$ are the individuals, we have, as a result, that the n individuals will represent the cases of effective instantiation. As n increases, the number of state descriptions in Carnap's style will grow exponentially.

As we have said, the Q-predicates represent kinds of individuals; they are the atoms of the algebra of monadic predicates. As atoms they constitute, at the same time, from an abstract point of view, possible worlds. In effect $(\pi \rightarrow 2) = 2^\pi = W$. But with regard to this, let us note that $L_1^2 = 2^{2^1}$. In this sense we have to distinguish carefully between the simple atoms, the pure conjunctions of primitive generators and the conjunctions referred to the first variable of the language and "open" to the action of quantifiers, the conjunctions, that is, which are expressed at the level of the realized partition. Here we may find a precise point of contact between the school of thought that has its roots in the analytical work done by von Wright and Hintikka and the original intuitions outlined by Montague and later on by the theorists of Boolean Semantics. In fact, also for Keenan and Faltz, in order to define an individual it is necessary to define, in a preliminary way, the atoms of the algebra. For these scholars an individual denotes the family of sets that contain the individual; in other words, it denotes the "functional role" played by a given atom on the entire chess-board concerning the formal operations in use and not at all a simple and static unitary "thing". Here, however, some immediate differences arise between the two schools at the theoretical level. When we refer to worlds, we have, in effect, to distinguish, according to the conceptual perspective outlined by Hintikka, between the different kinds of atoms-worlds, between the different layers through which these worlds can be expressed along the development of a precise vertical (and intensional) dimension. For example, we have to distinguish, first of all, the atoms being expressed at the level of the primitive logical space W from the atoms-worlds concerning the realized constituents at the monadic level.

Making this kind of distinction in a direct way is not so obvious in the logical system outlined by Montague. At the level of such a system the individuals, in effect, are still considered, in many respects, as the real primitives; but if we start with individuals as primitives, this involves the fact that the worlds (but not the contexts) are given at once and for ever with regard to their inner modal structure. They are, so to speak, at the back of the stage. From a genealogical point of view, they are the only worlds living in the "interior" of the coupled system, the only possible kind of possible worlds. In other words, they really are, in a preliminary way, the basic worlds (or protoworlds) which constitute the primitive logical space i.e., that particular space onto which the process of partition successively grafts itself. Then we place ourselves at a monadic level that reveals itself as closely correlated to the propositional case. Starting from this level, we successively build the polyadic structures without any possibility, however, of entering, in a complete way, the specific articulation of possible worlds that characterizes this latter type of expressive domain. This prevents us, for example, from referring to atoms-worlds directly as types or pseudo types, from seeing the inner generativity of the worlds as it expresses itself according to the successive development of the above mentioned vertical dimension and from taking into consideration the fact that, in general, the individuals, as defined before, when acting as generators of a given system, really perform their task with respect to the different levels of atoms previously established step by step.

Such a construction intimately depends on the propositional model. Therefore, it is not strong enough to cope with the complexities of human language. This forced many scholars in the sixties to introduce a multiplicity of possible worlds (and contexts) from the outside, in the form, for instance, of indices, without any possibility, however, of giving a precise account of the inner generativity and of the functional articulation proper to the different layers of possible worlds, without any possibility, for example, to distinguish the atoms-states from the atoms-urns or from the atoms-spaces. This is, precisely, the case of the theoretical construction outlined by Montague. In this way, a conceptually "meagre", but technically powerful formal architecture, within which it is not possible, as we have just pointed out, to draw the lines relative to the construction of a series of differentiated substrata along a vertical dimension, comes to connect itself to a nearly infinite multiplicity of successive possible variations.

In this manner, however, we shall not be able to follow, step by step, the entire map of constraints that progressively unfolds according to the development of the vertical dimension, particularly in consequence of that specific articulation of possible worlds which manifests extensively itself when we pass to the polyadic level and to the second-order level. With reference to the "genealogical" construction of possible worlds we have, therefore, from the beginning: $D<\pi, \iota> = W$ and $(W \rightarrow \mathrm{B}) = 2^W$. But recall that, according to Montague's first intuitions, $P = 2^U$. If we consider U as a set that is composed of individuals-atoms, we consequently have: $P = 2^{I_P}$. In effect, the set I_P consists of individuals-atoms considered as collections of properties, properties, however, that, in turn, represent, according to Keenan and Faltz, the elements of a given boolean algebra and cannot be considered as generators at all. For the theorists of Boolean Semantics who have revised Montague's conceptual formulation, the individual considered as an atom, really is the filter generated by the atom. On the other hand, in Hintikka's opinion, at the monadic level, the atoms express themselves as individuals when, exploiting, first of all, variables and quantifiers, they change into individual constituents, shaping themselves as kinds of individuals that are instantiated. This happens when we arrive to outline, starting from a shot universe of states-predicates, the passage to worlds-constituents populated by individuals through a constrained transformation of the old atoms.

At the level of the underlying algebraic structures the original intuitions by von Wright, as revisited by Hintikka, foreshadow, thus, in an abstract manner, the necessity of resorting to a specific series of power set operations that are linked, in an indissoluble way, to the successive transformation of the elements into atoms and of the atoms into generators. It is precisely at the level of this conceptual framework that the possible worlds appear to put on the clothes of types. Actually, once the necessary cylindrifications are effected, the primitive atoms of a given algebra B can change into generators, that is into instantiated individuals which are constrained by the action of quantifiers. So, we shall have the possibility of outlining new atoms-worlds (specific conjunctions, that is, of the new born generators) which will be in one-to-one correspondence with the atoms of the power set of B, 2^B.

As is obvious, the atoms that change into generators may refer, in turn, to a specific set of generators. Hence the possibility of the birth of a successive gain of power sets and of constrained transformations of elements into atoms; a gain that is characterized by the presence of a precise game of crossed correspondences. The constraints that are brought into play at the level of this particular chess-board of logical operations are very powerful and induce precise exponential growths.

With reference to this conceptual framework, it is convenient to underline again that when we write $(Ex)e_l$ (where e_l represents a monadic individual constituent) we certify that reference is made to entities which are characterized in a different manner with respect to the worlds arising from the primitive process of identification. We have moved away from the propositional level and we have entered the monadic level. $(Ex)e_l$ precisely means: there is at least one entity-individual x which exemplifies the individual constituent e_l with reference to the frame represented by the atoms-constituents C_i. The individual constituents indicate the possible kinds of individuals at play, they precisely point out that the partition has been delineated; the determiners, in turn, guarantee the existence in the different worlds of such kinds of individuals. In other words, they specify for each individual constituent whether or not it is instantiated in a given world. We can recognize, thus, the first skeleton (or the first roots) of a complex structured articulation within which linguistic information can begin to flow freely, so gaining, step by step, that variegated composition of differentiated forms which we can look at whenever we take into consideration the infinite multiplicity of expressions of natural language.

From an abstract point of view, according to the suggestions of the theorists of Boolean Semantics, the fundamental passage which constitutes the first stage of linguistic growth is given by the construction of $P^* = 2^{2^{I_P}}$. I_P represents, as we remember, the set of atoms of $P = T_N$. The elements of I_P are the individuals generated by the atoms. At the level of $P^* = T_{\bar{N}}$, however, the individuals constitute, in turn, a set of c-generators. We are really faced, now, with a free boolean ca-algebra. For instance, if there are 4 individuals we shall have 2^{2^4} sets of properties or elements of the resulting algebra. As a consequence of this particular one-to-one transformation we can easily realize that, while the atoms of the algebra P correspond to the principal ultrafilters, on the other hand, the generators of the algebra P^*, considered as a result of the application of precise transformation operators, correspond to principal filters that cannot, in any case, be expressed as ultrafilters. The old atoms "shape" themselves as instantiated individuals by exploiting specific closure procedures. Hence, in perspective, the possible construction, as we have just said, of a specific gain of power sets, a gain that involves the definition of ever new sets of generators according to the unfolding of a precise vertical dimension that possesses a very strong exponential character. This kind of sophisticated setting allows for the outline of a particular grid of constrained informational fluxes. It is with reference to such a complex grid that new boolean functions as well as new linguistic constructions can effectively appear.

So, when we consider the imposition of particular constraints on the atoms as well as the birth of new “shaped” generators, in other words, when we try to describe the constrained articulation concerning the successive gain of the power sets, we have to realize, according to the theoretical perspective first outlined by von Wright, that we are really faced with an entire series of concurrent passages:

1) there is, first of all, a process of functional partition through which the primitive atoms i.e., the primitive possible worlds-states become kinds of individuals; 2) there is a precise game of variables and quantifiers that governs the specific operations concerning the constrained closure of the different kinds of individuals; 3) specific processing procedures enter the stage operating with “shaped” elements-generators and not with bare ones; 4) at the level of the realized new conjunctions we can see the birth of atoms-worlds with respect to which the individuals come to play a role that is very different from the initial one. In fact, as a consequence of the performed closure operations, the individuals, instead of expressing themselven as simple collections of properties, come to assume the role of *elements* of atoms-worlds. The worlds are now characterized by the presence (or absence) of individuals-elements. It is in this sense that we have, finally, the possibility of speaking about the existence of empty domains of individuals. The urn concept, then, comes in, a concept that is borrowed from probability calculus. So we are now in the presence not of a shot universe of states, but of individuals populating different worlds (and with the possible additional presence of an observer who looks at these individuals from the outside, interacting possibly with them).

If we try to run again through the track of thoughts followed until now, step by step, and if we try to imagine the plot of the principal filters as well as the different algebraic structures that are in correspondence with the different entities generated by means of the aforesaid embedded transformations, it is, finally, easy to understand, for example, how the same first conceptual product of the aforesaid track (i.e. the individual considered as a member-element of an urn), reveals itself really as a much more complex “entity” than the pure collection of properties which Keenan and Faltz speak about.

The reality of generalized quantifiers presents itself to us, now, according to a much more variegated and dynamic conceptual perspective, a perspective, in particular, that is linked to the construction, in the course of time, of a series of worlds-atoms of increasing complexity. In this sense, the intensional relationship existing between linguistic expressions and worlds no longer appears as a sort of extrinsic link, as happens in Montague’s system, but as an intrinsic one. At the level of the linguistic fabric we are in the presence of a horizontal and a vertical dimension linked crosswise. The intensional constructions are, in particular, correlated to the development of the vertical dimension. Specific determinations of time, in this way, enter the stage from the interior of the system.

* * *

The track we have been referring to, despite its sophistication, still represents only the first part of our long journey. The linguistic structures that we have just examined are articulated according to an ideal grid whose essential parts are made up by monadic predicates, intransitive verbs and so on. As is evident, things are different and much more complex when we pass to the polyadic and second-order level. Actually, it is precisely at the polyadic level, for instance, that transitive verbs enter the stage. Further transformations occur. New kinds of models are necessary in order to understand the mathematical structures that support these transformations. New factors also appear, factors concerning, in particular, the supervenience of particular constraints (i.e., compatibility, symmetry, etc.) and of particular processes (i.e., processes of constitution of new kinds of individuals and constituents, etc.). An "open" land will reveal itself, placed exactly in front of us even if we are constitutive parts of it. In order to give some hints with respect to a possible exploration of the different aspects of this strange and mysterious realm let us first of all exploit (extending some original indications offered by R. Montague) some general ideas concerning the possible treatment of quantification at the polyadic level (and, in particular, at the dyadic level) with respect to the actual realization of the processes of constitution of new kinds of constituents.

We have just recalled the correspondence existing between the atoms of B and the ultrafilters. An ultrafilter J on an atomic boolean algebra B is a set of elements of B containing exactly one atom b and all the elements p such that $b \leq p$. If J is an ultrafilter on a boolean algebra B, then for all b and p in B

$b \vee p \in J$ iff either $b \in J$ or $p \in J$ (or both)

$b \wedge p \in J$ iff both $b \in J$ and $p \in J$

$b^{\perp} \in J$ iff $b \notin J$.

There is a one-to-one correspondence between the set of ultrafilters on B and the set of homomorphisms of B onto B, such that, if J is an ultrafilter on B and h_J is the corresponding homomorphism, then, for all b in B:

$h_J(b) = 1$ iff $b \in J$.

The homomorphisms are, in general, functions from a boolean algebra into another which preserve the boolean operations. Keenan and Faltz take T_{P_1} to be the set of (complete) homomorphisms from $T_{\overline{N}}$ into B. Thus, in a model of n individuals there are just 2^n such functions, the same as the number of (extensional) properties in a world of n individuals. In this sense the first-order predicates correspond one-to-one to sets of individuals or, equivalently, to the functions from individuals into B. From a general point of view, for every function f from the individuals on $T_{\overline{N}}$ into B, there is exactly one (complete) homomorphism from $T_{\overline{N}}$ into B whose values on the individuals are the same as f.

The type for P_1, T_{P_1}, is not, however, represented by $F_{B/P*}$, that is from the entire set of functions from $P*$ into B, but from $H_{B/P*}$: the set of complete homomorphisms from $T_{\overline{N}}$ into T_B. And, in fact, the elements of $H_{B/P*}$ are just the one-place predicates of standard first-order logic. In order to see in which way the elements of T_{P_1} are in a one-to-one correspondence with the elements of T_N, let us note that a complete homomorphism from $P*$ into B may be considered as a function which assigns a set of properties Q value 1 iff a fixed property p is a member of Q.

So, formally we have the following definition as stated by Keenan and Faltz[15]:

Definition 4. 1) For all $p \in P\ (=T_N)$ define a function f_p from P^* into B by: $f_p\ (Q) = 1$ if $p \in Q$ and $f_p\ (Q) = 0$ if $p \notin Q$.
2) $F_P =_{\text{def}} \{f_p : p \in P\}$

Consequently, we can prove that F_P and H_{B/P^*} are both subsets of F_{B/P^*}. T_{P_1} is a boolean *ca*-algebra.

We can exploit the method utilized just now in order to define also the type for P_2. The elements of T_{P_2} should simply be homomorphisms from $T_{\overline{N}} = P^*$ into T_{P_1}. In general, we have: $T_{P_{n+1}} = H_{T_{P_n}/P^*}$. Let us note, again, that in the same way that T_{P_1} is isomorphic to the power set of I_P, so T_{P_n} is isomorphic to the power set of $(I_P)^n$.

With regard to this, the essential problem, however, is constituted by the fact that in the passage from T_{P_1} to T_{P_2} there is not only the presence of a sort of mechanical passage from a given set of functions to another. At the level of T_{P_2} in fact, we are faced with a particular sort of algebras within which a specific series of constraints suddenly arises. These constraints concern, for instance, the action of particular factors: compatibility, symmetry and so on, as well as the action of well determined processes of constitution of individuals according to precise geometric guidelines. With regard to T_{P_2}, for example, once we come to use primitive binary relations, we must immediately make a series of choices and assumptions which are relative to the structural properties of such relations. In consequence of the structural properties that characterize precisely the dyadic predicates (i.e. that such predicates possess in an exclusively conceptual way), some specific conjunctions of these same predicates will be shown to be inconsistent. In this sense, at the T_{P_2} level, the process of construction and constitution of individuals turn out to be much more complex and articulated with respect to T_{P_1}. In the case, for instance, of T_{P_1}, an individual is, for the theorists of Boolean Semantics, basically a collection of properties, but, when we go into the realm of polyadic predicates, an individual, as well as being considered as a collection of properties, must also be defined as a chain or collection of relationships. This means that what must be joined together will no longer consist of simple entities or sets of properties but of configurations and graphs. Here we meet with a crucial point. In the light of the reconstruction by Keenan and Faltz the passage from the monadic case to the polyadic one (and in particular to the dyadic one) does not alter, from an essential point of view, the nature of generators. Both I_P and $(I_P)^2$ appear, at the beginning, as sets of atoms (more precisely as sets of specific "structures" generated by the atoms, structures that are in a one-to-one correspondence to the atoms) of the original algebra. These atoms (the corresponding generated structures) are then transformed into generators by means of well known algebraic operations.

But, according to the conceptual perspective advocated here, at the dyadic level the individual constituents are not represented, at the level of the denotational domain, by $(I_P)^2$ but by $2^{((I_P)^2)}$ (or from a specific set derived from it in accordance with the application of precise cancellation rules). Actually, we are faced with a double transformation.

Insofar as $2^{((I_P)^2)}$ presents itself as the real generator, the atoms-worlds that successively arise by means of the traditional algebraic operations must be considered as worlds or "modules" populated not by individuals-elements or by relations-elements but by specific relational systems.

From a traditional point of view, the type of Q in the schema $Qx, y \cdot A\ (x, y)$ is given by $\langle\langle e, \langle e, t\rangle\rangle, t\rangle$ and the size of the corresponding denotational domain is $2^{(2(n^2))}$. We are faced with a classical (Fregean) dyadic quantification. But utilizing the P transformation we may also define the following semantic equivalence: $(EyA(y))\ (x) \approx EyA(x, y)$. Also in this case we can introduce a double quantification obtaining a scheme such as: $Q_1\, x\, Q_2\, y \cdot A\ (x, y)$. And we know that according to Frege polyadic quantification is iterated unary quantification.

However, when we consider $2^{((I_P)^2)}$ as the real generator, the starting point of our logical construction is not given by the originary atoms (or more precisely by the sets of the specific structures generated by the atoms) but by specific "modules" considered as conjunctions of the atoms (atoms that have necessarily undergone a precise algebraic transformation). These modules, these centralized graphs are exactly the equivalent, at the polyadic level, of the individual constituents as defined at the monadic level. In this sense, the individual constituents considered at the monadic level as collections of properties, now present themselves also as specific collections of relations, as the branching points of well specified relational systems. We are inside a net that grows up in a continuous way, a net, moreover, that appears as closely linked to the development of a precise symbolic dynamics. With reference to this sophisticated net the role of compatibily factors becomes particularly essential. From here both the birth of complex cancellation procedures and the introduction by construction (a construction which, since articulated in the space "guided" by imagination, we could define as Kantian) of new constituents, in a potentially unlimited way, arise. Likewise, we would have, in a correlated way, the introduction of nested quantifiers. In addition, we would have an opening up of the logical second-order spaces with the consequent arrival on the scene of precise holistic issues.

It is in this framework, which is in itself already complicated enough, that the higher functions proper to the linguistic articulation come progressively to establish themselves: transitive verbs, determiners of higher level, etc. It is with reference to this very framework that a precise dynamics of graphs with the subsequent introduction of cycles, attractors, fixed points, etc. as well as the revelation of further constraints relative to problems of fitting, consistency, etc. will, finally, enter the stage. Precise forms of classification and, therefore, precise contexts of sense will appear. In this way, specific intensional structures will begin to emerge, in particular, intensional grammars defined with reference to orders-spaces of higher level.

From this comes the necessity of outlining, in the case of T_{P_n} (and in the case of second-order structures), the sophisticated dynamism of a great book of Language that presents itself, at the level of conscious representation, as an effective reality in action, a reality which emerges also through our thinking and which, at the same time, determines, first of all at the genetic level, our thinking. We no longer have before us a static book of Reality written in linguistic and mathematical characters. We have, on the contrary, a Language in action which makes itself the Word of reality, the book in progress of linguistic constructions. This language by reflecting itself in a simulation space (of which, what is more, we are as human beings the support), assumes its primary forms and represents itself to itself by means of the tools of a precise symbolic dynamics. Therefore we are no longer faced with concrete signs-symbols, but rather with complex conceptual structures which are inserted in the effective articulation of a coupled process. At the level of this process, alongside the aforesaid dynamics relative to configurations and graphs, specific, informational valuations proper to the subject and to the structures of reflection and cognition that characterize his activity, will enter the scene as well.

We have seen how, according to Putnam, the invention of new language represents the main tool to open up Reality, to discover new horizons of meaning. The awareness that comes from the intensional analysis of the semantic structures of natural language and the cognitive functions that underly these same structures, leads us to understand that the problem is not only that of extracting the information living deeply in things, but it is also of building simulation models able to bring out the information contained in the fibers of reality in such a way that this same information, irrupting into the neural circuits of elaboration proper to the subject, can, finally, induce and determine the emergence of new forms of conceptual order and linguistic construction. The problem is, likewise, that of supplying coagulum functions which are capable of leading the Source to nest deeply, according to stronger and more powerful moduli. The emergence process and the same creativity that has been progressively realized will present themselves as the "story" of the performed irruption and of the nesting carried out. They will articulate as forms of conceptual insight which spread out into a story, the story, in particular, of a biological realization. In order to "open" Reality, language must be embodied as an autonomous growth so that it will be possible, in perspective, to coagulate new linguistic constructions. Hence the importance of resorting to the outlining of recurrent processes and coupled processes in order to model the brain's functions. Likewise, this is the importance of that vertical (and intensional) dimension which grows upon itself, according to those exponential coefficients that we have previously considered and which appears indissolubly linked to the appearance and the definition of ever new forms of meaning, forms which necessarily spring up through the successive discovery-construction of new substrata and new dependency links according to Husserl's primitive intuitions.

Genealogical processes, recurrent processes, coupled structures, new measure spaces, new orders of acting imagination: such is the scenario from which new information can, finally, emerge. Here is the Language in action. Herein we may recognize the birth of new forms of seeing. Herein we can find the possibility to hear from a Source which, like a new "daimon", comes forth from within biological structures to dictate (in accordance with the constrained articulation of new generated linguistic resources) the message of its self-representation, of its "wild" autonomy and of its renewed creativity.

Prof. Arturo Carsetti
Dipartimento di Ricerche filosofiche
Università di Roma "Tor Vergata"
V. A. Cavaglieri, 6
00173 Roma, Italy

NOTES

(*) This work, which was done while the author was visiting professor at the Ecole Normale Supérieure, Paris, was supported by the National Research Council of Italy.

[1] Cfr. for further information: Barwise, J. and Perry J., Situations and Attitudes, Cambridge, MA, 1983.

[2] Cfr.: Montague, R., *Formal Philosophy*; Selected Papers of Richard Montague, edited with an introduction by R. Thomason, New Haven, 1974.

[3] Cfr. for more information: Dowty, D., Wall, R. and Peters, S., Introduction to Montague Semantics, Dordrecht, 1989; Thayse, A., (Ed.), From Natural Languages Processing to Logic for Expert Systems Introducing a Logic Based Approach to Artificial Intelligence, Vol. 3, New York, 1991.

[4] Cfr.: Keenan, E. and Faltz, L., Boolean Semantics for Natural Language, Dordrecht, 1985.

[5] Cfr.: Keenan, E. and Faltz, L., Boolean Semantics for Natural Language, Dordrecht, 1985, p. 48.

[6] Cfr.: Keenan, E. and Faltz, L., Boolean Semantics for Natural Language, Dordrecht, 1985, p. 77.

[7] Cfr.: (egr.): Barwise, J. and Cooper, R., "Generalized quantifiers and natural language", *Linguistic and Philosophy*, 4, (1981), pp. 159-219.

[8] Cfr.: Barwise, J. and Cooper, R. "Generalized quantifiers and natural language", *Linguistic and Philosophy*, 4, (1981) pp. 159-219.

[9] Cfr.: Keenan, E. and Faltz, L., Boolean Semantics for Natural Language, Dordrecht, 1985, p. 51.

[10] Cfr.: Quine, W. V., Word and Object, Cambridge, 1960, p. 179.

[11] Cfr.: Putnam, H., Representation and Reality, Cambridge, 1983, p. 137.

[12] Cfr.: Putnam, H., Representation and Reality, Cambridge, 1983, p. 134.

[13] Cfr.: for more information: Carsetti, A., "Meaning and complexity: a non-standard approach", *La Nuova Critica,* 19-20, (1992), pp. 109-127.

[14] Cfr.: Hintikka, J., "Distributive Normal Forms in First-order Logic" in Crossley, J.M. and Dummett, M.A. (Eds.), Formal Systems and Recursive Functions, Amsterdam, 1963, pp. 47-90.

[15] Cfr.: Keenan, E. and Faltz, L., Boolean Semantics for Natural Language, Dordrecht, 1985.

REFERENCES

Barwise, J. and Perry J., Situations and Attitudes, Cambridge, MA, 1983.

Bennett, M., "A variation and extension of a Montague fragment of English", in Partee B. (Ed.), Montague grammar, New York, 1976, pp. 119-163.

Carnap, R., Meaning and Necessity, 2 rd. Ed., Chicago, 1956.

Carsetti, A., "Meaning and complexity: a non-standard approach", "La Nuova Critica", 19-20, (1992), pp. 109.127.

Cresswell, M.J., Logics and Languages, London, 1973.

Dekker, P., "Dynamic interpretation, flexibility and monotonicity", in Stokhof M. and Torenvliet L. (Eds), Proceedings of the Seventh Amsterdam Colloquium, 1989, Amsterdam, 1990.

Dowty, D., Wall, R. and Peters, S., Introduction to Montague Semantics, Dordrecht, 1989.

Groenendijk, J. and Stokhof, M., "Dynamic predicate logic; towards a compositional, non-representational semantics of discourse", "First International Summer School on Natural Language Processing, Logic and Knowledge Representation", Groningen, June 5-17, 1989.

Hendriks, H., "Flexible Montague grammar", "First European Summer School on Natural Language Processing, Logic and Knowledge Representation", Groningen, 1989.

Henkin, L., "Completeness in the theory of types", "J. of Symb. Logic", 15, (1950), pp, 81-91.

Hintikka, J., "Distributive Normal Forms in First-order Logic" in Crossley, J.M. and Dummett, M.A. (Eds.), Formal Systems and Recursive Functions, Amsterdam, 1963, pp. 47-90.

Hintikka, J., Models for Modalities, Dordrecht, 1969.

Hintikka, J., "No Scope for Scope?", "Linguistics and Philosophy", 20, (1997) pp. 515-544.

Kamp, J., "A theory of truth and semantic representation", in Groenendijk, J. et al. (Eds), Truth, Interpretation and Information, Dordrecht, 1984.

Keenan, E. and Faltz, L., Boolean Semantics for Natural Language, Dordrecht, 1985.

Montague, R., "The proper treatment of quantification in ordinary English", in Hintikka I. et al. (Eds.), Approaches to Natural Language, Dordrecht, 1973, pp. 221-242.

Montague, R., Formal Philosophy; Selected Papers of Richard Montague, edited with an introduction by R. Thomason, New Haven, 1974.

Partee, B., (Ed.), Montague Grammar, New York, 1976.
Putnam, H., Representation and Reality, Cambridge, 1983.
Quine, W. V., Word and Object, Cambridge, 1960.
Quine, W. V., La Poursuite de la Verité, Paris, 1993.
Rantala, V., "Urn Models: a New Kind on Non-standard Models for First-order Logic", "J. Philos. Logic", 4, (1975), pp. 455-74.
Thayse, A., (Ed.), From Natural Languages Processing to Logic for Expert Systems Introducing a Logic Based Approach to Artificial Intelligence, Vol. 3, New York, 1991.
Tichy, P., "A New Approach to Intensional Analysis", "Nous", 5, (1971), pp. 275-98.
Thomason, R., "Some Extensions of Montague Grammar" in Partee B. (Ed.) Montague Grammar, New York, 1976.
Van Benthem, J. and Doets, K., "Higher-order logic" in Gabbay, D.I. and Guenthner, F. (Eds.), Handbook of Philosophical Logic I, Dordrecht, 1983, pp. 275-329.
Van Benthem, J., Exploring Logical Dynamics, Stanford, 1996.
von Wright, G.H., Logical Studies, London, 1957.

NAME INDEX

NAME INDEX

SUBJECT INDEX

THEORY AND DECISION LIBRARY

SERIES A: PHILOSOPHY AND METHODOLOGY OF THE SOCIAL SCIENCES

Editors: W. Leinfellner (*Vienna*) and G. Eberlein (*Munich*)

1. G. Zecha and P. Weingartner (eds.): *Conscience: An Interdisciplinary View*. Salzburg Colloquium on Ethics in the Sciences and Humanities (1984). 1987 ISBN 90-277-2452-0
2. R.W. Scholz: *Cognitive Strategies in Stochastic Thinking*. 1987 ISBN 90-277-2454-7
3. H. Nurmi: *Comparing Voting Systems*. 1987 ISBN 90-277-2600-0
4. M. Schmid and F.M. Wuketits (eds.): *Evolutionary Theory in Social Science*. 1987 ISBN 90-277-2612-4
5. C.L. Sheng: *A New Approach to Utilitarianism*. A Unified Utilitarian Theory and Its Application to Distributive Justice. 1991 ISBN 0-7923-1301-1
6. F. Forman: *The Metaphysics of Liberty*. 1989 ISBN 0-7923-0080-7
7. G. Bernard: *Principia Economica*. 1989 ISBN 0-7923-0186-2
8. J.W. Sutherland: *Towards a Strategic Management and Decision Technology*. Modern Approaches to Organizational Planning and Positioning. 1989 ISBN 0-7923-0245-1
9. C. Vlek and G. Cvetkovich (eds.): *Social Decision Methodology for Technological Projects*. 1989 ISBN 0-7923-0371-7
10. P. Hoyningen-Huene and F.M. Wuketits (eds.): *Reductionism and Systems Theory in the Life Sciences*. Some Problems and Perspectives. 1989 ISBN 0-7923-0375-X
11. K.G. Grunert and F. Ölander (eds.): *Understanding Economic Behaviour*. 1989 ISBN 0-7923-0482-9
12. G. Antonides: *The Lifetime of a Durable Good*. An Economic Psychological Approach. 1990 ISBN 0-7923-0574-4
13. G.M. von Furstenberg (ed.): *Acting under Uncertainty*. Multidisciplinary Conceptions. 1990 ISBN 0-7923-9063-6
14. R.L. Dukes: *Worlds Apart*. Collective Action in Simulated Agrarian and Industrial Societies. 1990 ISBN 0-7923-0620-1
15. S.O. Funtowicz and J.R. Ravetz: *Uncertainty and Quality in Science for Policy*. 1990 ISBN 0-7923-0799-2
16. J. Götschl (ed.): *Erwin Schrödinger's World View*. The Dynamics of Knowledge and Reality. 1992 ISBN 0-7923-1694-0
17. R. von Schomberg (ed.): *Science, Politics and Morality*. Scientific Uncertainty and Decision Making. 1993 ISBN 0-7923-1997-4
18. J. Baron: *Morality and Rational Choice*. 1993 ISBN 0-7923-2276-2
19. M. Allais and O. Hagen (eds.): *Cardinalism*. A Fundamental Approach. 1994 ISBN 0-7923-2398-X
20. E. Carlson: *Consequentialism Reconsidered*. 1995 ISBN 0-7923-3547-3
21. J. Götschl (ed.): *Revolutionary Changes in Understanding Man and Society*. 1995 ISBN 0-7923-3627-5
22. A. Camacho: *Division of Labor, Variability, Coordination, and the Theory of Firms and Markets*. 1996 ISBN 0-7923-3832-4

23. R. Hegselmann, U. Mueller and K.G. Troitzsch (eds.): *Modelling and Simulation in the Social Sciences from the Philosophy of Science Point of View.* 1996
ISBN 0-7923-4125-2
24. J. Nida-Rümelin: *Economic Rationality and Practical Reason.* 1997
ISBN 0-7923-4493-6
25. G. Barbiroli: *The Dynamics of Technology.* A Methodological Framework for Techno-Economic Analyses. 1997 ISBN 0-7923-4756-0
26. O. Weinberger: *Alternative Action Theory.* Simultaneously a Critique of Georg Henrik von Wright's Practical Philosophy. 1998 ISBN 0-7923-5184-3
27. A. Carsetti (ed.): *Functional Models of Cognition.* Self-Organizing Dynamics and Semantic Structures in Cognitive Systems. 2000 ISBN 0-7923-6072-9

KLUWER ACADEMIC PUBLISHERS – DORDRECHT / BOSTON / LONDON